Gaskinetic Theory is an introductory text on the molecular theory of gases and on modern transport theory. It is suitable for upper division undergraduates in physics and first year graduate students in aerospace engineering, upper atmospheric science and space research. The first part introduces basic concepts, including the distribution function, classical theory of specific heats, binary collisions, mean free path, and reaction rates. Elementary transport theory is used to express coefficients such as viscosity and heat conductivity in terms of molecular properties. The second part of the book covers advanced transport theory. Generalized transport equations are derived from the Boltzmann equation. The Chapman–Enskog and the Grad methods are discussed to obtain higher order transport equations for low density gases. The aerodynamics of solid bodies in rarefied gases is explored and the book concludes with the kinetic description of shock waves. Tamas Gombosi pursued a distinguished research career in Hungary, and moved to the University of Michigan in 1987. He is a corresponding Member of the International Academy of Astronautics. His papers in plasma physics and gaskinetic theory have been published widely in European and North American journals. His interests have encompassed astrophysics, geophysics and space research. He has edited four monographs in these fields.

Gaskinetic theory

Cambridge atmospheric and space science series

Editors

John T. Houghton
Michael J. Rycroft
Alexander J. Dessler

Titles in print in this series

M. H. Rees, *Physics and chemistry of the upper atmosphere*
Roger Daley, *Atmospheric data analaysis*
Ya. L. Al'pert, *Space plasma*, Volumes 1 and 2
J. R. Garratt, *The atmospheric boundary layer*
J. K. Hargreaves, *The solar–terrestial environment*
Sergei Sahzin, *Whistler-mode waves in a hot plasma*
S. P. Gary, *The theory of space plasma microinstabilities*
Tamas I. Gombosi, *Gaskinetic theory*
I. N. James, *Introduction to circulating atmospheres*

Gaskinetic Theory

Tamas I. Gombosi

Professor of Space Science
Professor of Aerospace Engineering
The University of Michigan, Ann Arbor

CAMBRIDGE
UNIVERSITY PRESS

Published by the Press Syndicate of the University of Cambridge
The Pitt Building, Trumpington Street, Cambridge CB2 1RP
40 West 20th Street, New York, NY 10011-4211, USA
10 Stamford Road, Oakleigh, Melbourne 3166, Australia

First published 1994

A catalogue record for this book is available from the British Library

Library of Congress cataloguing in publication data

Gombosi, Tamás I.
Gaskinetic theory / Tamás I. Gombosi.
p.   cm. – (Cambridge atmospheric and space science series)
Includes bibliographical references.
ISBN 0 521 43349 5. – ISBN 0 521 43966 3 (pbk.)
1. Kinetic theory of gases.   I. Title.   II. Series.
QC175.G66   1994
533'.7–dc20   93–1691 CIP

ISBN 0 521 43349 5 hardback
ISBN 0 521 43966 3 paperback

Transferred to digital printing 2003

To my wife Eszter
and my parents Magda and János

Contents

ix

Preface

This book provides a comprehensive introduction to the kinetic theory of gases for graduate students and interested researchers. The book is based on graduate level courses I taught in the Department of Aerospace Engineering and in the Department of Atmospheric, Oceanic and Space Sciences of the University of Michigan College of Engineering. These courses were intended to provide a broad introduction to the kinetic theory of gases.

The course on gaskinetic theory and real gas effects has been taught for a long period of time by Professors Thomas C. Adamson, Vi-Cheng Liu, Paul B. Hays, Martin Sichel and others before I was fortunate enough to teach it. I greatly benefited from their experience and lecture notes. The first part of the course (Phasespace distributions, Binary collisions and Elementary transport theory) generally follows the way this course was organized for the last decade or so. The second part is significantly revised and concentrates on the fundamentals of advanced transport theory. Although the course starts at a very elementary level, Chapters 5 and 6 are quite advanced.

This course was intended to provide a comprehensive background for students who eventually intend to carry out independent research in computational fluid dynamics, aerodynamics of high altitude flight, upper atmospheric science, planetary atmospheres and space physics. The reader is expected to have some familiarity with integral calculus, vector and tensor algebra, statistics and classical mechanics. It was my intent to produce a 'self contained' text and introduce all important concepts and definitions used in the book.

In the preparation of these lecture notes I have consulted other textbooks, monographs, review articles and research papers. As can be seen from the bibliographies most of the available textbooks are older than the students who take the course and therefore, throughout the text I tried to emphasize those aspects and techniques which are used by state of the art theories (especially in the fields of upper atmospheric research and space science).

I would like to express my thanks to Ms Lori Kwiecinski for typing the core of the manuscript and Dr Steven M. Guiter for finding a number of typos and some algebraic errors in the original manuscript. Special thanks are due to Mr Nathan Schwadron for his constructive comments and his help in preparing the manuscript.

Tamas I. Gombosi

Ann Arbor, Michigan
January, 1993

1

Introduction

The objective of gaskinetic theory is to explain and predict the macroscopic properties of gases from the properties of their microscopic constituents. Macroscopic properties include the equation of state, specific heats, transport coefficients (such as viscosity, diffusion coefficient, thermal conductivity), etc. The fundamental hypothesis of kinetic theory is that solids, liquids and gases are composed of finite particles (molecules) which are in various states of motion. The term molecule is taken generally to mean a regular arrangement of atoms, or it may refer to a free atom, an electron or an ion.

Statistical mechanics, like kinetic theory, also aims to explain and predict the macroscopic properties of matter in terms of the microscopic particles of which it is composed. The main difference between gaskinetic theory and statistical mechanics is the manner in which particle interactions are treated. In gaskinetic theory the details of the molecular interactions are taken into account. Statistical mechanics avoids these complicated and sometimes poorly known details and replaces them with certain powerful and general statistical ideas.

1.1 Brief history of gaskinetic theory

The history of the kinetic theory of gases can be traced back to ancient Greece, where around 400 B.C. Democritus hypothesized that matter is composed of small indivisible particles which he called atoms. The atoms of different materials were assumed to have different sizes and shapes, so that atoms represent the smallest quantity of a substance that retains its properties. Until about two hundred years ago the theory remained just one of several alternative speculations, essentially unsubstantiated by experiments.

The molecular hypothesis entered modern physical science in the eighteenth century. In 1738 Daniel Bernoulli (1700–1782) of Basel explained Boyle's law[1] (the observation that in an isothermal gas the product of the pressure and the vol-

1

ume remains constant) by assuming that at constant temperature the mean velocity of the gas particles remains constant. In this fundamental work Bernoulli derived a formula equivalent to $p = \rho u^2/3$, where p is the gas pressure, ρ is the gas mass density and u^2 is the mean square velocity of the gas molecules. The work of Bernoulli can be regarded as the beginning of the modern kinetic theory of gases.

It was not until the general acceptance of the mechanical theory of heat that the kinetic theory of gases was admitted into the mainstream of physical theory. Kinetic theorists also had to explain the specific heats, adiabatic change, and the velocity of sound in gases and provide an alternative theory of matter to replace the Newtonian static, corpuscular theory of gases.

Kinetic theory entered a new phase in 1858 when Rudolf Julius Emanuel Clausius (1822–1888) introduced the concept of mean free path.[2] Clausius' work greatly influenced James Clerk Maxwell (1831–1879) who is considered to be the founder of the modern kinetic theory of gases.

In 1859, Maxwell introduced the concept of the velocity distribution function of molecules for a gas in equilibrium and recognized the equipartition principle of the mean molecular energy for molecules of different mass in a gas mixture.[3] Combining the mean free path and his own concept of velocity distribution function, Maxwell also derived formulae for the transport coefficients of the gas (viscosity, thermal conductivity and diffusion coefficient).[4] Today these papers are regarded as one of the cornerstones of gaskinetic theory.

Another cornerstone is the fundamental integro-differential equation of particle transport derived in 1872 by Ludwig Boltzmann (1844–1906). In his classic paper Boltzmann[5] established the H-theorem, which shows that molecular collisions cause any initial distribution function to relax to a Maxwellian velocity distribution as long as the gas is not influenced by external effects. In the same paper Boltzmann studied non-uniform gases and derived his famous equation for the spatial and temporal evolution of the distribution function which must be satisfied under all conditions.

In 1905 Hendrik Antoon Lorentz (1853–1928), professor of theoretical physics at the University of Leiden, applied Boltzmann's transport theory to the problem of electrical conduction in an electron gas.[6] In fact, this treatment described the electron component of gaseous plasmas, although the significance of the theory was not recognized for a long time.

In 1917 David Enskog (1887–1947) published his Ph.D. dissertation,[7] in which he developed the detailed mathematical theory of non-equilibrium gases. Very similar results were independently published by Sydney Chapman (1888–1970) in a series of papers.[8,9] This highly successful method is the basis of contemporary gaskinetic theory.

In 1949 Harold Grad (1923–1986) developed a new method of solving the Boltzmann equation by expanding the solution into a series of orthogonal polynomials.[10] This method leads to Grad's 13 moment approximation which was widely used from the late 1950s. More recently with the improvement of computer technology the 20 moment approximation is being used for several applications. It can also be shown that the Chapman–Enskog and the Grad methods are mathematically equivalent.

The history of the kinetic theory of gases is full of fascinating discoveries, understandings and misunderstandings. We refer the interested reader to science history books specifically dealing with the emergence of kinetic theory.[11,12]

Today, gaskinetic theory is a widely used tool in various areas of science and engineering. This book concentrates on the kinetic description of transport phenomena with special emphasis on applications in aerospace engineering and space science.

1.2 The road to gaskinetic theory in science and engineering

1.2.1 Hydrostatics

The practical study of fluid behavior (here the term 'fluid' means both gases and liquids) started a very long time ago with hydrostatics. Hydrostatics is the science of crude engineering approximations to simple fluid phenomena, which was primarily used to solve problems, such as the discharge rate of rivers, hydrostatic pressure on a dam, or wind pressure on a wall, etc. These problems were usually treated with very primitive analysis, which generally goes little beyond very simple principles, such as Bernoulli's law:

$$p = p_0 + \rho g(z_0 - z) \tag{1.1}$$

where p is pressure, ρ is the constant mass density, g is the gravitational acceleration, and z is altitude (or depth). Problems involving more complicated flows were usually circumvented by the introduction of empirical constants after some form of dimensional analysis.

A typical example for such empirical constants is the discharge coefficient used to describe the discharge rate through an orifice. For instance, consider a reservoir filled with fluid (see Figure 1.1). There is a small orifice with area A at the bottom of the reservoir. The fluid discharges through the orifice with a flow velocity of $v=(2gh)^{1/2}$. The discharge rate of the fluid is $Q=CAv$, where C is the discharge coefficient of the reservoir. This discharge coefficient, C, has to be determined as an empirical constant from experimental calibration of the orifice. The functional dependence of Q on A, h, etc., can be formulated by applying either physical intuition or dimensional analysis.

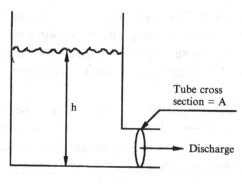

Figure 1.1

In different problems we need different empirical coefficients, i.e., we need drag coefficients to calculate drag, lift coefficients to calculate lift, etc. Things would not be too bad if the empirical coefficients were really constant over a wide range of experimental conditions and for various fluids. This is not the case because in using a coefficient we are taking gross averages of quantities characterizing physical phenomena. The physical process is usually too complicated to be represented by such a simple parameter.

1.2.2 Hydrodynamics (fluid mechanics)

Hydrodynamics represents a more sophisticated approach to the description of fluids than hydrostatics. The main difference is the much more rigorous and refined use of physical and mathematical principles. The basic idea of hydrodynamics is to take a small fluid element in the flow field and consider the forces acting on it and to calculate the resulting velocities and the changes of the element as a function of space and time. The basic relations are appropriately formulated conservation laws of mechanics, such as the conservation of mass, momentum and energy.

The internal state of the fluid element is determined by laws of thermodynamics and the equation of state. The ensuing relations combined with initial and boundary conditions define the mathematical problem which is usually a set of differential equations. The solutions 'almost' describe the flow field: almost, because some empirical coefficients are still needed. In hydrodynamics one still needs a few basic empirical coefficients or relations describing some internal properties of the fluid. These empirical coefficients include: (i) the viscosity coefficient, which relates the viscous stress to the rate of strain of the fluid element (velocity gradient); (ii) the heat conduction coefficient, which relates the heat flow to the temperature gradient; (iii) the diffusion coefficient, which relates the mass transfer rate to the gradients of concentration, temperature and pressure.

These fundamental coefficients and the equation of state depend on the internal structure of the fluid: their values can be determined through experimental calibrations of different fluids and have a wide range of validity.

Compared with hydrostatics, fluid dynamics is a big step forward. Instead of empirical coefficients for each individual flow problem we need only a few basic coefficients which apply to a large family of problems. Hydrodynamics also provides much more information about the flow field. This is an advantage only if we can solve the hydrodynamic equations for a given flow problem.

Hydrodynamics is not a new science. It had been in existence for a long time before physicists and engineers discovered it. In the 18th century it was a favorite field for mathematicians since it involved interesting aspects of the theory of differential equations. A number of fundamental differential equations were solved for flow problems.

In the 19th century, hydrodynamics played with paradoxes, such as the interpretation of drag force on a body in a stream of perfect fluid. Hydrodynamics was not considered physically realistic until Ludwig Prandtl[13] worked out the boundary layer theory and resolved many paradoxes. His contribution transformed hydrodynamics from a subject of pure academic interest to a discipline of enormous physical and engineering significance.

Compressible fluid dynamics deals with gases and concentrates on the description of subsonic, transonic, supersonic and hypersonic flows. This is generally called gas dynamics. As more and more problems are created by science and technology, new physical phenomena and problems are encountered. These new challenges include the increasing speed and energy range in flows, flows at very low density (such as high altitude flight, spacecraft re-entry, interplanetary flight, space science), regions of very large gradients (shock waves), flows with chemical reactions, flows of ionized species (plasma physics), and non-equilibrium flows.

In these cases we repeatedly encounter the problem that the classical gas transport coefficients (viscosity, heat conduction) lose their meaning, because the internal state of the gas becomes too complicated and the coefficients cease to be constant. In addressing these new problems, it is not sufficient to consider the continuum model generally used by gas dynamics. We must build the flow theory from a more basic point of view. This new approach is the molecular approximation, which leads to gaskinetic theory.

1.2.3 Gaskinetic theory

The next step in the description of the fluid phenomena is kinetic theory which is capable of deriving transport coefficients from the fundamental properties of the gas molecules. Gaskinetic theory is based on the concept that a 'gas' is a collection of very many particles which are acted upon by their surroundings and by mutual encounters. The molecules are too small to be seen, but they obey the classical laws of matter (possibly modified to describe some quantum effects, too).

The classical kinetic theory of gases emerged from a combination of mechanics and statistics. The motions of the molecules are described by probability and not by their individual paths. This approach is the kinetic theory of gases, which is based on first principles and has led to an improved description of the pressure, temperature and equation of state, as well as the description of transport properties (viscosity, thermal conductivity, diffusion coefficients).

This book is an introduction to the theory of kinetic theory of gases, that is the molecular theory of compressible fluids. The foundations of molecular theory will be used to interpret those flow quantities which cannot be determined from the continuum theory of fluids.

1.3 Basic assumptions of gaskinetic theory

The kinetic theory of gases is based on three fundamental approximations. These are the molecular hypothesis, the assumption that classical conservation laws can be applied, and the application of statistical methods.

1.3.1 Molecular hypothesis

The fundamental assumption that all matter is composed of small elementary building blocks can be traced to ancient Greek philosophers. The molecular hypothesis assumes that (i) matter is composed of small discrete units (called molecules), (ii) the molecules are the smallest quantity of a substance that retains its chemical properties, (iii) all molecules of a given substance are alike, and (iv) the states of matter (solids, liquid, gas) differ essentially in the arrangement and state of motion of the molecules.

Gases differ from liquids and solids by completely filling any container in which they are placed, by being extraordinarily compressible, by rapidly diffusing into one another, and by having very low densities. These facts suggest that the molecules in a gas are widely separated from one another (meaning that the average distance between molecules is much greater than the size of the molecules) and that the molecules move about throughout the entire space occupied by the gas.

In a gas, the molecules move about freely and most of the time are separated by distances that are large compared to their dimensions. They move in straight lines until two (or more) happen to come so close together that they act strongly upon each other and something like a collision occurs, after which they separate and move off in new directions, generally with different velocities.

This book will primarily apply kinetic theory to ideal (perfect) gases. We define a perfect gas using the following assumptions:

(i) molecules are point-like with no internal structure or internal degrees of freedom (perfect gases are monatomic);

(ii) molecules exert forces on each other only when their centers are within a distance which is very small compared to their average separation (the region in which the particles interact is called the sphere of influence);

(iii) outside each other's sphere of influence molecules obey the laws of classical mechanics (their motion is described by a non-relativistic equation of motion).

In perfect gases heat energy is interpreted as random translational energy of the molecules. However, molecules which possess internal structure may store energy in their internal degrees of freedom (kinetic energy of rotation and vibration or potential energy of vibration). Although classical kinetic theory makes specific predictions of how the kinetic energy is partitioned among all degrees of freedom, the results do not agree with observations except in the cases of the simplest molecules. This failure clearly shows the limits of classical gaskinetic theory.

Friction is entirely absent in molecular interactions. In kinetic theory the conversion of mechanical work into heat by friction is interpreted as a conversion of bulk translational energy into random molecular motion. After the body has been brought to rest by friction, there may be as much motion as before, only the formerly organized motion now becomes part of the completely disorganized heat motion.

1.3.2 Classical conservation laws

It should be repeatedly emphasized that relativistic corrections, quantum mechanical and non-perfect gas effects are usually not considered in the classical kinetic theory of gases. The notable exceptions are the calculation of specific heats and collision cross sections, where quantum mechanical effects usually must be taken into consideration.

The internal structure of atoms and molecules enters into kinetic theory in two essentially different ways. Quantum mechanical effects are relatively unimportant in those parts of gaskinetic theory which treat only the translational motion of molecules (such as viscosity and diffusion). In these problems the internal molecular structure (which is governed by quantum mechanics) matters only so far as it determines the exact nature of the intermolecular forces. On the other hand it should be recognized that the intermolecular force laws in gaskinetic theory are taken as semi-empirical expressions. Quantum mechanical theory of atomic and molecular structure provides a method for obtaining these semi-empirical force laws.

Quantum mechanical effects must be taken into account in any gaskinetic problem in which the energy stored in internal degrees of freedom is considered. The molecular rotation and vibration and the motion of electrons must all be treated by methods of quantum theory, and the classical methods are correct only for sufficiently high temperatures.

1.3.3 *Statistical nature of the theory*

The kinetic theory of gases is a statistical theory: in addition to dynamical methods it also uses a statistical approach. The dynamical description of the motion of a single particle is given by the non-relativistic equation of motion, which is a second order differential equation with two integration constants for each translational degree of freedom. If the values of these integration constants are specified (by the initial conditions) and all the forces acting on the particle are known at all spatial locations and at all times, its entire future dynamical behavior is fully described by the equation of motion.

Regarding a gas as a system of N particles (each with three translational degrees of freedom), there would be altogether N vector equations of motion with 6N integration constants. The problem is that for typical gases N is a prohibitively large number. There are a very large number of molecules in every 'macroscopically infinitesimal' sample.

In order to illustrate this point Figure 1.2 shows the gas concentration as a function of altitude[14] in the atmosphere of Earth between the surface and 1000 km. It can be seen that there are about 10^{25} molecules in a cubic meter at the surface. To solve $\geq 10^{25}$ equations is not easy. Besides, it is not needed. Complete knowledge of the history of each molecule in the gas is not what we are seeking. We need to know only so much about the molecular motions as is necessary to under-

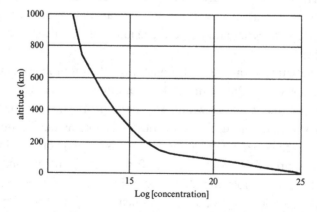

Figure 1.2

stand and predict the observable, macroscopic properties of the gas. The measurement of these quantities involves taking a time average over the instantaneous states of motion of the individual particles in 'macroscopically infinitesimal' volumes.

These average quantities might be the density, bulk velocity, pressure, etc. Gaskinetic theory involves the description of a very large number of individual molecules by statistical methods. This description is meaningful, because we have a large number of molecules in any macroscopically infinitesimal volume.

1.4 Notations

Throughout this book vectorial quantities either are denoted by coordinate subscripts (such as x_i, where i=1, 2, 3) or are printed in boldface type (**x**). Throughout the book coordinate indices are denoted by i,j,k,l,m and n. When using Cartesian coordinates this notation means x_1=x, x_2=y, and x_3=z. In expressions written in component notation, summation must be applied with respect to coordinate indices which occur twice in one term. This means that the dot product of vectors **x** and **v** can be denoted either as **x·v** or as $x_i v_i$.

The differential operator vector, ∇, is sometimes also written in component notation, $\nabla=(\partial/\partial x_1, \partial/\partial x_2, \partial/\partial x_3)$. The velocity space del operator is denoted by ∇_v, and it can be written in component notation as $\nabla_v=(\partial/\partial v_1, \partial/\partial v_2, \partial/\partial v_3)$.

The subscripts s and t refer to different particle species. Summation with respect to these subscripts is always indicated explicitly (i.e. $\Sigma_t f_t$).

The Kronecker symbol, δ_{ij}, is equal to +1 when the two indices are equal, and it is equal to zero when the two indices are different. For instance, δ_{11}=1 and δ_{12}=0.

The permutation tensor, ε_{ijk}, is equal to zero when at least two indices are equal, it has the value of +1 when the three subscripts are all different and form an even permutation of 1,2 and 3, and finally it takes the value of –1 when the three subscripts are all different and form an odd permutation. For instance, ε_{312}=1, ε_{113}=0, and ε_{213}=–1. The permutation tensor is necessary to express vector products in component notation. For instance the i-th component of the vector product of vectors **x** and **v** can be denoted either as $(\mathbf{x}\times\mathbf{v})_i$ or as $\varepsilon_{ijk} x_j v_k$.

Using our index notation the trace of tensor A can be written as A_{ii}. The i,j element of the unit tensor is obviously $I_{ij}=\delta_{ij}$. The trace of the unit tensor is δ_{ii}=3. Tensors with two or more indices can be contracted to obtain lower order tensors. For instance, by contracting the two indices of a regular matrix we get a scalar quantity, $a=A_{ii}$. By contracting two indices of a three index tensor we get a vector, $q_i=Q_{ijj}$.

1.5 Solid angles and curvilinear coordinates

1.5.1 Spherical coordinates and solid angles

In gaskinetic theory one usually has to deal with problems involving seven independent variables: time, three components of the position vector and three components of the velocity vector. Every symmetry of the problem can help us to reduce the number of independent variables and thus make the problem somewhat more tractable. The most commonly used symmetry assumptions in gaskinetic theory are those of spherical and cylindrical geometries, therefore the use of spherical or cylindrical coordinates can result in significant mathematical simplifications. In this section we briefly outline the basic concepts of solid angles, spherical and cylindrical coordinate systems.

The spherical coordinates (r, θ, ϕ) of a point, P, denote the distance of the point from the origin of the coordinate system, the polar angle measured between the +z axis and the radius vector, and the azimuth angle measured between the +x axis and the projection of the radius vector onto the (x,y) plane. The spherical coordinates, r, θ, and ϕ, vary from 0 to ∞, from 0 to π, and from 0 to 2π, respectively (the range of the angle, θ, is constrained to avoid redundancy). This coordinate system is shown in Figure 1.3. Figure 1.3 also shows the new orthogonal unit vectors, $\mathbf{e}_r$, $\mathbf{e}_\theta$, and $\mathbf{e}_\phi$. The relation between the spherical and Cartesian coordinates can be expressed as

$$x = r \sin\theta \cos\phi$$
$$y = r \sin\theta \sin\phi \qquad (1.2)$$
$$z = r \cos\theta$$

If r and θ are fixed and ϕ is varied, P moves along a parallel of latitude, which maps out a circle of radius $r\sin\theta$. If the azimuth angle changes by $d\phi$, the arc length

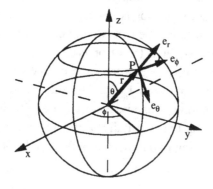

Figure 1.3

along this circle changes by rsinθdφ. If r and φ are fixed and θ is varied, P moves along a meridian. If the polar angle changes by dθ, the arc length along the meridian changes by rdθ. If r is fixed and θ and φ are varied, the point, P, moves on the surface of a sphere of radius r. If the azimuth angle changes by dφ and the polar angle changes by dθ, the corresponding area element on the surface of the sphere is

$$dS = r^2 \sin\theta \, d\theta \, d\phi \tag{1.3}$$

If the radius vector is also allowed to change, i.e. the radius changes by dr, the azimuth angle changes by dφ and the polar angle changes by dθ, the corresponding volume element can be expressed as

$$dV = d^3r = r^2 \sin\theta \, d\theta \, d\phi \, dr \tag{1.4}$$

Here and throughout this book three dimensional volume elements will be denoted by d^3r. It is obvious that in Cartesian coordinates $d^3r=dxdydz$.

The solid angle is a generalization to three dimensions of the radian definition of angle in a plane. If s is the length of arc along a circle of radius r, then the central angle, α, is defined as $\alpha=s/r$. In three dimensions the circle is replaced by a sphere of radius r, and the arc is replaced by a surface area on the sphere, S, bounded by an arbitrary closed curve C. The definition of the solid angle covered by the surface area, S, as seen from the center of the sphere, is $\Omega=S/r^2$. Since the largest value of S is $4\pi r^2$, one can immediately see that Ω can vary from 0 to a maximum value of 4π. In analogy to plane angles the solid angle is measured in units of steradian.

Using the above definition of solid angle one can obtain the infinitesimal solid angle element occupied by the surface element $dS=r^2\sin\theta \, d\theta d\phi$:

$$d\Omega = \frac{dS}{r^2} = \sin\theta \, d\theta \, d\phi \tag{1.5}$$

Sometimes it is more convenient to replace the polar angle, θ, by its cosine, $\mu=\cos\theta$. In this case the spherical coordinates are r, μ and ϕ, with μ changing from 1 ($\theta=0$) to $\mu=-1$ ($\theta=\pi$). In this case the infinitesimal surface element can be expressed as $dS=-r^2d\mu d\phi$. The negative sign can be eliminated by interchanging the integration boundaries: in this case the integral over μ goes from -1 to $+1$. When one integrates over the entire solid angle the integrals go from $\mu=-1$ to $\mu=+1$ and from $\phi=0$ to $\phi=2\pi$.

Some fundamental differential operators in spherical coordinates are given in the Appendix.

1.5.2 Cylindrical coordinates

In some cases problems show rotational symmetry about a specific axis. In such cases the use of cylindrical coordinates might offer great mathematical simplifications.

The cylindrical coordinates (ρ,ϕ,z) of a point P denote the projection of the radius vector to the (x,y) plane, the azimuth angle in the (x,y) plane measured counterclockwise from the x axis, and the z coordinate, respectively. This coordinate system is shown in Figure 1.4 together with the curvilinear orthogonal unit vectors, e_ρ, e_ϕ, and e_z. It can be seen from Figure 1.4 that the cylindrical coordinates, ρ, ϕ, and z vary from 0 to ∞, from 0 to 2π, and from $-\infty$ to $+\infty$, respectively. Using elementary geometric considerations one can readily express the relation between the cylindrical and Cartesian coordinates in the following form:

$$x = \rho\cos\phi$$
$$y = \rho\sin\phi \qquad (1.6)$$
$$z = z$$

The three dimensional infinitesimal volume element can be expressed in cylindrical coordinates as

$$d^3r = \rho\,d\phi\,dr\,dz \qquad (1.7)$$

Some fundamental differential operators in cylindrical coordinates are given in the Appendix.

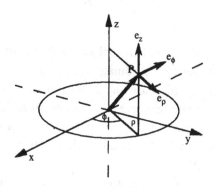

Figure 1.4

1.6 Problems

1.1 Using index notation show that $\nabla \cdot (\nabla \times \mathbf{B}) = 0$, where the vector, $\mathbf{B}$, is a function of location, $\mathbf{r}$.

1.2 In a vacuum the vector form of Maxwell's equations is the following:

$$\nabla \cdot \mathbf{E} = \frac{\rho}{\varepsilon_0} \qquad \nabla \times \mathbf{E} = -\frac{\partial \mathbf{B}}{\partial t}$$

$$\nabla \cdot \mathbf{B} = 0 \qquad \nabla \times \mathbf{B} = \varepsilon_0 \mu_0 \frac{\partial \mathbf{E}}{\partial t} + \mu_0 \mathbf{j}$$

where $\mathbf{E}$ and $\mathbf{B}$ are the electric and magnetic field vectors, ρ is the electric charge density, $\mathbf{j}$ is the electric current density, and μ_0 and ε_0 are the magnetic permeability and electric permittivity of free space, respectively. Write these equations using index notation.

1.3 Comet Halley has a total surface area of 500 km^2. According to the images made by the Giotto spacecraft, cometary activity was concentrated in a 50 km^2 area on the sunlit side. Approximate the nucleus by a sphere and calculate the solid angle covered by the active area.

1.7 References

[1] Bernoulli, D., *Hydrodynamica, sive de viribus et motibus fluidorum commentarii*, Argentoria, Strasbourg, 1738.

[2] Clausius, R., Über die mittlere Länge der Wege, welche bei der Molecularbewegung gasförmigen Körper von den einzelnen Molecülen zurückgelegt werden, nebst einigen anderen Bemerkungen über die mechanischen Wärmetheorie, *Ann. Phys. [2]*, **105**, 239–258, 1858.

[3] Maxwell, J.C., Illustrations of the dynamical theory of gases. I. On the motions and collisions of perfectly elastic spheres, *Phil. Mag. [4]*, **19**, 19–32, 1860 (read 1859).

[4] Maxwell, J.C., Illustrations of the dynamical theory of gases. II. On the process of diffusion of two or more kinds of moving particles among one another, *Phil. Mag. [4]*, **20**, 21–37, 1860.

[5] Boltzmann, L., Weitere Studien über das Wärmegleichgewicht unter Gasmolekülen, *Sitz. Math.-Naturwiss. Cl. Akad. Wiss. Wien*, **66**, 275–370, 1872.

[6] Lorentz, F.A., The motions of electrons in metallic bodies, *Proc. Amsterdam Acad.*, **7**, 438, 1905.

[7] Enskog, D., The kinetic theory of phenomena in fairly rare gases, Ph.D. Thesis, University of Uppsala, Sweden, 1917.

[8] Chapman, S., On the law of distribution of velocities, and on the theory of viscosity and thermal conduction, in a non-uniform simple monatomic gas, *Phil. Trans. Roy. Soc., A*, **216**, 279, 1916.

[9] Chapman, S., On the kinetic theory of a gas, Part II, A composite monatomic gas, diffusion, viscosity and thermal conduction, *Phil. Trans. Roy. Soc., A*, **217**, 115, 1917.

[10] Grad, H., On the kinetic theory of rarefied gases, *Comm. Pure Appl. Math.*, **2**, 331, 1949.

[11] Brush, S.G., *The kind of motion we call heat*, North-Holland Publ. Co., New York, 1976.

[12] Garber, E., Brush, S.G., and Everitt, C.W.F., *Maxwell on molecules and gases*, MIT Press, Cambridge, Mass., 1986.

[13] Prandtl, L., Flüssigkeiten bei sehr kleimer Reibung, *III. Int. Math. Kongress*, Heidelberg, Teubner, Leipzig, p.484, 1905.

[14] Banks, P.M., and Kockarts, G., *Aeronomy*, Academic Press, New York, 1973.

2

Equilibrium kinetic theory

Maxwell's velocity distribution function[1,2] was the key element which made it possible to develop connections between the motions of microscopic molecules and the macroscopic properties of gases (observable by the methods of classical physics).

In this chapter we introduce the concept of phase space velocity distribution functions and derive a specific functional form for the distribution of molecular velocities for gases in equilibrium. Using this statistical law for the velocity distribution we derive some simple relations between molecular statistics and thermodynamics. For additional reading we refer the reader to a series of articles and books on classical kinetic theory.[3-12]

2.1 Distribution functions

First we briefly summarize the elementary ideas involved in the calculation of average values. As a very simple example, let us calculate the average number of children per family in a particular community. Let N_k denote the number of families in the community having k children (k=0, 1, 2, ...). The total number of families is $N=\Sigma N_k$, so the average number of children, $<k>$, will be equal to the total number of children divided by the total number of families:

$$<k> = \frac{1}{N}\left[0 \times N_0 + 1 \times N_1 + 2 \times N_2 + ...\right] = \frac{\sum_{k=0}^{\infty} k N_k}{\sum_{k=0}^{\infty} N_k} \qquad (2.1)$$

One may also define the 'normalized' distribution numbers, $f_k=N_k/N$, which specify the fraction of families having k children. It should be noted that f_k is normalized to unity, i.e. $\Sigma f_k=1$. Using this definition equation (2.1) can be written as

$$<k> = \sum_{k=0}^{\infty} k f_k \qquad (2.2)$$

In this example the quantity to be averaged (the number of children) can take only discrete values. As a matter of fact possible values of k are only positive integers or zero.

The problem is somewhat more complicated in the case when k can take on a continuous range of values. Suppose we wish to know the average weight of cars owned by the people of our community. In this case we cannot use the distribution of integers as we did in the previous problem, because the number of cars with the exact weight of w (for example exactly 1000 kg) is certainly zero.

However, if we ask for the number of cars whose weights lie within a certain interval (such as cars with weights between 950 kg and 1000 kg), a definite and meaningful answer can be given. Information about the distribution of car weights in the city can then be obtained by dividing the weight range into equal intervals and determining the number of cars whose weights fall into each interval, Δw. Let ΔN_k denote the number of cars with weights between $w_k = k \Delta w$ and $w_{k+1} = (k+1) \Delta w$. This way the problem is converted into a discrete distribution.

It is obvious that the number of cars with weights between w_k and w_{k+1} is proportional to Δw. The smaller the value chosen for Δw, the more details we know about the distribution. The most meaningful information about the distribution is given by the fraction of cars per unit interval of weight,

$$f_k = \frac{\Delta N_k}{N \Delta w} \qquad (2.3)$$

As we choose Δw to be smaller and smaller the statistical information contained by f_k becomes more and more detailed. When Δw becomes infinitesimally small, f_k becomes a continuous function:

$$f(w) = \lim_{\Delta w \to 0} f_k = \lim_{\Delta w \to 0} \frac{\Delta N_k}{N \Delta w} \qquad (2.4)$$

This point is illustrated by Figure 2.1, which shows histograms of f_k as the weight interval, Δw, decreases.

The function f(w) is referred to as the normalized distribution of car weights. From equation (2.4) it is obvious that f(w) is normalized to unity:

$$\int_0^{\infty} dw\, f(w) = \lim_{\Delta w \to 0} \sum_{k=0}^{\infty} f_k\, \Delta w = \lim_{\Delta w \to 0} \sum_{k=0}^{\infty} \frac{\Delta N_k}{N \Delta w} \Delta w = \lim_{\Delta w \to 0} \sum_{k=0}^{\infty} \frac{\Delta N_k}{N} = 1 \qquad (2.5)$$

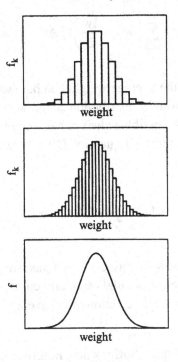

Figure 2.1

The distribution function can be interpreted as a probability density function: $f(w)dw$ is the probability that a randomly selected car will have its weight between w and $w+dw$.

With the help of the weight distribution, f_k, one can calculate the average weight of a car in the community. First we multiply the average weight in each interval by the number of cars in that particular weight interval over the total number of cars. In the second step we sum these quantities over all intervals. This procedure can be mathematically written in the following form:

$$<w> = \sum_{k=0}^{\infty} \frac{w_k + w_{k+1}}{2} \frac{\Delta N_k}{N}$$

$$= \sum_{k=0}^{\infty} \left(w_k + \frac{\Delta w}{2} \right) \frac{\Delta N_k}{N \Delta w} \Delta w = \sum_{k=0}^{\infty} \left(w_k + \frac{\Delta w}{2} \right) f_k \Delta w \qquad (2.6)$$

As Δw becomes smaller and smaller, more and more accurate values are obtained for $<w>$. As $\Delta w \to 0$, equation (2.6) becomes the following:

$$< w >= \lim_{\Delta w \to 0} \sum_{k=0}^{\infty} \left(w_k + \frac{\Delta w}{2} \right) f_k \, \Delta w = \int_0^{\infty} dw \, w \, f(w) \qquad (2.7)$$

Equation (2.7) tells us that the average weight can be obtained by multiplying the weight by the normalized weight distribution function (probability density function) and integrating over all possible values of car weights.

In general, the average value of a quantity, Q(w) (where Q is a function of car weights), can be obtained by

$$< Q >= \int_0^{\infty} dw \, Q(w) f(w) \qquad (2.8)$$

In what follows we apply the same principle to a gas, where the number of molecules is much larger than the car population of any city, therefore statistical methods yield an even more accurate description of the average quantities.

2.2 Phase space distributions and macroscopic averages

2.2.1 Phase space

Let us start with a very simple case: consider a single particle moving in one dimension under the influence of a well defined force field. This system can be completely described in terms of the particle's spatial coordinate, x, and its velocity, v_x. This description is complete, because the laws of classical mechanics are such that the knowledge of x and v_x at any instant permits prediction of the values of x and v_x at any other time. The particle can be represented by a single point in the Cartesian coordinate system (x, v_x). Specification of location and velocity is equivalent to specifying a point in this two dimensional space, called phase space (Figure 2.2). As the location and the velocity of the particle change in time, the point representing the particle moves around this two dimensional phase space.

In gaskinetic theory the three dimensional motion of individual particles between collisions is governed by the equation of motion which is a set of three second order differential equations. In each dimension we have two integration constants. These six constants are independent of each other and they determine the particle orbit between collisions. Between collisions each particle can be represented by a single point in the six dimensional phase space (x,y,z,v_x,v_y,v_z). In vector form the six dimensional phase space (sometimes called µ-space) can be denoted by $(\mathbf{r},\mathbf{v})$, where $\mathbf{r}$ refers to the configuration space location, while $\mathbf{v}$ is the three dimensional velocity space location. Just as in our previous example, the

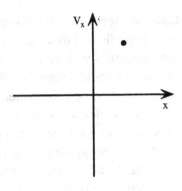

Figure 2.2

particle is represented by a single point in this six dimensional phase space. The vectors **r** and **v** are the Eulerian coordinates of the phase space. As the location and the velocity of the particle change in time, the point representing the particle moves around this six dimensional phase space.

2.2.2 Phase space distribution

Let us consider a gas which is non-uniformly distributed in a container. Let $\Delta^3 N$ be the number of molecules contained in the small configuration space volume element, $\Delta V = \Delta^3 r$, located between x and x+Δx, y and y+Δy, z and z+Δz. The superscript 3 refers to the three dimensional nature of the small volume element. The mean density of the gas in the small volume element is $\Delta^3 N / \Delta^3 r$. Next we take the limit where the volume element becomes macroscopically infinitesimal (in other words $\Delta^3 r$ approaches zero but it remains large on the scale of intermolecular distances). In this way we define the particle number density (or particle concentration):

$$n(\mathbf{r}) = \lim_{\Delta^3 r \to 0} \frac{\Delta^3 N}{\Delta^3 r} \qquad (2.9)$$

The local number density, $n(\mathbf{r})$, gives the number of molecules per unit volume in the infinitesimal vicinity of location **r**. It is thus the measure of the distribution of molecules in configuration space, in short it is the configuration space distribution function. If the configuration space distribution function is known at a particular location, **r**, the number of molecules, $d^3 N$, in the macroscopically infinitesimal volume element, $d^3 r$, is given by

$$d^3 N = n(\mathbf{r})\, dx\, dy\, dz = n(\mathbf{r})\, d^3 r \qquad (2.10)$$

It should be noted that the configuration space distribution function characterizes the distribution of particles in space, regardless of their velocity.

At any instant, each of the gas molecules will have a velocity, which can be specified by a velocity space point, v. One could define the velocity space concentration function of all molecules in a macroscopic container. The velocity space concentration defines the number of molecules in an infinitesimal velocity space volume element around a given velocity vector independently of their spatial location in the container. However, such a velocity space concentration is not a particularly useful physical quantity. A statistically useful physical quantity should characterize the distribution of molecular velocities at any given configuration space location.

One can generalize the concept of configuration space particle distribution function to the six dimensional phase space. The phase space distribution function, $F(\mathbf{r},\mathbf{v})$, characterizes the number of particles in a six dimensional infinitesimal volume element, $d^3r\,d^3v$. If the phase space distribution function, F, is known at a particular phase space location, $(\mathbf{r},\mathbf{v})$, the number of molecules, d^6N, to be found in the configuration space volume element between x and x+dx, y and y+dy, z and z+dz with velocity components between v_x and v_x+dv_x, v_y and v_y+dv_y, v_z and v_z+dv_z is given by

$$d^6N = F(\mathbf{r},\mathbf{v})\,dx\,dy\,dz\,dv_x\,dv_y\,dv_z = F(\mathbf{r},\mathbf{v})\,d^3r\,d^3v \qquad (2.11)$$

Equation (2.11) shows that the phase space distribution function, $F(\mathbf{r},\mathbf{v})$, is the density of particles in the $d^3r\,d^3v$ phase space volume element around the phase space point $(\mathbf{r},\mathbf{v})$.

$F(\mathbf{r},\mathbf{v})$ and $n(\mathbf{r})$ are closely related. The configuration space distribution function, $n(\mathbf{r})$, describes the number of particles in a small volume element, d^3r, around **r**, regardless of their velocity. This means that if we integrate the phase space distribution function, $F(\mathbf{r},\mathbf{v})$, over all possible velocity values, we must get the configuration space distribution function, $n(\mathbf{r})$:

$$n(\mathbf{r}) = \int_{-\infty}^{\infty} dv_x \int_{-\infty}^{\infty} dv_y \int_{-\infty}^{\infty} dv_z\, F(\mathbf{r},\mathbf{v}) = \iiint_{\infty} d^3v\, F(\mathbf{r},\mathbf{v}) \qquad (2.12)$$

Equation (2.12) shows that the phase space distribution function, $F(\mathbf{r},\mathbf{v})$, is normalized to the particle number density, n. One can also introduce the normalized phase space distribution function, $f(\mathbf{r},\mathbf{v})$:

$$f(\mathbf{r}, \mathbf{v}) = \frac{F(\mathbf{r}, \mathbf{v})}{n(\mathbf{r})} \qquad (2.13)$$

This definition means that $f(\mathbf{r},\mathbf{v})d^3rd^3v$ is the fraction of all particles in the spatial volume element d^3r around $\mathbf{r}$ which have their velocity vectors between $\mathbf{v}$ and $\mathbf{v}+d^3v$. In other words $f(\mathbf{r},\mathbf{v})d^3v$ is the conditional probability of finding a particle in the velocity space volume element d^3v around $\mathbf{v}$, provided that the particle is located in the configuration space volume element d^3r around $\mathbf{r}$. The normalized phase space distribution function, $f(\mathbf{r},\mathbf{v})$, is the probability density of finding a particle at the velocity space point $\mathbf{v}$ at the configuration space location $\mathbf{r}$. It is easy to see that $f(\mathbf{r},\mathbf{v})$ is normalized to 1:

$$\iiint_{\infty} d^3v \ f(\mathbf{r}, \ \mathbf{v}) = \frac{1}{n(\mathbf{r})} \iiint_{\infty} d^3vF(\mathbf{r},\mathbf{v}) = 1 \qquad (2.14)$$

2.2.3 Macroscopic averages

In gaskinetic theory macroscopic physical quantities usually characterize the gas at a given location. Such quantities are the concentration of molecules, mass density, temperature, pressure, flow velocity, etc. These macroscopic quantities might vary with time or configuration space location, but they do not reflect the distribution of molecular velocities at the given time and location. These macroscopic parameters are averages of quantities that depend on molecular velocities. These velocity space averages can be defined in a way which is analogous to the method we used to calculate the average weight of cars.

Let $Q(\mathbf{v})$ be any molecular quantity that is only a function of the velocity of the given molecule. Examples of such quantities include the molecular mass (which is independent of the molecular velocity in our non-relativistic approximation), momentum, or energy. The average value of Q for molecules in a macroscopically infinitesimal configuration space volume element, d^3r, located at position $\mathbf{r}$ is given by

$$\overline{Q}(\mathbf{r}) = <Q> = \iiint_{\infty} d^3v \ Q(\mathbf{v}) \ f(\mathbf{r},\mathbf{v}) \qquad (2.15)$$

This definition is analogous to equation (2.7), deduced for the average weight of cars in our simple example.

In order to demonstrate the use of this definition let us consider the following example. Assume that the gas is uniformly distributed in a container of volume $\mathcal{V}$, that there are N molecules in the container and all molecules are alike and have mass m. In this case the gas concentration is uniformly $n=N/\mathcal{V}$ everywhere in the container. Let us also assume that the distribution of molecular velocities is the same at every location inside the container, and that this velocity distribution can be described as follows. The velocity components of each molecule (v_x, v_y, v_z) are independent of each other and they can assume any value between $-v_0$ and $+v_0$ with equal probability. There are no particles with velocity components outside of the $(-v_0,+v_0)$ range. Mathematically such a velocity distribution can be described as

$$F(\mathbf{v}) = G(v_x)G(v_y)G(v_z) \tag{2.16}$$

since v_x, v_y, and v_z are independent. In expression (2.16) the function, G, is defined as

$$G(v_i) = A\,H(v_0 - v_i)H(v_0 + v_i) \tag{2.17}$$

where A is a constant value and H(x) is Heaviside's step function (H(x)=1 if x>0 and H(x)=0 if x<0, see Appendix 9.5.2). This distribution is shown in Figure 2.3 (the Figure shows dimensionless coordinates v_x/v_0, v_y/v_0, and v_z/v_0).

The value of A can be obtained from the normalization of the distribution function:

$$\iiint_\infty d^3v F(\mathbf{v}) = \frac{N}{\mathcal{V}} \tag{2.18}$$

Substituting the distribution function given by (2.16) into this equation yields the following relation:

$$\int_{-v_0}^{v_0} dv_x A \int_{-v_0}^{v_0} dv_y A \int_{-v_0}^{v_0} dv_z A = 8v_0^3 A^3 = \frac{N}{\mathcal{V}} \tag{2.19}$$

This means that $A^3=N/(8\mathcal{V}v_0^3)$. Using this value of the normalization constant one can readily derive the normalized phase space distribution function:

$$f(\mathbf{v}) = \frac{H(v_0 - v_x)H(v_0 + v_x)H(v_0 - v_y)H(v_0 + v_y)H(v_0 - v_z)H(v_0 + v_z)}{8v_0^3}$$

$$\tag{2.20}$$

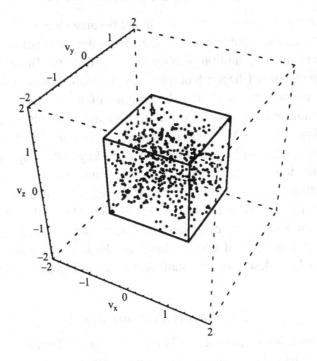

Figure 2.3

Using the normalized phase space distribution function we can calculate averaged molecular quantities for the gas in the container. For instance, one can use equation (2.15) to calculate the average translational energy of molecules:

$$< E_{kin} >= \iiint_{\infty} d^3v \tfrac{1}{2} mv^2 f(\mathbf{v}) = \frac{m}{16v_0^3} \int_{-v_0}^{v_0} dv_x \int_{-v_0}^{v_0} dv_y \int_{-v_0}^{v_0} dv_z (v_x^2 + v_y^2 + v_z^2) = \frac{mv_0^2}{2}$$

$$(2.21)$$

2.3 The Maxwell–Boltzmann distribution

In this chapter we follow James Clerk Maxwell's original derivation of the velocity distribution function[1] which was carried out in 1859. In retrospect we know that Maxwell's original derivation is only valid for gases in equilibrium (so is his second attempt to derive the distribution function[2]). In the 1859 paper, however, he used his distribution of molecular velocities to obtain transport coefficients for spatially non-uniform gases.[1] It can be shown that the mathematical assumptions Maxwell used in the derivation of the distribution function exclude all spatial nonuniformity in the gas (however, this fact did not became clear until much later). In 1877 Ludwig Boltzmann published a more rigorous derivation of the same distribution function using the condition of dynamical equilibrium[3]. Boltzmann's derivation will be presented in Chapter 5 of this book. In appreciation of Maxwell's and Boltzmann's fundamental work the distribution of molecular velocities under equilibrium conditions is called the Maxwell–Boltzmann distribution. This section presents a derivation of the Maxwell–Boltzmann distribution function.

Maxwell's velocity distribution function was the key to developing connections between the motions of microscopic molecules and macroscopic, measurable physical quantities. In the middle of the 19th century statistical methods had already been widely used in physical sciences, and Maxwell's distribution law turned out to have a mathematically identical form to the normal distribution introduced earlier by Laplace and Gauss. However, the idea that physical processes themselves should be described by a statistical function was a novel one.

2.3.1 *Maxwell's assumptions*

In this section we follow the outline of Maxwell's original derivation of the mathematical form of the velocity distribution function.[1] Maxwell's basic assumptions could be summarized as follows.

(i)   The gas is composed of an indefinite number of small, hard, and perfectly elastic spheres acting on one another only during impact.

(ii) When two spheres collide, their direction after the collision is distributed with equal probability over all solid angles. This means that the probability of finding a particle moving within a solid angle element $d\Omega$ after its last collision can be expressed as $d\Omega/4\pi$.

(iii) At any spatial location the distribution of molecular velocities is independent of time.

(iv) After a large number of collisions the orthogonal components of the molecular velocities are statistically independent. In Cartesian coordinates the probability of a molecule having a velocity component in the v_x direction with a magnitude between v_x and v_x+dv_x is $f_x(\mathbf{r},v_x)dv_x$. Similarly, the probability that the same molecule has a velocity component in the v_y direction with a magnitude between v_y and v_y+dv_y is $f_y(\mathbf{r},v_y)dv_y$. Finally, the probability that the same molecule has a velocity component in the v_z direction with a magnitude between v_z and v_z+dv_z is $f_z(\mathbf{r},v_z)dv_z$. Thus, the average number of particles to be found with velocities between $\mathbf{v}$ and $\mathbf{v}+d^3v$ is given by

$$d^3N = n\,f_x(\mathbf{r},v_x)f_y(\mathbf{r},v_y)f_z(\mathbf{r},v_z)dv_x\,dv_y\,dv_z \qquad (2.22)$$

where n is the particle concentration.

(v)   The distribution function is isotropic. This means that there is no preferred direction and that the distribution function is independent of the orientation of the coordinate system. Mathematically this assumption means that the normalized distribution function of molecular velocities, $f(\mathbf{r},\mathbf{v})$, must be constant on all v^2=constant surfaces, or $f(\mathbf{r},\mathbf{v})=f(\mathbf{r},v^2)$. This condition can also be written as

$$df(\mathbf{r},\mathbf{v}) = 0 \qquad (2.23)$$

on a surface where $(v_x^2+v_y^2+v_z^2)$=constant.

It should be noted that assumption (v) is only valid for spatially homogeneous (i.e. constant density) gases where the distribution of molecular velocities does not change with location. This point can be illustrated with a very simple example. Let us consider a one dimensional gas composed of molecules with mass m moving with a given speed, u, in either the +x or the –x direction. When two of these molecules collide their velocities reverse, but their speeds remain the same. In this one dimensional example local isotropy means that at that particular location the average numbers of molecules moving in the +x and –x directions are the same. Figure 2.4 shows three nearby locations along the x axis, so that the distances between points A, B and C are much smaller than the average distance traveled by the molecules between two consecutive collisions. At a given location the fluxes of molecules moving in the +x and –x directions are given by $uN^+(x)$ and $uN^-(x)$, respectively. Here $N^{\pm}(x)$ is the number density of the left or right moving population. Isotropy at A and C means that $N^+(A)=N^-(A)$ and $N^+(C)=N^-(C)$.

The distances AB and BC are very short compared to the average distance between collisions, therefore there is very little change in the flux moving left between points C and B. Similarly, there is very little change in the flux moving right between points A and B. This can be summarized as $N^+(B)=N^+(A)$ and $N^-(B)=N^-(C)$. Isotropy at point B requires that the condition, $N^+(B)=N^-(C)$, must be satisfied. Isotropy at all three points requires $N=N^+(A)=N^-(A)$ and $N=N^+(C)=N^-(C)$.

This simple example shows that velocity space isotropy is physically equivalent to requiring spatial homogeneity in the macroscopic gas parameters. This means that in gases that satisfy Maxwell's assumptions the macroscopic parameters (interpreted as velocity space statistical averages over macroscopically infinitesimal samples of molecules) do not change with time or spatial location. Our kine-

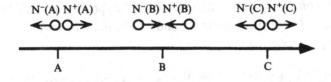

Figure 2.4

matic definition of equilibrium is based on this result: in equilibrium the distribution of molecular velocities is independent of time and configuration space location, consequently $f=f(\mathbf{v})$. Later in this book we shall present a dynamical definition of equilibrium. It should be mentioned that particle transport is intimately related to spatial variations of macroscopic quantities, therefore strictly speaking Maxwell's assumptions exclude all transport phenomena. This paradox will be resolved later in this book.

2.3.2 *The distribution of molecular velocities under equilibrium conditions*

Equations (2.22) and (2.23) represent a well defined mathematical problem which can be solved. Here we present a solution by using the method of Lagrange multipliers. First let us calculate the change of the distribution function, f, when v_x, v_y and v_z are changed by dv_x, dv_y, and dv_z, respectively:

$$df = \frac{\partial f}{\partial v_x} dv_x + \frac{\partial f}{\partial v_y} dv_y + \frac{\partial f}{\partial v_z} dv_z \qquad (2.24)$$

In the next step we make use of the fact that the velocity components are statistically independent and therefore the normalized velocity distribution function can be written as $f(\mathbf{v})=f_x(v_x)f_y(v_y)f_z(v_z)$:

$$df = f_x(v_x)f_y(v_y)f_z(v_z)\left(\frac{1}{f_x} \frac{df_x}{dv_x} dv_x + \frac{1}{f_y} \frac{df_y}{dv_y} dv_y + \frac{1}{f_z} \frac{df_z}{dv_z} dv_z \right) \quad (2.25)$$

The condition for the existence of a non-trivial solution of equation (2.25) is that the terms in the parenthesis should cancel, i.e.

$$\frac{1}{f_x} \frac{df_x}{dv_x} dv_x + \frac{1}{f_y} \frac{df_y}{dv_y} dv_y + \frac{1}{f_z} \frac{df_z}{dv_z} dv_z = 0 \qquad (2.26)$$

Next we recall that according to our physical considerations equation (2.26) must be satisfied on the surface of all velocity space spheres centered at $\mathbf{v}=0$. This means that both velocity space points, $\mathbf{v}$ and $\mathbf{v}+d^3v$, must be on the surface of the same velocity space sphere, i.e.

$$v_x^2 + v_y^2 + v_z^2 = (v_x + dv_x)^2 + (v_y + dv_y)^2 + (v_z + dv_z)^2 \qquad (2.27)$$

Neglecting the second order terms in the infinitesimal quantities, dv_x, dv_y, and dv_z, equation (2.27) can also be written in a much simpler form:

$$v_x \, dv_x + v_y \, dv_y + v_z \, dv_z = 0 \tag{2.28}$$

This result means that on the surface of any given velocity space sphere centered at v=0 the velocity differentials are not independent. They cannot be given any arbitrary values, but must satisfy equation (2.28), a so-called condition equation. Equation (2.26) has to be solved while taking into account the condition equation, (2.28).

Using a method invented by Lagrange, known as the method of undetermined multipliers, one can combine equation (2.26) and the condition equation, (2.28), and obtain a single equation in which dv_x, dv_y, and dv_z can be considered to be independent. Let us multiply the condition equation, (2.28), by a constant, 2β, whose value we shall determine later (β is the undetermined Lagrange multiplier) and add the resulting equation to equation (2.26). The result is the following:

$$\left(\frac{1}{f_x} \frac{df_x}{dv_x} + 2\beta v_x \right) dv_x + \left(\frac{1}{f_y} \frac{df_y}{dv_y} + 2\beta v_y \right) dv_y + \left(\frac{1}{f_z} \frac{df_z}{dv_z} + 2\beta v_z \right) dv_z = 0$$

$$\tag{2.29}$$

In this equation all three infinitesimal quantities are independent, so the condition for non-trivial solution is that the multipliers of dv_x, dv_y, and dv_z must be identically zero. This means that

$$d\left[\ln(f_x)\right] = -2\beta v_x \, dv_x \quad \Rightarrow \quad f_x(v_x) = A_x \, e^{-\beta v_x^2} \tag{2.30a}$$

$$d\left[\ln(f_y)\right] = -2\beta v_y \, dv_y \quad \Rightarrow \quad f_y(v_y) = A_y \, e^{-\beta v_y^2} \tag{2.30b}$$

$$d\left[\ln(f_z)\right] = -2\beta v_z \, dv_z \quad \Rightarrow \quad f_z(v_z) = A_z \, e^{-\beta v_z^2} \tag{2.30c}$$

These solutions can be combined to obtain the mathematical form of the distribution function:

$$f(\mathbf{v}) = A_x A_y A_z \, e^{-\beta\left(v_x^2 + v_y^2 + v_z^2\right)} = A \, e^{-\beta v^2} \tag{2.31}$$

where $A = A_x A_y A_z$. Equation (2.31) is the Maxwell–Boltzmann distribution which is one of the fundamental tools of gaskinetic theory.

The distribution function that we have just derived has two unknown constants, A and β. Next we find an expression for the normalization constant, A. The determination of the other parameter, β, is more complicated and it involves some considerations of basic thermodynamics. For the time being we treat β as an unknown constant and will determine its physical meaning later.

First we recall that the normalized velocity distribution function, $f(\mathbf{v})$, can also be interpreted as the probability density of finding a randomly selected particle with velocity $\mathbf{v}$. We know that probability density functions must be normalized to 1, therefore

$$\iiint_\infty d^3v \, f(\mathbf{v}) = 1 \tag{2.32}$$

The Maxwell–Boltzmann distribution function is spherically symmetric (it depends only on the magnitude of the velocity, v), therefore equation (2.32) can most easily be evaluated using spherical coordinates:

$$\iiint_\infty d^3v f(\mathbf{v}) = A \int_0^\infty dv \, v^2 \, e^{-\beta v^2} \int_0^\pi d\Theta \, \sin\Theta \int_0^{2\pi} d\varphi = 4\pi \, A \int_0^\infty dv \, v^2 \, e^{-\beta v^2} \tag{2.33}$$

The remaining definite integral can be integrated by parts:

$$\int_0^\infty dv \, v^2 e^{-\beta v^2} = -\left[\frac{v}{2\beta} \, e^{-\beta v^2} \right]_0^\infty + \frac{1}{2\beta} \int_0^\infty dv \, e^{-\beta v^2} = \frac{1}{4\beta} \sqrt{\frac{\pi}{\beta}} \tag{2.34}$$

Substituting expressions (2.33) and (2.34) into the normalization relation given by equation (2.32) yields the following value for the normalization constant, A:

$$A = \left(\frac{\beta}{\pi} \right)^{3/2} \tag{2.35}$$

This value for the normalization constant results in the following form of the normalized distribution of molecular velocities under equilibrium conditions:

$$f(\mathbf{v}) = \left(\frac{\beta}{\pi} \right)^{3/2} e^{-\beta v^2} \tag{2.36}$$

The unknown parameter, β, has the dimensions of $1/v^2$. The distribution given by expression (2.36) has the mathematical form of the well known normal distribution (or Gaussian distribution) with a half width of $(2/\beta)^{1/2}$. Figure 2.5 shows a one dimensional cut (along the v_x axis) of the dimensionless form of the normalized Maxwell–Boltzmann distribution function, f/f_{max}, in terms of the dimensionless variable $\beta^{1/2}v_x$ (for $v_y=0$ and $v_z=0$ values). Inspection of Figure 2.5 reveals that this distribution is symmetric around its maximum value at $v_x=0$ with an exponential decrease in both directions.

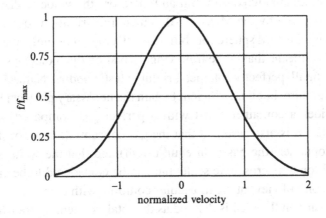

Figure 2.5

Next we examine the physical interpretation of the remaining free parameter, β. This can be obtained by calculating the average translational kinetic energy of the molecules in a gas in equilibrium. The average translational kinetic energy of the molecules can be obtained by

$$< E_{kin} >= \iiint_{\infty} d^3v \, \tfrac{1}{2} mv^2 f(\mathbf{v}) = 2\pi m \left(\frac{\beta}{\pi}\right)^{3/2} \int_0^\infty dv \, v^4 e^{-\beta v^2} = \frac{3m}{4\beta} \qquad (2.37)$$

where m is the molecular mass. Equation (2.37) shows that β is directly related to the average kinetic energy of the molecules. In the next section we examine how the average kinetic energy of the molecules can be expressed in terms of macroscopic parameters.

2.4 Determination of the Lagrange multiplier

In this section we will relate the unknown Lagrange multiplier, β, to known thermodynamic quantities. This will be done in two steps. First we express the gas pressure on a wall with the help of the newly derived Maxwell–Boltzmann distribution of molecular velocities. In the second step we express the same pressure with the help of the thermodynamic equation of state for perfect gases. Finally, the two results will be compared and thus an expression will be obtained for the β parameter.

2.4.1 Gas pressure on a wall

We start with the kinetic determination of the gas pressure on a wall. It is assumed that the gas is in kinematic equilibrium and therefore the velocity distribution of gas molecules is given by the Maxwell–Boltzmann distribution function. For the sake of simplicity the hard sphere (or 'billiard ball') model of molecular collisions will be used. This means that in the following derivation the molecules are considered to be very small, perfectly spherical, rigidly elastic bodies. Naturally this is an oversimplified picture, but it is sufficient to gain the necessary physical insight.

Let us consider a container filled with a perfect gas composed of our hard sphere molecules. It is also assumed that the gas concentration is constant everywhere in the container (the gas is in equilibrium) and that the walls of the container are rough on a macroscopic scale, but can be considered to be smooth surfaces from the point of view of the molecules colliding with the wall.

The gas pressure on the wall is defined as the total momentum transferred from the gas molecules to a unit wall area in unit time. In what follows we will calculate this physical quantity.

We first consider a small, locally planar surface area, dS. We define a coordinate system with the z axis normal to this surface element and pointing inward to the container (see Figure 2.6). The mass of the wall is so much larger than the mass of the individual molecules that the colliding hard sphere molecules are reflected elastically from the surface. In this elastic collision the perpendicular velocity component (perpendicular to the surface element) of the reflected particle is reversed, while the parallel component remains unchanged (see Figure 2.6).

One can readily calculate the change of the particle velocity vector due to the collision with the wall. If the Cartesian velocity components of a molecule before the collision with the wall were $\mathbf{v}=(v_x, v_y, -v_z)$, then after the collision the velocity can be expressed as $\mathbf{v}'=(v_x, v_y, v_z)$. The change of the velocity vector due to the collision is

$$\Delta \mathbf{v} = \mathbf{v}' - \mathbf{v} = (0, 0, 2v_z) \qquad (2.38)$$

In the next step we calculate the number of particles with a given velocity vector which collide with the wall surface element in unit time and change their momentum. Consider the group of molecules with velocity vectors in an infinitesi-

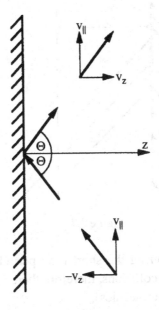

Figure 2.6

mal neighborhood of **v** which hit the surface element, dS, during the infinitesimal time interval, dt. These molecules would all be contained at the beginning of the infinitesimal time interval, dt, in an oblique cylinder with base dS and altitude $v_z$dt (see Figure 2.7 for a schematic representation of this situation). The volume of this 'impact' cylinder is $v_z$dt dS (called impact cylinder because it contains all particles with velocities between **v** and **v**+d^3v which impact the surface element dS during the time interval dt). At the same time we know that the number of molecules per unit configuration space volume in an infinitesimal velocity space volume element, d^3v, around velocity **v** is nf(**v**)d^3v. Combining these two results one can obtain the number of particles with initial velocity between **v** and **v**+d^3v which hit dS in dt time:

$$d^6N = n f(\mathbf{v}) v_z d^3v \, dt \, dS \qquad (2.39)$$

The momentum transferred to the wall surface element, dS, during the time interval, dt, can be calculated by multiplying d^6N by the momentum change of particles with pre-collision velocities, **v**, and integrating over the entire velocity space. However, it is obvious from Figure 2.7 that only molecules with $v_z<0$ are able to hit the wall (molecules with $v_z>0$ are moving away from the wall), therefore we only integrate over these velocities. With these considerations one obtains the following expression for the perpendicular momentum component transferred

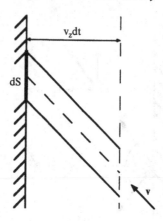

Figure 2.7

to the surface element, dS, during the short time period, dt (the present model takes into account only elastic collisions, therefore there is no change in the parallel momentum of the colliding molecules):

$$d^3P_z = \int_{-\infty}^{\infty} dv_x \int_{-\infty}^{\infty} dv_y \int_{-\infty}^{0} dv_z (2mv_z) nv_z f(\mathbf{v}) dS \, dt \qquad (2.40)$$

Under equilibrium conditions the distribution of molecular velocities is given by the Maxwell-Boltzmann distribution function. Substituting expression (2.36) into equation (2.40) yields the following:

$$d^3P_z = 2mn \left[\sqrt{\frac{\beta}{\pi}} \int_{-\infty}^{\infty} dv_x e^{-\beta v_x^2} \right] \left[\sqrt{\frac{\beta}{\pi}} \int_{-\infty}^{\infty} dv_y e^{-\beta v_y^2} \right] \left[\sqrt{\frac{\beta}{\pi}} \int_{-\infty}^{0} dv_z v_z^2 e^{-\beta v_z^2} \right] dS \, dt$$

$$(2.41)$$

It can be easily seen that the values of the first and second brackets are 1 (because of the normalization of the distribution function), while the definite integral in the third bracket gives $1/4\beta$. Using these values equation (2.41) becomes the following:

$$d^3P_z = \frac{mn}{2\beta} \, dS \, dt \qquad (2.42)$$

At this point we must recall that the gas pressure is defined as the total perpendicular momentum transferred from the gas molecules to unit wall area in unit time. This means that the pressure can be written as $p = d^3P_z/(dS \, dt)$, therefore we obtain the following expression for the gas pressure:

$$p = \frac{mn}{2\beta} \qquad (2.43)$$

Equation (2.43) expresses the gas pressure in terms of the molecular mass density, mn, and the still unknown Lagrange multiplier, β.

2.4.2 Equation of state

Next we want to express the gas pressure on the container's wall in terms of thermodynamic quantities. We start from Boyle's law for perfect gases:

$$p\mathcal{V} = \nu R T \qquad (2.44)$$

where $\mathcal{V}$ is the volume of the container, R is the universal gas constant (R=8.3145 joule mole^{-1} K^{-1}), ν is the number of moles in the container (1 mole is the amount of gas which weighs m/m_0 grams, where m_0 is the atomic mass unit) and T is the gas temperature. It should be emphasized that this is an experimental law which has fundamental significance. It relates universal constants to theoretical quantities introduced by gaskinetics. One can combine equations (2.43) and (2.44) to obtain an expression for the Lagrange multiplier:

$$\beta = \frac{mn\mathcal{V}}{2\nu R T} \qquad (2.45)$$

On the other hand we know that the number of molecules in a mole of gas is always the same. This universal constant, N_A, is called Avogadro's number and its value is $N_A = 6.0221 \times 10^{23}$ molecules. The number of moles in the container can also be expressed as $\nu = n\mathcal{V}/N_A$. With the help of this relation equation (2.45) can be rewritten as

$$\beta = \frac{m}{2kT} \qquad (2.46)$$

where we have denoted the ratio of the universal gas constant and Avogadro's number by k. This ratio, $k = R/N_A$, is also a universal constant. This quantity was first introduced by Ludwig Boltzmann and it is called Boltzmann's constant ($k = 1.3807 \times 10^{-23}$ jouleK^{-1}).

Equation (2.46) is our desired relation between the unknown Lagrange multiplier and macroscopic gas properties. The final form of the Maxwell–Boltzmann distribution is obtained by substituting the recently derived value of β into expression (2.36):

$$f(\mathbf{v}) = \left(\frac{m}{2\pi kT}\right)^{3/2} e^{-\frac{mv^2}{2kT}} \tag{2.47}$$

Finally, one can substitute the newly derived expression for the Lagrange multiplier into equation (2.43) to obtain the following equation of state for perfect gases in kinematic equilibrium:

$$p = nkT \tag{2.48}$$

2.5 Elementary properties of the Maxwell–Boltzmann distribution

2.5.1 Distribution of molecular speeds

In some problems we are not interested in the distribution of molecular velocity vectors. Under certain symmetry conditions the distribution of molecular speeds (the magnitude of molecular velocity is called speed) gives a perfectly adequate description of the gas.

Let $f^*(v)dv$ denote the probability of finding a particle with speed between v and $v+dv$. Here f^* is the distribution function of molecular speeds. One can readily find an expression for f^* if the distribution of molecular velocity vectors is given by the Maxwell–Boltzmann distribution function, $f(\mathbf{v})$. We start by using the fact that f^* is a probability density function, therefore it is normalized to unity:

$$\int_0^\infty dv\, f^*(v) = 1 \tag{2.49}$$

The Maxwell–Boltzmann distribution function of molecular velocity vectors not only contains statistical information about the distribution of molecular speeds, but also specifies the probability that a given particle with speed v moves in the direction defined by the infinitesimal solid angle element, $d\Omega = \sin\theta\, d\theta\, d\phi$. Using velocity space spherical coordinates one can write the normalization condition of the Maxwell–Boltzmann distribution in the following form:

$$\iiint_\infty d^3v \left(\frac{m}{2\pi kT}\right)^{3/2} e^{-\frac{mv^2}{2kT}} = \int_0^{2\pi} d\phi \int_0^{\pi} d\theta \sin\theta \int_0^\infty dv\, v^2 \left(\frac{m}{2\pi kT}\right)^{3/2} e^{-\frac{mv^2}{2kT}}$$

$$= 4\pi \int_0^\infty dv\, v^2 \left(\frac{m}{2\pi kT}\right)^{3/2} e^{-\frac{mv^2}{2kT}} = 1 \tag{2.50}$$

The last remaining integral in equation (2.50) can be compared with expression (2.49). This comparison immediately yields the following functional form for the speed distribution of molecules under equilibrium conditions:

$$f^*(v) = 4\pi v^2 \left(\frac{m}{2\pi\, kT}\right)^{3/2} e^{-\frac{mv^2}{2kT}} \tag{2.51}$$

The function f^* is plotted in Figure 2.8. Inspection of Figure 2.8 reveals that the speed distribution function starts from $f^*=0$ at zero speed, reaches its only maximum at a speed v_m and then decreases with increasing speed. The most probable speed, v_m, can be obtained by recognizing that f^* has its only extremum at this point, therefore the condition $df^*(v_m)/dv=0$ must be satisfied. This means that

$$\left(\frac{2}{v_m} - \frac{mv_m}{kT}\right) f^*(v_m) = 0 \tag{2.52}$$

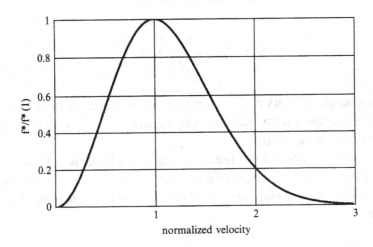

Figure 2.8

Equation (2.52) yields the solution $v_m=(2kT/m)^{1/2}$ for the most probable particle speed (the maximum of f^*). As we shall see later the most probable particle speed is different than the average particle speed.

2.5.2 Mixture of gases

So far we have considered only the simplest case when all molecules in the container were identical. When the gas is in equilibrium and more than one molecular species occupies the same volume (each species is characterized by a different type of molecule), there is no reason for the molecular velocities to depart from the Maxwell–Boltzmann distribution. As we discussed before, under kinematic equilibrium conditions all macroscopic parameters are time independent and homogeneous (there are no spatial gradients). One specific macroscopic property of the gas is its temperature. A natural consequence of the kinematic equilibrium condition is that the temperatures of all gas components are independent of time and location and that the temperatures of all gas components are the same.

One of the fundamental assumptions of gaskinetic theory is that the spheres of influence of the individual molecules are much smaller than the average intermolecular distance. This assumption means that the presence of more than one molecular species only negligibly reduces the volume available for a given gas component, consequently the concentrations of the individual species remain unmodified. The gas mixture is in equilibrium, therefore the velocity distribution of each molecular component is a Maxwell–Boltzmann velocity distribution with a common temperature, T. The total (unnormalized) velocity distribution function of the gas mixture can be written in the following form:

$$F(\mathbf{v}) = \sum_s n_s \left(\frac{m_s}{2\pi kT} \right)^{3/2} e^{-\frac{m_s v^2}{2kT}} \qquad (2.53)$$

Here the subscript, s, refers to the different molecular species present in the gas mixture. The distribution function given by equation (2.53) is normalized to the total gas concentration, $n = \Sigma n_s$.

It is interesting to note that this result can readily yield the well known Dalton's law (which states that the pressure of a gas mixture is the sum of partial pressures of the gas components). Using equation (2.48) the total pressure of the gas mixture can be written

$$p = nkT = \sum_s n_s kT = \sum_s p_s \qquad (2.54)$$

Equation (2.54) is the well known Dalton's law. It should be emphasized again that we were able to explain this empirical law by using first principles and the basic assumptions of equilibrium gaskinetic theory.

The normalized distribution function of molecular velocities in the gas mixture is obtained by dividing the total distribution function (given by equation (2.53)) by the total particle concentration:

$$f(\mathbf{v}) = \sum_s \frac{n_s}{n} \left(\frac{m_s}{2\pi kT} \right)^{3/2} e^{-\frac{m_s v^2}{2kT}} \tag{2.55}$$

2.5.3 Moments of the Maxwell–Boltzmann distribution

Next we examine some of the fundamental macroscopic properties of the Maxwell–Boltzmann function which describes the equilibrium distribution of molecular velocities in a gas containing a single molecular species. The results to be determined below can readily be extended to multispecies gases with the help of the distribution function discussed in Subsection 2.5.2.

Let us start by calculating the average molecular velocity vector, $\mathbf{u}$. According to our earlier results the average of the molecular velocity vector, $\mathbf{v}$, is given by the following integral (see equation (2.8)):

$$<\mathbf{v}> = \mathbf{u} = \iiint_\infty d^3 v \, \mathbf{v} \, f(\mathbf{v}) \tag{2.56}$$

Equation (2.56) is a vector equation with three components. Let us first examine one of the components in detail. The other components can be obtained in an analogous manner. The v_x component of the average molecular velocity is the following:

$$u_x = \left(\sqrt{\frac{\beta}{\pi}} \int_{-\infty}^{\infty} dv_x \, v_x \, e^{-\beta v_x^2} \right) \left(\sqrt{\frac{\beta}{\pi}} \int_{-\infty}^{\infty} dv_y \, e^{-\beta v_y^2} \right) \left(\sqrt{\frac{\beta}{\pi}} \int_{-\infty}^{\infty} dv_z \, e^{-\beta v_z^2} \right) \tag{2.57}$$

The values of the second and the third integrals are 1 (remember, this is the way we determined the normalization of the Maxwell–Boltzmann distribution), therefore equation (2.57) simplifies to

$$u_x = \sqrt{\frac{\beta}{\pi}} \int_{-\infty}^{\infty} dv_x \, v_x \, e^{-\beta v_x^2} \tag{2.58}$$

There is an important remark to be made in connection with equation (2.58). Equation (2.58) can be solved immediately by examining the nature of the integrand. The results can be used to trivially evaluate a number of molecular integrals.

Let us consider an arbitrary odd function of x, g(x). We know that odd functions satisfy the following relation:

$$g(x) = -g(-x) \qquad\qquad (2.59)$$

In other words, if we replace the argument of an odd function with its negative, the value of the function also changes its sign. Some simple examples for odd functions are $g(x)=x$ or $g(x)=\sin x$.

Even functions retain the same value when the argument is replaced by its negative:

$$h(x) = h(-x) \qquad\qquad (2.60)$$

Next we take the integrals of an even function, $h(x)$ and an odd function, $g(x)$, from $-\infty$ to $+\infty$. Let us start with the even function, $h(x)$. In this case we get

$$\int_{-\infty}^{\infty} dx\, h(x) = \int_{-\infty}^{0} dx\, h(x) + \int_{0}^{\infty} dx\, h(x) = -\int_{0}^{-\infty} dx\, h(x) + \int_{0}^{\infty} dx\, h(x)$$

$$= -\int_{0}^{\infty} d(-x)\, h(-x) + \int_{0}^{\infty} dx\, h(x) = 2\int_{0}^{\infty} dx\, h(x) \qquad (2.61)$$

In the case when the integrand is an odd function, $g(x)$, the integration yields the following result:

$$\int_{-\infty}^{\infty} dx\, g(x) = \int_{-\infty}^{0} dx\, g(x) + \int_{0}^{\infty} dx\, g(x) = -\int_{0}^{-\infty} dx\, g(x) + \int_{0}^{\infty} dx\, g(x)$$

$$= -\int_{0}^{\infty} d(-x)\, g(-x) + \int_{0}^{\infty} dx\, g(x) = 0 \qquad (2.62)$$

This is a very important and frequently used result in kinetic theory. Therefore, if one takes the integral of an odd function on a symmetric domain around zero (for instance from $x=-a$ to $x=+a$) then the value of the integral is identically zero.

Let us apply our result to the remaining integral in equation (2.58). The integrand is an odd function of the integration variable, v_x, because the even function $\exp(-\beta v_x^2)$ is multiplied by an odd function, v_x. The integration boundaries are symmetric, ($-\infty$ and $+\infty$), therefore we conclude that the integral in equation (2.58) vanishes:

$$<v_x> = u_x = 0 \qquad\qquad (2.63)$$

The average values of the v_y and v_z velocity components can be calculated in a similar manner, yielding $<v_y>=0$ and $<v_z>=0$. These results can be summarized by the following vector equation:

$$<\mathbf{v}> = \mathbf{u} = 0 \tag{2.64}$$

This result tells us that a gas in equilibrium has no organized (bulk) motion in any direction (remember, we assumed that there was no preferred direction in the system, therefore we are absolutely free to choose the v_x axis to be in any direction). However, the particle speed (magnitude of velocity) is a positive semidefinite scalar quantity ($v \geq 0$), and its average value is not zero:

$$\bar{v} = <v> = \iiint_\infty d^3v\, v\, f^*(v) = 4\pi \left(\frac{\beta}{\pi}\right)^{3/2} \int_0^\infty dv\, v^3 e^{-\beta v^2} = \sqrt{\frac{8kT}{\pi m}} \tag{2.65}$$

In our non-relativistic case the same volume of gas can be observed from different coordinate systems. Equation (2.64) was obtained in a frame of reference where the container was stationary. The physical meaning of equation (2.64) is that under equilibrium conditions the gas as a whole does not move with respect to the container. However, the same volume of gas can also be described in a coordinate system moving with a constant non-relativistic velocity, $\mathbf{u}_0$, with respect to the container. In this case an elementary Galilean transformation has to be carried out, i.e. the particle velocity vector, $\mathbf{v}$, has to be replaced by the velocity relative to the container, $\mathbf{v}-\mathbf{u}_0$. In this case the Maxwell–Boltzmann distribution becomes the following:

$$f(\mathbf{v}) = \left(\frac{\beta}{\pi}\right)^{3/2} \exp\left\{-\beta\left[\left(v_x - u_{0x}\right)^2 + \left(v_y - u_{0y}\right)^2 + \left(v_z - u_{0z}\right)^2\right]\right\} \tag{2.66}$$

One can calculate the components of the average molecular velocity vector in this moving frame of reference using the so-called 'drifting' distribution function given by expression (2.66). In this case the v_x component of the average velocity is the following:

$$u_x = \left[\sqrt{\frac{\beta}{\pi}} \int_{-\infty}^\infty dv_x\, v_x\, e^{-\beta\left(v_x - u_{0x}\right)^2}\right]\left[\sqrt{\frac{\beta}{\pi}} \int_{-\infty}^\infty dv_y\, e^{-\beta\left(v_y - u_{0y}\right)^2}\right]\left[\sqrt{\frac{\beta}{\pi}} \int_{-\infty}^\infty dv_z\, e^{-\beta\left(v_z - u_{0z}\right)^2}\right]$$

$$\tag{2.67}$$

In order to evaluate expression (2.67) we introduce a new variable, $c = v - u_0$. With the help of this new variable the v_x component of the average molecular velocity can be written as

$$u_x = \left[\sqrt{\frac{\beta}{\pi}} \int_{-\infty}^{\infty} dc_x (c_x + u_{x0}) e^{-\beta c_x^2} \right] \left[\sqrt{\frac{\beta}{\pi}} \int_{-\infty}^{\infty} dc_y \, e^{-\beta c_y^2} \right] \left[\sqrt{\frac{\beta}{\pi}} \int_{-\infty}^{\infty} dc_z \, e^{-\beta c_z^2} \right]$$

$$= \sqrt{\frac{\beta}{\pi}} \int_{-\infty}^{\infty} dc_x (c_x + u_{x0}) e^{-\beta c_x^2} = \sqrt{\frac{\beta}{\pi}} \int_{-\infty}^{\infty} dc_x \, c_x \, e^{-\beta c_x^2} + u_{x0} \sqrt{\frac{\beta}{\pi}} \int_{-\infty}^{\infty} dc_x \, e^{-\beta c_x^2} = u_{x0}$$

$$(2.68)$$

One can calculate the values of the other components of the average molecular velocity vector. This exercise results in $u_y = u_{0y}$ and $u_z = u_{0z}$, so finally we conclude that in a frame of reference moving with velocity u_0 with respect to the container the average molecular velocity of a gas in equilibrium is

$$u = u_0 \qquad\qquad (2.69)$$

Finally let us calculate the mean translational kinetic energy of the molecules in the volume of gas (in the frame of reference of the container). This can be obtained by averaging the translational energy of a molecule, $E_{kin} = mv^2/2$:

$$< E_{kin} > = \frac{1}{2} m \left(\frac{\beta}{\pi} \right)^{3/2} \int_{-1}^{1} d\mu \int_{0}^{2\pi} d\varphi \int_{0}^{\infty} dv \, v^4 e^{-\beta v^2} = \frac{3}{4} \frac{m}{\beta} = \frac{3}{2} kT \qquad (2.70)$$

This is a very important result which will be extensively used later in this book. Equation (2.70) explicitly relates a macroscopic quantity, the gas temperature, to an average property of the molecules in the gas, namely the average translational energy of the molecules.

2.6 Specific heats of gases

The classical equilibrium gaskinetic theory is capable of predicting some fundamental thermodynamic properties of gases, including specific heats at constant volume, C_V, and at constant pressure, C_p. However, these predictions are accurate only for the simplest molecules. Classical gaskinetic theory clearly fails to predict the specific heats of gases composed of complicated molecules. In this case quantum mechanical effects are important and classical mechanics loses validity.

In this section we will outline the basic concepts behind the derivation of specific heats in equilibrium gases. In discussing the specific heats of diatomic molecules we will refer to some basic quantum mechanical concepts without going into specific details. The interested reader is referred to textbooks on atomic and molecular physics.[13]

2.6.1 Equipartition of translational energy

We start by calculating the average kinetic energy of the motion of the particle in the v_x, v_y and v_z directions. It is again assumed that the gas is in kinematic equilibrium and therefore the Maxwell–Boltzmann distribution function can be used to describe the distribution of molecular velocities. First let us calculate the average translational energy in the v_x direction:

$$\overline{E}_{kin,x} = \frac{1}{2}m<v_x^2> = \frac{1}{2}m\left[\sqrt{\frac{\beta}{\pi}}\int_{-\infty}^{\infty}dv_x v_x^2 e^{-\beta v_x^2}\right]\left[\sqrt{\frac{\beta}{\pi}}\int_{-\infty}^{\infty}dv_y e^{-\beta v_y^2}\right]\left[\sqrt{\frac{\beta}{\pi}}\int_{-\infty}^{\infty}dv_z e^{-\beta v_z^2}\right]$$

$$(2.71)$$

The second and third integrals in equation (2.71) yield 1, therefore the equation simplifies to the following expression:

$$\overline{E}_{kin,x} = \frac{m}{2}\sqrt{\frac{\beta}{\pi}}\int_{-\infty}^{\infty}dv_x v_x^2 e^{-\beta v_x^2} = \frac{m}{2}\sqrt{\frac{\beta}{\pi}}2\int_{0}^{\infty}dv_x v_x^2 e^{-\beta v_x^2} = \frac{m}{4\beta} = \frac{1}{2}kT \quad (2.72)$$

Similar derivations can be carried out for the mean energy in v_y and v_z directions as well. The results are

$$\overline{E}_{kin,y} = \frac{1}{2}m\left[\sqrt{\frac{\beta}{\pi}}\int_{-\infty}^{\infty}dv_x e^{-\beta v_x^2}\right]\left[\sqrt{\frac{\beta}{\pi}}\int_{-\infty}^{\infty}dv_y v_y^2 e^{-\beta v_y^2}\right]\left[\sqrt{\frac{\beta}{\pi}}\int_{-\infty}^{\infty}dv_z e^{-\beta v_z^2}\right] = \frac{1}{2}kT$$

$$(2.73)$$

$$\overline{E}_{kin,z} = \frac{1}{2}m\left[\sqrt{\frac{\beta}{\pi}}\int_{-\infty}^{\infty}dv_x e^{-\beta v_x^2}\right]\left[\sqrt{\frac{\beta}{\pi}}\int_{-\infty}^{\infty}dv_y e^{-\beta v_y^2}\right]\left[\sqrt{\frac{\beta}{\pi}}\int_{-\infty}^{\infty}dv_z v_z^2 e^{-\beta v_z^2}\right] = \frac{1}{2}kT$$

$$(2.74)$$

Equations (2.72) through (2.74) show that under equilibrium conditions the average translational energy is the same in the v_x, v_y and v_z directions. At the same time we recall that there are no preferred directions in the entire phase space when the gas is in equilibrium, consequently we conclude that the average translational energy is the same in every direction.

It can be easily seen that the sum of the average translational kinetic energies in the orthogonal directions, v_x, v_y and v_z, is the previously calculated mean translational energy (see equation (2.70)):

$$\overline{E}_{kin} = \overline{E}_{kin,x} + \overline{E}_{kin,y} + \overline{E}_{kin,z} \qquad (2.75)$$

The three orthogonal velocity space directions, v_x, v_y, and v_z, represent three translational degrees of freedom of the molecule. Our results have shown that a monatomic gas in equilibrium has translational energy which is equally shared between these three degrees of freedom. This result is called the equipartition of energy between the translational degrees of freedom.

This is a very interesting result that has important consequences for the thermodynamic properties of gases. First we examine the consequences of the equipartition of translational energy in the case of perfect gases which have no internal degrees of freedom.

2.6.2 Specific heats of monatomic gases

As indicated previously, perfect gases are assumed to be composed of monatomic molecules with no internal structure (and consequently they have no internal degrees of freedom). As mentioned above, the total mean translational kinetic energy of a monatomic molecule is $<E_{kin}>=3kT/2$. This mean energy is closely related to a macroscopic thermodynamic quantity, the internal energy of perfect gases. Here the word 'internal' refers to a macroscopically infinitesimal volume of gas containing a very large number of particles and not to a single molecule. The internal energy per unit mass of the perfect gas, e, can be expressed in the following form:

$$e = \frac{\overline{E}_{kin}}{m} = \frac{3}{2}\frac{kT}{m} = \frac{3}{2}\frac{kN_A}{mN_A}T = \frac{3}{2}\frac{R}{M}T \qquad (2.76)$$

where M is the mass of 1 mole of gas, $M = mN_A$.

The specific heat of a gas at constant volume, C_V, is the heat per unit mass which is needed to raise the gas temperature by 1 K, while the volume occupied by the gas is kept constant:

$$\delta Q|_v = C_v \, dT \tag{2.77}$$

Here $\delta Q|_v$ is the heat added per unit mass to raise the gas temperature by dT, while the volume occupied by the gas is kept constant.

Next we want to express $\delta Q|_v$ in terms of molecular quantities and compare the result with expression (2.77). This goal can be achieved by using equation (2.76) to calculate the change of internal energy per unit mass when the temperature of a monatomic perfect gas is increased by 1 K. In this process the volume of the container remains constant, therefore the change of internal energy is identical to $\delta Q|_v$:

$$\delta Q|_v = de = \frac{3}{2}\frac{k}{m}\, dT = \frac{3}{2}\frac{R}{M}\, dT \tag{2.78}$$

Comparing equations (2.77) and (2.78) yields the following expression for the specific heat at constant volume, C_v:

$$C_v = \frac{3}{2}\frac{k}{m} = \frac{3}{2}\frac{R}{M} \tag{2.79}$$

Expression (2.79) is a fundamentally important result, because it directly relates the specific heat of a perfect gas to molecular quantities. This result tells us that the specific heat of a perfect gas at constant volume depends only on the mass of the gas molecules and the universal constant, k.

Next we turn our attention to the derivation of the specific heat at constant pressure. It is well known in elementary thermodynamics that the gas has to be allowed to expand if the pressure is kept constant while the gas is heated. This means that some of the heat energy is used to expand the gas (to increase the volume) and only the rest of the heat energy is available to increase the internal energy of the gas:

$$\delta Q|_p = de + p\, d\vartheta|_p = C_p \, dT \tag{2.80}$$

In equation (2.80) $\vartheta = 1/\rho$ is the specific volume of the gas (ρ being the mass density, $\rho = mn$). From the kinetic equation of state one can readily obtain an expression for the product of the pressure and the specific volume:

$$p\vartheta = \frac{nkT}{mn} = \frac{k}{m}T \tag{2.81}$$

With the help of this definition one can express the work done by the expanding gas while the pressure is kept constant:

$$p\,d\vartheta\big|_p = p\frac{\partial}{\partial T}\left(\frac{kT}{mnkT}\right)\bigg|_p dT = \frac{k}{m}p\frac{\partial}{\partial T}\left(\frac{T}{p}\right)\bigg|_p dT = \frac{k}{m}dT \qquad (2.82)$$

Next we substitute this result into equation (2.80). This step leads to the following result:

$$\delta Q\big|_p = de + \frac{k}{m}\,dT = \frac{3}{2}\frac{k}{m}\,dT + \frac{k}{m}\,dT = \frac{5}{2}\frac{k}{m}\,dT = C_p\,dT \qquad (2.83)$$

Comparing the sides of equation (2.83) yields the final expression for the specific heat of monatomic gases at constant pressure:

$$C_p = \frac{5}{2}\frac{k}{m} \qquad (2.84)$$

Expression (2.84) directly relates the specific heat of a monatomic perfect gas to the mass of the molecules. Using equations (2.79) and (2.84) one can readily calculate the specific heat ratio for monatomic gases:

$$\gamma = \frac{C_p}{C_v} = \frac{5}{3} \qquad (2.85)$$

Equations (2.79), (2.84) and (2.85) represent a very important result. In classical thermodynamics the specific heats of various gases are empirical quantities which have to be determined experimentally. Here specific expressions were derived for these parameters from first principles using the basic assumptions of kinetic theory and assuming that the gas was in equilibrium.

The agreement of this prediction with room temperature measurements on all the noble gases is remarkably good. For instance, C_p/C_v=1.67 for argon and 1.64 for neon, and the variation with temperature is slight. Figure 2.9 shows the observed values of C_v for argon gas, which is close to the predicted value of 3/2.

On the other hand it should also be noted that the specific heats of polyatomic gases are not in agreement with the values derived above, indicating that the energy associated with internal degrees of freedom of the molecules cannot be neglected.

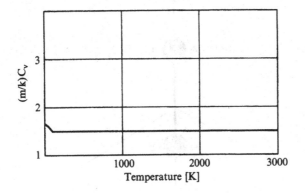

Figure 2.9

2.6.3 Specific heats of diatomic gases

In the previous section we were able to express the specific heats of perfect (therefore monatomic) gases with the help of the universal Boltzmann constant, k, and the mass of the molecules. In this section we examine the specific heats of the somewhat more complicated but still quite simple diatomic molecules. As can be seen from the schematics shown in Figures 2.10 and 2.11 these molecules are capable of storing energy in forms of motion other than the translational motion of the center of mass of the entire molecule.

In addition to its translational energy, part of the energy of a diatomic molecule is stored in the internal degrees of freedom of rotation and vibration. This is a different situation from that of the monatomic perfect gas in which only the three translational kinetic energies were considered. Monatomic molecules were considered to be point-like hard spheres without any internal degrees of freedom.

Now consider a diatomic molecule. A diatomic molecule possesses six degrees of freedom, meaning that six coordinates are needed to specify the positions of the two atoms in the molecule. One can choose three of these coordinates to specify the location of the center of mass of the diatomic molecule. The velocity components corresponding to these three coordinates specify the translational motion of the molecule as a whole, therefore they correspond to the translational degrees of freedom. The internal degrees of freedom are associated with the relative motion of the molecules.

The line joining the nuclei of the diatomic molecule is called the longitudinal axis of the molecule. The orientation of this longitudinal axis in the three dimensional space can be described by two independent coordinates. The simplest model of the diatomic molecule assumes that the distance between the two nuclei

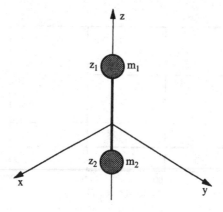

Figure 2.10

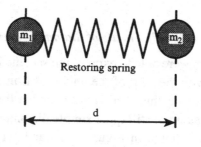

Figure 2.11

in the molecule remains constant at all times. This model (shown in Figure 2.10) is referred to as the 'rigid rotator' or 'dumbbell' model. In the case of a rigid rotator diatomic molecule the rotation of the molecule is fully described by the two independent coordinates (such as the polar and the azimuth angles) characterizing the orientation of the longitudinal axis of the molecule.

Since in gaskinetic theory the nuclei are treated as point masses, the energy associated with the rotation around the longitudinal axis is negligible (the moment of inertia with respect to the longitudinal axis is zero). Rotation around any two orthogonal axes in the plane perpendicular to the longitudinal axis (for instance around the x and y axes in Figure 2.10) is associated with finite energy.

The origin of the coordinate system shown in Figure 2.10 is at the center of mass of the molecule. The longitudinal axis is chosen to be in the z direction. Let z_1 and z_2 denote the distances from the center of mass to each nucleus. In the rigid rotator model the two nuclei are separated by a fixed distance, $d = z_1 + z_2$. The distances, z_1 and z_2 can be easily expressed in terms of the nuclear masses, m_1 and m_2, and the distance, d:

$$z_1 = \frac{m_2}{m_1 + m_2} d \qquad z_2 = \frac{m_1}{m_1 + m_2} d \qquad\qquad (2.86)$$

Using these expressions for the distances z_1 and z_2, one can easily obtain the moments of inertia with respect to the x, y, and z axes:

$$I_x = m_1 z_1^2 + m_2 z_2^2 = m^* d^2$$
$$I_y = m_1 z_1^2 + m_2 z_2^2 = m^* d^2 \qquad\qquad (2.87)$$
$$I_z = 0$$

where we have introduced the reduced mass of the molecule, $m^* = m_1 m_2/(m_1 + m_2)$. The total rotational energy of the molecule is

$$E_{rot} = \frac{1}{2} I_x \omega_x^2 + \frac{1}{2} I_y \omega_y^2 + \frac{1}{2} I_z \omega_z^2 = \frac{1}{2} m^* d^2 (\omega_x^2 + \omega_y^2) = \frac{1}{2} I (\omega_x^2 + \omega_y^2) \quad (2.88)$$

where ω_x, ω_y, and ω_z are the angular velocity components and $I = m^* d^2$ is the moment of inertia around any axis in the (x, y) plane.

This result means that the diatomic molecule has two rotational degrees of freedom. Such a molecule has a total of five degrees of freedom: three due to the translational motion of the center of mass and two due to the rotation of the longitudinal axis of the molecule.

Inspection of equation (2.88) reveals that the rotational energy of a rigid rotator molecule depends only on the two perpendicular (to the longitudinal axis) components of the angular velocity and it is independent of the location or velocity of the center of mass (in other words it is independent of the translational coordinates and velocities). It is also obvious that the rigid rotator model of diatomic molecules exhibits rotational symmetry around the longitudinal axis, therefore there is no preferred direction in the (x,y) plane.

Following Maxwell assumptions (made for the translational degrees of freedom) we assume that in kinematic equilibrium the angular velocity components are statistically independent of each other. This means that the normalized distribution function of angular velocities, $g(\omega_x, \omega_y)$, must be a separable function, i.e. $g(\omega_x, \omega_y) = g_x(\omega_x) g_y(\omega_y)$. Also, the rotational symmetry means that the distribution of angular velocities depends only on the magnitude of the perpendicular angular velocity, $\omega = (\omega_x^2 + \omega_y^2)^{1/2}$. The specific mathematical form of the distribution function, g, can be obtained by applying the previously used method of Lagrange multipliers:

$$g(\omega) = \left(\frac{\beta_r}{\pi}\right) e^{-\beta_r\omega^2} \tag{2.89}$$

where β_r is the Lagrange multiplier for the rotational motion. It should be noted that equation (2.89) describes a two dimensional distribution. Instead of the Lagrange multiplier one can formally introduce the rotational temperature, T_r, by analogy to the distribution of translational velocities:

$$T_r = \frac{I}{2k\beta_r} = \frac{m^*d^2}{2k\beta_r} \tag{2.90}$$

Using this rotational temperature the distribution of angular velocities can be written in the following form:

$$g(\omega) = \left(\frac{I}{2\pi kT_r}\right) \exp\left[-\frac{I\omega^2}{2kT_r}\right] \tag{2.91}$$

Under general conditions the translational temperature, T, and the rotational temperature, T_r, can be different. However, under kinematic equilibrium conditions the gas is characterized by a single temperature, therefore all temperatures must be the same, i.e. $T_r=T$. This means that the average rotational energy of the molecules in the gas is the following:

$$<E_{rot}> = \frac{1}{2}I\left(\frac{I}{2\pi kT}\right)\int\limits_{-\infty}^{\infty} d\omega_x \int\limits_{-\infty}^{\infty} d\omega_y (\omega_x^2 + \omega_y^2) \exp\left[-\frac{I\omega^2}{2kT}\right] = kT \tag{2.92}$$

One can also calculate the average energy of rotation around the x and y axes. This calculation leads to the following results:

$$<E_{x,rot}> = \frac{1}{2}kT \qquad <E_{y,rot}> = \frac{1}{2}kT \tag{2.93}$$

These results mean that under equilibrium conditions the average energy per rotational degree of freedom is kT/2, the same as the average energy per translational degree of freedom. This indicates that the principle of energy equipartition holds for the rotational degrees of freedom, which contribute to the total average energy of a rigid rotator molecule. Hence, the total average energy of a rigid rotator molecule is:

$$< E_{total} >=< E_{translational} > + < E_{rotational} >= \frac{5}{2}kT \qquad (2.94)$$

If the distance, d, between the two nuclei in the diatomic molecule does not remain constant at all times, then they may vibrate along the longitudinal axis. This vibrational motion introduces new degrees of freedom into the system. The vibrational coordinate is the variable distance between the nuclei, d. The vibrational motion is described by putting the origin of the coordinate system at the center of mass and keeping the orientation of the longitudinal axis fixed (see Figure 2.11). The distances to the center of mass from each of the nuclei are again denoted by z_1 and z_2 (with $d=z_1+z_2$) while the time derivatives of the distances z_1, z_2 and d are denoted by q_1, q_2 and q, respectively ($\dot{z}_1 =q_1$, $\dot{z}_2 =q_2$, $\dot{d} =q$). Equation (2.86) relates the distances z_1 and z_2 to d. The vibrational velocities of the nuclei can be similarly expressed in terms of q:

$$E_{vib} = \frac{1}{2}m_1 q_1^2 + \frac{1}{2}m_2 q_2^2 = \frac{1}{2}m^* q^2 \qquad (2.95)$$

Here m^* is again the reduced mass of the molecule.

The vibrational motion of the nuclei can be approximated as harmonic oscillation around the equilibrium distance between the two nuclei, d_0. The governing equation of this harmonic oscillation is the following:

$$m^* \frac{d^2 z}{dt^2} = -Kz \qquad (2.96)$$

where $z=d-d_0$ is the deviation from the equilibrium distance and K is the so-called spring coefficient which characterizes the attracting or repelling force between the molecules. The solution of this second order harmonic differential equation is the following:

$$z=\delta \cos\left(\omega_v t + \Psi_0\right) \qquad (2.97)$$

where δ is the amplitude of vibration, Ψ_0 is the initial phase of the oscillation, and the natural frequency of the oscillator (the frequency of vibration), ω_v, is related to the restoring force constant, K, and the reduced mass, m^*, and is given by $\omega_v=(K/m^*)^{1/2}$. The velocity of the relative motion of the two nuclei can be obtained by taking the time derivative of z:

$$q = \frac{dz}{dt} = -\omega_v \, \delta \sin\left(\omega_v t + \Psi_0\right) \tag{2.98}$$

It should be noted that the governing equation of the molecular vibration is a second order differential equation (equation (2.96)), consequently the solution has two integration constants which fully specify the vibrational motion. In the present solution (equations (2.97) and (2.98)) these independent parameters were chosen to be δ and Ψ_0; however, one could use other independent variables.

The kinetic energy of the vibrational motion can be obtained by substituting equation (2.98) into (2.95):

$$E_{kin} = \frac{1}{2} m^* \omega_v^2 \delta^2 \sin^2\left(\omega_v t + \Psi_0\right) \tag{2.99}$$

The relative motion of the two nuclei in the diatomic molecule takes place in the potential field of the restoring force, consequently the vibrating molecules possess potential as well as kinetic energy. This means that the vibrational motion of the molecule introduces two degrees of freedom, corresponding to the potential and kinetic energies of vibration.

The restoring force (given in equation (2.96)) is the negative gradient of the potential, therefore the potential energy of vibration can be written in the following form:

$$E_{pot} = \frac{1}{2} K z^2 = \frac{1}{2} m^* \omega_v^2 \delta^2 \cos^2\left(\omega_v t + \Psi_0\right) \tag{2.100}$$

In the next step we calculate the vibration averaged values of the kinetic and potential energies of a single molecule. These averages refer to a single molecule and are denoted by $<E_{kin}>_{vib}$ and $<E_{pot}>_{vib}$, respectively. Taking the vibration period averages of equations (2.99) and (2.100), one obtains the following result:

$$< E_{kin} >_{vib} = < E_{pot} >_{vib} = \frac{1}{4} m^* \omega_v^2 \delta^2 \tag{2.101}$$

Equation (2.101) reveals that the average values of the kinetic and potential energies of a given molecule are identical.

Using our earlier results one can derive the normalized distribution function of the independent quantities describing molecular vibration, z and q. Let us denote the normalized distribution function by h(z,q). The two parameters are independent of each other; therefore the distribution function can be separated: $h = h_1(z)h_2(q)$. One can express the total particle energy in terms of these variables:

$$E_{tot} = E_{kin} + E_{pot} = \frac{1}{2}m^*q^2 + \frac{1}{2}Kz^2 = \frac{1}{2}m^*\omega_v^2\delta^2 \qquad (2.102)$$

which means that in the absence of external excitations $q^2+\omega_v^2z^2=\omega_v^2\delta^2$ is constant for every individual particle. One can use our well proven method of Lagrange multipliers to obtain the normalized distribution function, h:

$$h(z,q) = \frac{\beta_v\omega_v}{\pi}\exp\left[-\beta_v(q^2 + \omega_v^2z^2)\right] \qquad (2.103)$$

where β_v is the Lagrange multiplier for the vibrational motion. One can again replace the Lagrange multiplier by a formally defined vibrational temperature, T_v:

$$T_v = \frac{m^*}{2k\beta_v} \qquad (2.104)$$

In kinematic equilibrium the gas is characterized by a single temperature, therefore all temperatures must be the same, i.e. $T_v=T$. The distribution can also be expressed in terms of δ and Ψ_0:

$$h(\delta,\Psi_0) = \delta\frac{m^*\omega_v^2}{2\pi kT}\exp\left[-\frac{m^*\omega_v^2\delta^2}{2kT}\right] \qquad (2.105)$$

Equation (2.102) expresses the vibration averaged potential and kinetic energies of a single molecule in terms of the vibration amplitude, δ. One can use the distribution function given in equation (2.105) to calculate the mean value of kinetic and potential vibrational energies for the entire gas:

$$<E_{kin}> = \int_0^{2\pi}d\Psi_0\int_0^\infty d\delta <E_{kin}>_{vib}\delta\frac{m^*\omega_v^2}{2\pi kT}\exp\left[-\frac{m^*\omega_v^2\delta^2}{2kT}\right]$$

$$= \frac{1}{4}m^*\omega_v^2\frac{m^*\omega_v^2}{kT}\int_0^\infty d\delta\,\delta^3\exp\left[-\frac{m^*\omega_v^2\delta^2}{2kT}\right] = \frac{1}{2}kT \qquad (2.106a)$$

$$<E_{pot}> = \int_0^{2\pi}d\Psi_0\int_0^\infty d\delta <E_{pot}>_{vib}\delta\frac{m^*\omega_v^2}{2\pi kT}\exp\left[-\frac{m^*\omega_v^2\delta^2}{2kT}\right]$$

$$= \frac{1}{4}m^*\omega_v^2\frac{m^*\omega_v^2}{kT}\int_0^\infty d\delta\,\delta^3\exp\left[-\frac{m^*\omega_v^2\delta^2}{2kT}\right] = \frac{1}{2}kT \qquad (2.106b)$$

This result means that under equilibrium conditions the average energy per vibrational degree of freedom is kT/2, the same as the average energy per translational or rotational degree of freedom. This implies that the principle of energy equipartition holds for all degrees of freedom, translational and internal.

We conclude that the total average energy of a diatomic molecule is

$$<E_{total}>=<E_{translational}>+<E_{rotational}>+<E_{kinetic}>+<E_{potential}>=\frac{7}{2}kT$$

$$(2.107)$$

The mean energy associated with each degree of freedom in the diatomic molecule is kT/2. It should be noted that vibration along the longitudinal axis of the molecule represents two degrees of freedom, because there is a potential energy term (associated with the restoring potential) with the same mean value as the kinetic energy of vibration.

One can generalize the results discussed above as the energy equipartition theorem. Let the total energy of the system (e.g. one polyatomic molecule) be expressed in terms of the position coordinates and velocity components for all degrees of freedom. Under equilibrium conditions the system (molecule) possesses a mean energy of kT/2 for each quadratic term appearing in this energy equation.

This theorem can be used to obtain the specific heat of any molecule. The internal energy per unit mass, e, is kT/2m per degree of freedom (see equation (2.76)) and the specific heat at constant volume is given by C_v=de/dT. If the number of internal degrees of freedom of a molecule is n, then C_v is the following:

$$C_V = \frac{n+3}{2}\frac{k}{m} \qquad (2.108)$$

For instance, perfect gases have n=0, and C_v=3k/2m. A rigid rotator diatomic molecule has n=2 and C_v=5k/2m. A rotating and vibrating diatomic molecule has n=4 and C_v=7k/2m.

We showed earlier that the specific heat at constant pressure was C_p=C_v+k/m (see equation (2.83)), therefore

$$C_p = \frac{n+5}{2}\frac{k}{m} \qquad (2.109)$$

Equations (2.108) and (2.109) can be used to obtain the specific heat ratio for a gas composed of molecules with n internal degrees of freedom:

$$\gamma = \frac{C_p}{C_V} = \frac{n+5}{n+3} \qquad (2.110)$$

Conversely, if one knows the specific heat ratio of a gas the number of internal degrees of freedom can be expressed as

$$n = \frac{5 - 3\gamma}{\gamma - 1} \qquad (2.111)$$

Our results, based on the classical kinetic theory of gases, make a very specific prediction for the specific heats of vibrating diatomic gases. We derived a specific expression for C_V, which is constant at all temperatures and depends only on the molecular mass. This result is clearly contradicted by experiments.

Figure 2.12 shows the measured values of the specific heat at constant volume, C_V, as a function of gas temperature, T, for hydrogen molecules. The specific heats of other diatomic molecules exhibit similar behavior. The main feature of the observed temperature dependence of C_V is its step-like behavior. At the lowest temperatures C_V is $3k/2m$, then the curve rises to $5k/2m$ and stays nearly constant at this value over a limited temperature range (which includes room temperature for many gases). Finally, there is a second rise toward the predicted value of $7k/2m$. This transition takes place at different temperatures for different gases.

This plot clearly shows that classical kinetic theory and statistical mechanics fail to account for the energy associated with the internal degrees of freedom of the molecules. This failure is the consequence of the quantum mechanical nature of molecular rotation and vibration.

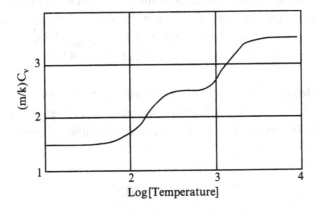

Figure 2.12

According to classical mechanics, a rigid body can rotate with any value of angular momentum. On a molecular scale, however, angular momentum can have only discrete values, i.e.,

$$L = n_r \frac{h}{2\pi} \qquad (2.112)$$

where h is the Planck constant, $h = 6.6261 \times 10^{-34}$ Joule·s, and $n_r = 1,2,3,....$ Planck's constant, h, represents a very small amount of angular momentum. For instance, a bicycle wheel rotating at 3 r.p.s. has an angular momentum of around 1 Joule·s, some 34 orders of magnitude larger. The kinetic energy of rotation is

$$E_{rot} = \frac{L^2}{2I} = n_r^2 \frac{h^2}{8\pi^2 I} \qquad (2.113)$$

The smallest possible rotational kinetic energy is clearly $E_{rot,min} = h^2/8\pi^2 I$. This minimum energy is needed to start rotation, i.e. to excite the molecule to its first rotational energy level. At temperatures so low that

$$kT << \frac{h^2}{8\pi^2 I} \qquad (2.114)$$

the chance of rotational excitation in a collision is very small and only a negligible fraction of molecules will possess rotational energy. This means that at very low temperatures molecular rotation does not contribute to the internal energy of the molecules and consequently to the specific heat of the gas.

When kT is comparable to the energy of the lowest rotational level a significant fraction of the molecules become rotationally excited and the specific heat of the gas starts to increase. At temperatures where $kT >> E_{rot,min}$, the discreteness of the rotational energy states becomes unimportant and classical kinetic theory gives the correct result.

A similar explanation can be given for the vibrational degrees of freedom. The only difference is that the molecule possesses vibrational energy even in the lowest energy state. The possible quantum mechanical values of vibrational energy are

$$E_{vib} = \left(n_v + \frac{1}{2} \right) h \omega_v \qquad (2.115)$$

where ω_v is the natural vibrational frequency of the molecule and $n_v=0,1,2,3,....$ The minimum vibrational energy of the molecule is obviously

$$E_{vib,min} = \frac{1}{2} h \omega_v \qquad (2.116)$$

This residual 'zero point energy' does not vanish even at the lowest temperatures. If kT is much less than the separation between consecutive vibrational energy levels, the chance of vibrational excitation in a collision is very small and nearly all molecules are in the lowest vibrational state. At these temperatures vibration does not make any significant contribution to the specific heat of the gas. When the temperature is so high that $kT>>E_{vib,min}$, the quantization is unimportant and the vibrational specific heat is given correctly by the equipartition theorem. At intermediate temperatures the contribution of molecular vibration to the specific heat increases with increasing temperature.

The vibrational energy level spacing is always greater than the minimum energy needed to excite rotation, the vibrational transition takes place at higher temperatures than the rotational one. Therefore the first rise in Figure 2.12 is attributed to rotation and the second one to vibration.

Except in the high temperature limit, classical kinetic theory gives incorrect results when applied to internal degrees of freedom. This difficulty does not occur when classical kinetic theory is applied to translational degrees of freedom. Since gaskinetic theory is mainly concerned with translation (perfect gases are assumed to be monatomic molecules with no internal degrees of freedom), it is practically unaffected by quantum mechanical effects.

2.7 Problems

2.1 All molecules in a volume have their velocity vectors uniformly distributed inside a velocity space sphere with radius v_0. What is the normalized velocity distribution function? Calculate the average velocity, $<v>$, the mean speed, $<|v|>$, and the mean energy, $<mv^2/2>$.

2.2 All molecules in a volume have their velocity vectors uniformly distributed on the surface of a negligibly thin velocity space spherical shell with radius v_0. What is the normalized velocity distribution function? Calculate the average velocity, $<v>$, the mean speed, $<|v|>$, and the mean energy, $<mv^2/2>$. [Hint: use the Dirac delta function, $\delta(x)$. The main features of the Dirac delta function are discussed in Appendix 9.5.1].

2.3 A volume of gas is in equilibrium at temperature T. Determine $<mv_x v_y>$ and $<mv_x^3/2>$.

2.4 A volume of gas is in equilibrium at temperature T. Calculate the average value of the reciprocal of the particle speed, $<1/v>$, and the average value of the reciprocal of the particle energy, $<2/mv^2>$.

2.5 For a gas in equilibrium find the ratio between the characteristic speeds of molecules, $<v>:<v^2>^{1/2}:v_m$, where v_m is the most probable speed.

2.6 The spatial density of air molecules (average molecular weight = 28.7 amu, 1 amu = 1.67×10^{-27} kg) at room temperature (T=300 K) is $n_0=2.6\times10^{25}$ molecules/m^3. How many molecules in a m^3 of gas are moving within a cone of half angle 5 degrees toward the ceiling with speeds between 200 m/s and 250 m/s? (There are at least two ways to solve this problem. One of them will lead to a special function, called the error function. The error function is discussed in Appendix 9.5.3 of this book.)

2.7 For a gas in equilibrium derive the distribution function of translational kinetic energy, $f_E(E)$, where $f_E(E)dE$ is the fraction of molecules whose kinetic energy lies in the range between E and E+dE. Calculate the most probable energy, E_m.

2.8 All molecules in a polyatomic gas are capable of independently rotating around the x, y and z axes, and vibrating along the x and y axes. What is the specific heat ratio of the gas?

2.9 The distribution function of molecular velocities of a flowing single species gas (composed of molecules with mass m) is given by

$$f(\mathbf{v}) = \left(\frac{m}{2\pi kT}\right)^{3/2} \exp\left\{-\frac{m}{2kT}\left[(v_x - u_x)^2 + (v_y - u_y)^2 + (v_z - u_z)^2\right]\right\}$$

where T is the gas temperature, $\mathbf{u}=(u_x,u_y,u_z)$ is the average flow velocity, while v_x, v_y, and v_z represent molecular velocity components in a stationary coordinate system. Find $<mv_x^2/2>$, $<mv^2/2>$.

2.10 The normalized velocity distribution in a low density anisotropic gas is given by

$$f(\mathbf{v}) = \left(\frac{m}{2\pi kT_1}\right)^{1/2}\left(\frac{m}{2\pi kT_2}\right)\exp\left(-\frac{mv_x^2}{2kT_1} - \frac{mv_y^2}{2kT_2} - \frac{mv_z^2}{2kT_2}\right)$$

Calculate $<mv_x^2/2>$, $<mv_y^2/2>$, $<mv_z^2/2>$, and $<mv^2/2>$.

2.11 Consider the following normalized velocity distribution function:

$$f(\mathbf{v}) = \left(\frac{\beta}{\pi}\right)^{3/2} e^{-\beta v^2}\left[1 + Av_x v_y + Bv_z\left(v^2 - \frac{5}{2\beta}\right)\right]$$

where A, B, and β are constants. Calculate $<v>$, $<mv_x^2/2>$, $<mv_y^2/2>$, $<mv_z^2/2>$, and $<mv^2v/2>$.

2.12 A cylinder contains gas in equilibrium at temperature T. In a reversible, adiabatic expansion of the gas, a piston moves out slowly with velocity u, where u is much smaller than the average thermal speed (see Figure 2.13). What is the gas pressure on the piston? What is the energy loss of a molecule (with angle of incidence θ) rebounding from the piston?

2.13 What specific heat is predicted by the classical theory of specific heats for the linear triatomic molecule CO_2 and the nonlinear triatomic molecule H_2O?

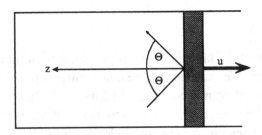

Figure 2.13

2.8 References

[1]   Maxwell, J.C., Illustrations of the dynamical theory of gases. I. On the motions and collisions of perfectly elastic spheres, *Phil. Mag.* [4], **19**, 19–32, 1860 (read 1859).

[2]   Maxwell, J.C., Illustrations of the dynamical theory of gases. II. On the process of diffusion of two or more kinds of moving particles among one another, *Phil. Mag.* *[4]*, **20**, 21, 1860.

[3]   Boltzmann, L., Über die Natur der Gasmolecüle, *Sitz. Math.-Naturwiss. Cl. Acad. Wiss., Wien II.*, **74**, 55, 1877.

[4]   Jeans, J., *An introduction to the kinetic theory of gases*, Cambridge University Press, London, 1940.

[5]   Kennard, E.H., *Kinetic theory of gases*, McGraw-Hill, New York, 1938.

[6]   Knudsen, M., *The kinetic theory of gases*, Methuen and Co. Ltd., London, 1934.

[7]   Loeb, L.B., *The kinetic theory of gases*, McGraw-Hill, New York, 1934.

[8]   Patterson, G.N., *Molecular flow of gases*, John Wiley & Sons, New York, 1956.

[9]   Present, R.D., *Kinetic theory of gases*, McGraw-Hill, New York, 1958.

[10] Sears, F.W., *An introduction to thermodynamics, the kinetic theory of gases, and statistical mechanics*, Addison-Wesley, Cambridge, Mass., 1950.

[11] Vincenti, W.G., and Kruger, C.H., *Introduction to physical gas dynamics*, John Wiley, New York, 1965.

[12] Hirschfelder, J.O., Curtiss, C.F., and Bird, R.B., *The molecular theory of gases and liquids*, John Wiley, New York, 1954.

[13] Fowler, R.H., *Statistical mechanics*, Cambridge University Press, London, 1936.

3

Binary collisions

So far our discussion of the kinetic theory of gases has been limited to kinematic equilibrium properties. We shall see later that molecular collisions are responsible for establishing the equilibrium condition (see Chapter 5). In the absence of equilibrium, intermolecular interactions result in transport of macroscopic gas quantities, such as mass, momentum and energy. Under equilibrium conditions the distribution of molecular velocities is the same Maxwell–Boltzmann distribution at every configuration space location. In other words the effects of molecular collisions cancel each other (the distribution function is constant in time and configuration space) and therefore the details of individual collisions do not play a role in determining the distribution of molecular velocities.

The situation is entirely different if we allow even the slightest deviation from equilibrium. In this case molecular collisions result in the transport of macroscopic quantities (such as mass, momentum and energy) accompanied by a gradual approach to the equilibrium velocity distribution. The details of the macroscopic transport and change of the distribution function are controlled by the specific nature of the molecular collision process. Molecular collisions represent the microscopic process governing all macroscopic transport phenomena.

This chapter lays part of the necessary groundwork for the dynamical description of transport phenomena in gases by examining the details of the most fundamental physical process in the gas: molecular collisions. We begin the discussion of collisional effects by considering in detail the process of two particle (or binary) collisions. For the sake of simplicity it will be assumed throughout this chapter that the gas is composed of monatomic molecules which do not possess any internal degrees of freedom (or if the molecules are not monatomic their states of internal motion are assumed to be unaffected by the collisions).

3.1 Kinematics of two particle collisions

3.1.1 Center of mass and relative position coordinates

This subsection reviews the classical dynamics of two particle central force encounters and the basic dynamical formulae to be used in this and later chapters of this book. Detailed discussion of this problem can be found in almost all elementary textbooks on classical mechanics[1,2] and in most books on kinetic theory.[3–9]

It will be assumed that the molecules can be represented as point centers of force and they interact via conservative forces directed along the line connecting the two molecules.

Let us consider two molecules with masses m_1 and m_2, position vectors $\mathbf{r}_1$ and $\mathbf{r}_2$, and velocities $\mathbf{v}_1$ and $\mathbf{v}_2$. It can be shown that the interaction between these two molecules depends only on their relative position and velocity. We introduce the radius vector of center of mass, $\mathbf{r}_c$:

$$\mathbf{r}_c = \frac{m_1\mathbf{r}_1 + m_2\mathbf{r}_2}{m_1 + m_2} \tag{3.1}$$

The other quantity of physical interest is the relative position vector of the two particles, $\mathbf{r}$:

$$\mathbf{r} = \mathbf{r}_1 - \mathbf{r}_2 \tag{3.2}$$

The two molecules move under each other's influence and the two equations of motion can be written in the following form:

$$m_1 \frac{d^2\mathbf{r}_1}{dt^2} = \mathbf{F}_{12} \qquad m_2 \frac{d^2\mathbf{r}_2}{dt^2} = \mathbf{F}_{21} \tag{3.3}$$

Here $\mathbf{F}_{12}$ and $\mathbf{F}_{21}$ are forces acting on molecules 1 and 2 due to the presence of the other molecule, respectively. These forces depend only on the relative position of the molecules, $\mathbf{r}$, and the fact that the forces acting on the two particles are of equal magnitude and point in opposite directions, $\mathbf{F}_{12}(\mathbf{r})=-\mathbf{F}_{21}(\mathbf{r})$. It can be easily seen that there is no force acting on the center of mass and consequently it does not accelerate (moves with constant velocity):

$$\frac{d^2\mathbf{r}_c}{dt^2} = \frac{1}{m_1+m_2}\left[m_1\frac{d^2\mathbf{r}_1}{dt^2} + m_2\frac{d^2\mathbf{r}_2}{dt^2}\right] = \frac{\mathbf{F}_{12}(\mathbf{r})+\mathbf{F}_{21}(\mathbf{r})}{m_1+m_2} = 0 \tag{3.4}$$

One can also readily calculate the relative acceleration of the two molecules with respect to each other:

$$\frac{d^2\mathbf{r}}{dt^2} = \frac{d^2\mathbf{r}_1}{dt^2} - \frac{d^2\mathbf{r}_2}{dt^2} = \frac{\mathbf{F}_{12}(\mathbf{r})}{m_1} - \frac{\mathbf{F}_{21}(\mathbf{r})}{m_2} = \left(\frac{1}{m_1} + \frac{1}{m_2}\right)\mathbf{F}_{12}(\mathbf{r}) \qquad (3.5)$$

Equation (3.5) can be written as an equation of motion for a single particle with mass m* (where m* is the reduced mass of the two molecules) in a central field of force:

$$m^* \frac{d^2\mathbf{r}}{dt^2} = -\frac{dU(r)}{dr}\mathbf{e}_r \qquad (3.6)$$

where $\mathbf{e}_r$ represents the unit vector along the relative position vector of the two molecules and U(r) is the potential of the conservative intermolecular force, $\mathbf{F}_{12}$:

$$\mathbf{F}_{12}(\mathbf{r}) = -\frac{dU(r)}{dr}\mathbf{e}_r \qquad (3.7)$$

These results show that one may introduce a new set of independent variables which simplify the description of the collision. These new variables refer to the center of mass of the two molecules and to their relative position, velocity and acceleration. It was shown that the center of mass velocity remains constant during the interaction of the two molecules, while the relative motion of the molecules can be described as the motion of a single particle with mass m* under the influence of a conservative central field of force characterized by potential U(r).

Next we examine the motion of a single point-like particle under the influence of a conservative central field of force.

3.1.2 Particle motion in a central field of force

A central force is always directed toward a fixed point. A conservative central force can be derived from a scalar potential that is only a function of the distance between the particle and the center of force. We choose the center of force as the origin of the coordinate system. In this case the force acting on the particle is always parallel (or antiparallel) to the radius vector.

At any instant the radius vector and the velocity vector of the particle are included in a plane. Since the force vector lies in this plane the particle has no acceleration normal to this plane. This means that the particle velocity and the force vector will always remain in the same plane and the particle trajectory will also be confined to this plane. In other words a particle moving under the influence of a central field of force exhibits a planar motion.

Let r and ϕ be polar coordinates of the particle in the plane of motion. As is obvious from Figure 3.1 the Cartesian coordinates can simply be written as

$$x = r \cos\phi \qquad y = r \sin\phi \qquad (3.8)$$

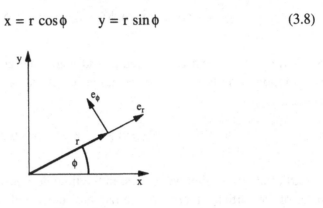

Figure 3.1

As the particle moves along its trajectory, the polar coordinates specifying its location in the plane vary with time. This means that the particle's motion is fully described by the pair of time dependent functions, r(t) and ϕ(t).

The Cartesian velocity components can be obtained by taking the time derivative of equation (3.8):

$$\dot{x} = \dot{r} \cos\phi - r \dot{\phi} \sin\phi \qquad \dot{y} = \dot{r} \sin\phi + r \dot{\phi} \cos\phi \qquad (3.9)$$

In equation (3.9) we have introduced a simple notation for the time derivative, $\dot{x} = dx/dt$. The second time derivatives of the Cartesian coordinates of the particle can also be readily calculated:

$$\ddot{x} = \ddot{r} \cos\phi - 2 \dot{r} \dot{\phi} \sin\phi - r \dot{\phi}^2 \cos\phi - r \ddot{\phi} \sin\phi \qquad (3.10a)$$

$$\ddot{y} = \ddot{r} \sin\phi + 2 \dot{r} \dot{\phi} \cos\phi - r \dot{\phi}^2 \sin\phi + r \ddot{\phi} \cos\phi \qquad (3.10b)$$

In equations (3.10) we have introduced a simple notation for the second time derivatives for the particle coordinates, $\ddot{x} = d^2x/dt^2$.

In Cartesian coordinates the relative velocity vector of the particles, $\mathbf{g} = \mathbf{v}_1 - \mathbf{v}_2$, can be expressed as $\mathbf{g} = g_x \mathbf{e}_x + g_y \mathbf{e}_y$, where $\mathbf{e}_x$ and $\mathbf{e}_y$ represent the orthonormal vectors (orthogonal unit vectors) in the x and y directions. The velocity vector can also be expressed as a linear combination of the orthonormal vectors, $\mathbf{e}_r$ and $\mathbf{e}_\phi$, $\mathbf{g} = g_r \mathbf{e}_r + g_\phi \mathbf{e}_\phi$ (see Figure 3.1). The velocity components, g_r and g_ϕ, can readily be expressed in terms of the time derivatives of the Cartesian coordinates of the particle:

$$g_r = \dot{x} \cos\phi + \dot{y} \sin\phi \qquad g_\phi = -\dot{x} \sin\phi + \dot{y} \cos\phi \qquad (3.11)$$

Similarly, one can express the radial and azimuthal components of the particle acceleration, **a**, in terms of the second time derivatives of the Cartesian components:

$$a_r = \ddot{x} \cos\phi + \ddot{y} \sin\phi \qquad a_\phi = -\ddot{x} \sin\phi + \ddot{y} \cos\phi \qquad (3.12)$$

Earlier in this section we expressed the time derivatives of the x and y coordinates of the particle in terms of the radial distance and azimuth angle. In the next step we substitute these results (equations (3.9) and (3.10)) into equations (3.11) and (3.12). After some manipulation we obtain the following results expressing the radial and azimuthal velocity components in terms of r and ϕ:

$$g_r = \dot{r} \qquad g_\phi = r\dot{\phi} \qquad (3.13)$$

$$a_r = \ddot{r} - r\dot{\phi}^2 \qquad a_\phi = 2\dot{r}\dot{\phi} + r\ddot{\phi} \qquad (3.14)$$

The physical interpretation of the extra terms in equation (3.14) is well known from classical mechanics: $r\dot{\phi}^2$ and $2\dot{r}\dot{\phi}$ correspond to the centrifugal and Coriolis accelerations, respectively.[1,2]

Substituting the newly derived radial and azimuthal acceleration components into the governing equation of the relative motion of the two particles (equation (3.6)), the equation of motion can be written in the following form:

$$m^* a_r = m^*(\ddot{r} - r\dot{\phi}^2) = -\frac{dU(r)}{dr} \qquad (3.15a)$$

$$m^* a_\phi = m^*(2\dot{r}\dot{\phi} + r\ddot{\phi}) = 0 \qquad (3.15b)$$

This result shows that in a central field of force the acceleration of the particle is always in the radial direction and the azimuthal acceleration is identically zero everywhere.

The fact that there is no azimuthal acceleration implies that angular momentum is conserved. The angular momentum vector of the relative motion, **L**, can be expressed as follows:

$$\mathbf{L} = \mathbf{r} \times m^* \mathbf{g} \tag{3.16}$$

Both **r** and **g** are in the plane of motion, therefore the angular momentum vector is perpendicular to this plane. We showed above that the plane of motion does not change with the relative motion of the two particles, therefore the direction of the angular momentum vector is independent of time. This means that **L** can be expressed as $\mathbf{L} = L\mathbf{e}_n$, where $\mathbf{e}_n$ is the time independent normal vector of the plane of relative motion. The scalar coefficient of the angular momentum, L, can be obtained as

$$L = m^* r^2 \dot{\phi} \tag{3.17}$$

It is easy to see that the time derivative of L is identically zero:

$$\dot{L} = 2m^* r \dot{r} \dot{\phi} + m^* r^2 \ddot{\phi} = m^* r a_\phi = 0 \tag{3.18}$$

At the same time we know that $\mathbf{e}_n$ remains constant, so we finally conclude that

$$\dot{\mathbf{L}} = 0 \tag{3.19}$$

Equation (3.19) is a fundamentally important result with very important implications for gaskinetic theory. It says that in a binary collision there are no external torques applied to the system and therefore the angular momentum vector of the relative motion is conserved.

Next we turn our attention to the radial component of the equation of motion. We start by writing equation (3.15a) in the following form:

$$m^* a_r + \frac{dU}{dr} = 0 \tag{3.20}$$

Equation (3.20) can be rewritten by using the definition of the magnitude of the angular momentum vector, L, and by recalling that L is a conserved quantity throughout the motion of the particles:

$$m^* (\ddot{r} - r\dot{\phi}^2) + \frac{dU}{dr} = m^* \ddot{r} - \frac{L^2}{m^* r^3} + \frac{dU}{dr} = m^* \ddot{r} + \frac{d}{dr}\left(\frac{L^2}{2\,m^* r^2} + U\right) = 0 \tag{3.21}$$

This result yields the following equation of motion for the distance between the two particles:

$$m^* \ddot{r} = -\frac{d}{dr}\left(U + \frac{L^2}{2m^* r^2}\right)$$

(3.22)

This equation leads to another conservation law. This can be seen by multiplying equation (3.22) by dr/dt and integrating the resulting relation with respect to time:

$$\int dt\, \dot{r}\left[m^* \ddot{r} + \frac{d}{dr}\left(\frac{L^2}{2\,m^* r^2} + U\right)\right] = \int d\dot{r}\, m^* \dot{r} + \int dr\, \frac{d}{dr}\left(\frac{L^2}{2\,m^* r^2} + U\right) = 0$$

(3.23)

The two integrals can be easily evaluated and one can use equation (3.17) to express L in terms of the polar coordinates. This manipulation yields the following conservation law:

$$\frac{1}{2}m^* \dot{r}^2 + \frac{1}{2}m^* r^2 \dot{\phi}^2 + U = \text{cons}\tan t$$

(3.24)

This equation expresses the conservation of energy for the relative motion of the two particles. It is no surprise that the sum of radial and tangential kinetic energies and the potential energy is constant.

3.1.3 Angle of deflection

In this subsection we will investigate the angle of deflection of the relative velocity vector, **g**, during an intermolecular collision.

As discussed previously the relative motion of two interacting particles is confined to a plane. In the absence of any interaction the relative velocity vector of the molecules, **g**, would retain its initial value, $\mathbf{g}_0$, at all times during the motion. This means that the trajectory of the relative motion would be a straight line, as shown in Figure 3.2.

In this section we consider a coordinate system which has its origin at molecule 2 and the x axis antiparallel to the initial relative velocity vector, $\mathbf{g}_0$ (see Figure 3.2). The radius vector pointing to particle 1 naturally varies with time as the particles move with respect to each other. The trajectory is fully described by the potential of the interparticle force, U(r), the initial relative velocity vector, $\mathbf{g}_0$, and by the impact parameter, b, characterizing the 'miss distance' between the particles.

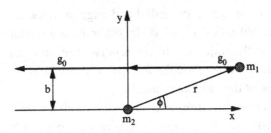

Figure 3.2

The impact parameter, b, is the smallest distance between the trajectories of the molecules if there were no interaction between them (see Figure 3.2). For a given particle trajectory the angle of deflection will be determined by the magnitude of the relative velocity at infinity (before collision), $\mathbf{g}_0$, and by the newly defined impact parameter, b.

It is obvious that when the two molecules are very far apart their influence on each other is negligible and the relative motion is practically a straight line as shown in Figure 3.2. In this case the angular momentum of the relative motion (with respect to molecule 2) can be expressed as follows (see equation (3.16)):

$$L = m^* g_0 \, r \sin \phi = m^* g_0 \, b \tag{3.25}$$

We know from our earlier discussions that the angular momentum is conserved throughout the trajectory, therefore equation (3.25) expresses the conserved quantity, L, in terms of the magnitude of the initial velocity, g_0, and the impact parameter, b. One can also express the total energy of the relative motion, E, in the limiting case of $r \to \infty$:

$$E = \lim_{r \to \infty} \left\{ \frac{1}{2} m^* \dot{r}^2 + \frac{L^2}{2 m^* r^2} + U(r) \right\} = \frac{1}{2} m^* g_0^2 \tag{3.26}$$

When evaluating the $r \to \infty$ limit in equation (3.26) we used equation (3.17) expressing the azimuthal velocity in terms of the conserved quantity, L, and used the fact that the intermolecular potential describes a short range force, which rapidly vanishes for large values of the intermolecular distance, r. Now the relative motion of the particles is described by the following two equations:

$$m^* r^2 \dot{\phi} = m^* g_0 \, b \tag{3.27a}$$

$$\frac{1}{2} m^* \dot{r}^2 + \frac{m^* g_0^2 b^2}{2 r^2} + U(r) = \frac{1}{2} m^* g_0^2 \tag{3.27b}$$

These two equations describe the radial and angular velocities, $\dot{r}$ and $\dot{\phi}$, as a function of polar coordinates, r and ϕ. In other words, equations (3.27a) and (3.27b) fully describe the trajectory of the relative motion of the two molecules. The parameters of the trajectory are the interparticle potential, U(r), and two independent initial values of the motion, g_0 and b.

Equations (3.27) provide a complete but fairly cumbersome description of the trajectory of the relative motion. A much simpler way to describe the trajectory is to give the radial distance as a function of azimuth angle, r(ϕ). This can be done by eliminating the time dependence and combining the two equations into one single differential equation.

First we rewrite equations (3.27) in the following form:

$$(d\phi)^2 = \frac{g_0^2 \, b^2}{r^4}(dt)^2 \tag{3.28a}$$

$$(dr)^2 = \frac{g_0^2}{r^2}\left[r^2 - b^2 - \frac{2r^2 U(r)}{m^* g_0^2}\right](dt)^2 \tag{3.28b}$$

These two equations can now be combined to yield the differential equation describing the function r(ϕ):

$$\frac{dr}{d\phi} = \pm \frac{r}{b}\sqrt{r^2 - b^2 - \frac{2r^2 U(r)}{m^* g_0^2}} \tag{3.29}$$

Equation (3.29) describes a double valued function, meaning that two ϕ values correspond to each value of r. The (−) sign corresponds to the first portion of the trajectory in which particle 1 approaches particle 2 and the interparticle distance decreases with increasing azimuth angle. The (+) sign corresponds to the second portion of the trajectory in which particle 1 moves away from particle 2 and r increases with increasing ϕ.

Equation (3.29) can be formally integrated to yield an implicit equation for the function r(ϕ):

$$\phi - \phi_m = \pm \int_{r_m}^{r} dr \, \frac{b}{r}\left[r^2 - b^2 - \frac{2r^2 U(r)}{m^* g_0^2}\right]^{-1/2} \tag{3.30}$$

The integration constants in equation (3.30), r_m and ϕ_m, refer to the radial distance and azimuth angle of closest approach, respectively. The solution is schematically shown in Figure 3.3 for a repelling force (positive interparticle potential).

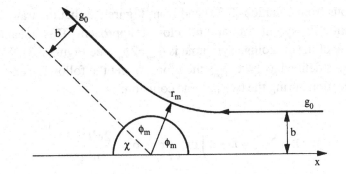

Figure 3.3

It can be seen from equation (3.30) (and also in Figure 3.3) that the trajectory of the relative motion has two asymptotes, one along the initial direction as the particles approach each other from infinite distance and the other along the final, post-encounter direction of motion, when the particles leave each other. The angle of deflection, χ, is defined as the angle between these two asymptotes measured from the initial to the final direction of motion.

Equation (3.30) has two unknown parameters, the radial distance of closest approach and the azimuth angle of closest approach. These parameters must be expressed in terms of b and g_0 if we want to get a full description of the trajectory of the relative motion.

Let us start with r_m. It is obvious from the definition of closest distance that, for $\phi < \phi_m$, r is decreasing with increasing ϕ, dr/dϕ<0, while, for $\phi > \phi_m$, r is increasing with increasing ϕ, dr/dϕ>0. This means that at closest approach dr/dϕ=0, therefore from equation (3.29),

$$r_m^2 - b^2 - \frac{2r_m^2 U(r_m)}{m^* g_0^2} = 0 \qquad (3.31)$$

Equation (3.31) is a complicated equation for r_m, but it can be solved for every specific intermolecular potential. This solution can be used to obtain the azimuth angle of closest approach, ϕ_m. We know that, well before the collision, the particles are infinitely far apart. At this point the azimuth angle is 0 (see Figure 3.2), therefore we conclude that $r(\phi=0)=\infty$. Substituting this condition into equation (3.30) yields the following expression for ϕ_m:

$$\phi_m = \int_{r_m}^{\infty} dr \frac{b}{r} \left[r^2 - b^2 - \frac{2r^2 U(r)}{m^* g_0^2} \right]^{-1/2} \qquad (3.32)$$

It is obvious from equation (3.29) and from Figure 3.3 that the trajectory of the relative motion is symmetric around closest approach. This means that the azimuth angle of the outgoing asymptote is $\phi_{out}=2\phi_m$ (see Figure 3.3). The angle of deflection, χ, is defined as $\chi=\pi-\phi_{out}$, therefore we get the following expression for the total deflection during the two particle collision:

$$\chi = \pi - 2\phi_m = \pi - 2\int_{r_m}^{\infty} dr\, \frac{b}{r}\left[r^2 - b^2 - \frac{2r^2 U(r)}{m^* g_0^2}\right]^{-1/2} \tag{3.33}$$

This is a very important result which shows that the impact parameter, b, and the deflection angle, χ, are physically equivalent quantities. Equation (3.33) provides a unique transformation between these two parameters for a given intermolecular potential and initial relative velocity. This result will play a major role later in this chapter, when we discuss collision cross sections.

It is very interesting to note that the deflection angle described by equation (3.33) depends on the ratio of the potential energy and the kinetic energy of the relative motion, $2U/m^* g_0^2$. This means that a fast relative motion in the same potential field will result in small deflection. This point can be seen by considering a very fast relative motion, so fast that the term $2U/m^* g_0^2$ can be neglected everywhere along the relative trajectory. In this case equation (3.31) becomes $r_m=b$ (the relative motion is a straight line and the impact parameter is the closest distance) and therefore

$$\chi = \pi - 2\int_{b}^{\infty} dr\, \frac{b}{r\sqrt{r^2 - b^2}} = \pi - 2\left[\text{Arc}\cos\left(\frac{b}{r}\right)\right]_{b}^{\infty} = 0 \tag{3.34}$$

In other words very fast moving particles pass each other without any noticeable change in their trajectories. Slow moving particles, on the other hand, have plenty of time to interact while in the vicinity of the other particle, therefore their trajectory is more significantly influenced.

In order to illustrate the physical process behind equation (3.33) let us evaluate it in the simplest possible case. Let us consider the collision of two hard, elastic spheres. These spheres are perfectly impenetrable with diameters of d_1 and d_2, respectively. Figure 3.4 shows a schematic representation of this collision. It is assumed that the spheres have no influence on one another unless their surfaces are in direct contact. When the spheres collide the distance between their centers is $d_{12}=(d_1+d_2)/2$. It is obvious from Figure 3.4 that the interparticle potential is zero when the distance between the molecular centers is larger than d_{12}. The particles cannot penetrate into each other, therefore $U=\infty$ for $r<d_{12}$ (corresponding to infi-

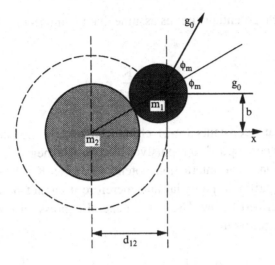

Figure 3.4

nite repelling force at contact). In the case of a collision the distance of closest approach is d_{12}, therefore in the case of hard sphere collisions equation (3.33) becomes the following:

$$\chi = \pi - 2 \int_{d_{12}}^{\infty} dr \frac{b}{r\sqrt{r^2 - b^2}} = \pi - 2 \left[\text{Arc} \cos\left(\frac{b}{r}\right) \right]_{d_{12}}^{\infty} = 2 \left[\text{Arc} \cos\left(\frac{b}{d_{12}}\right) \right] \quad (3.35)$$

This means that in the case of hard sphere collisions the relation between the impact parameter, b, and the deflection angle, χ, is given by

$$b = d_{12} \cos\left(\frac{\chi}{2}\right) \quad (3.36)$$

Note that this result is independent of the initial relative speed, g_0. This should be intuitively reasonable, because we know that in hard sphere collisions the geometry determines the deflection angle. By the way, this result can be seen immediately in Figure 3.4. In this simple case our general result given by equation (3.33) seems to be trivial. However, for more complicated molecular interaction models equation (3.33) becomes a fundamental tool of gaskinetic theory.

3.1.4 Inverse power interactions

The expression for the angle of deflection, χ, can be greatly simplified for a special class of interparticle potentials. Let us assume that the intermolecular force can be written as

$$\mathbf{F}_{12}(\mathbf{r}) = \frac{K_a}{r^a}\mathbf{e}_r \qquad (3.37)$$

where K_a and a are constants (a>1) characterizing the molecular interaction. Positive K_a values correspond to repulsive forces, while negative values describe attractive interactions. The interaction potential is $U(r)=K_a r^{-(a-1)}/(a-1)$. The a=1 case leads to a logarithmic potential and therefore it cannot be treated with the method to be described below. Using the general expression for U(r), equation (3.33) becomes the following:

$$\chi = \pi - 2\int_{r_m}^{\infty} dr \frac{b}{r}\left[r^2 - b^2 - \frac{2K_a}{(a-1)m^*g_0^2 r^{a-3}}\right]^{-1/2} \qquad (3.38)$$

The evaluation of this integral can be made somewhat simpler after expressing it in terms of dimensionless variables, therefore we introduce the following new quantities:

$$x = \frac{b}{r} \qquad x_m = \frac{b}{r_m} \qquad x_0 = b\left(\frac{m^*g_0^2}{K_a}\right)^{\frac{1}{a-1}} \qquad (3.39)$$

It should be noted that for attracting potentials (when K_a is negative) one has to take the absolute value of K_a when calculating these dimensionless quantities. Using these quantities equation (3.38) can be written in the following form:

$$\chi = \pi - 2\int_0^{x_m} dx\left[1 - x^2 - \frac{2}{(a-1)}\left(\frac{x}{x_0}\right)^{a-1}\right]^{-1/2} \qquad (3.40)$$

The upper integration boundary, x_m, can be obtained by solving the following equation:

$$x_m^2 = 1 - \frac{2}{(a-1)}\left(\frac{x_m}{x_0}\right)^{a-1} \qquad (3.41)$$

It can be seen that the deflection angle is only a function of the dimensionless impact parameter, x_0, and the power of the intermolecular force, a.

The integral in equation (3.40) can be evaluated in some specific cases. Let us start with the lowest order case, when a=2. This corresponds to gravitational or electrostatic interactions, when the force decreases as the square of the distance between the particles.

In the a=2 case the integration boundary, x_m, can be obtained as the positive root of a quadratic equation:

$$x_m = \frac{\sqrt{1 + x_0^2} - 1}{x_0} \tag{3.42}$$

The integral specifying the relation between the deflection angle and the dimensionless impact parameter can now be carried out:

$$\chi = \pi - 2 \int_0^{x_m} dx \, \frac{1}{\sqrt{1 - x^2 - \frac{2x}{x_0}}} = \pi - 2 \operatorname{Arcsin}\left[\frac{x_0 x + 1}{\sqrt{1 + x_0^2}} \right]_0^{x_m} = 2 \operatorname{Arcsin}\left[\frac{1}{\sqrt{1 + x_0^2}} \right] \tag{3.43}$$

Finally, one can rewrite this result in terms of the impact parameter, b:

$$\tan\left(\frac{\chi}{2} \right) = \frac{b_0}{b} \tag{3.44}$$

where $b_0 = K_2/m^* g_0^2$. Expression (3.44) also helps us to understand the physical meaning of the parameter b_0. It is obvious that $\chi = \pi/2$ for $b = b_0$, therefore b_0 is the impact parameter value separating forward and backward scattering.

Equation (3.44) reveals another fundamental result. It is obvious that $\chi(b=0) = \pi$ and $\chi(b=\infty) = 0$. Clearly, b=0 describes a 'head on' collision in which the colliding particle simply reverses direction after impact. The other extreme, b=∞, describes the situation in which the particles experience no collision whatsoever: the two particles pass at a very large distance moving along straight lines without influencing each other.

In the case where a=3, the integral becomes even more simple. First of all, the maximum value of x becomes

$$x_m = \frac{x_0}{\sqrt{1 + x_0^2}} \tag{3.45}$$

and the integral itself does not yield inverse trigonometric functions:

$$\chi = \pi - 2 \int_0^{x_m} dx \frac{1}{\sqrt{1 - x^2 \left(1 + \frac{1}{x_0^2}\right)}}$$

$$= \pi - \frac{2}{\sqrt{1 + \frac{1}{x_0^2}}} \text{Arc sin} \left[x \sqrt{1 + \frac{1}{x_0^2}} \right]_0^{x_m} = \pi \frac{\sqrt{1 + x_0^2} - x_0}{\sqrt{1 + x_0^2}} \qquad (3.46)$$

The relation between χ and b in the case where a=3 can be given as

$$\chi = \pi \frac{\sqrt{b_0^2 + b^2} - b}{\sqrt{b_0^2 + b^2}} \qquad (3.47)$$

where $b_0 = (K_3 / m^* g_0^2)^{1/2}$.

The case a=5 is of particular interest, because for this special inverse power interaction the transport coefficients can be evaluated analytically. Molecules interacting with this interparticle potential are called Maxwell molecules (this molecular interaction will be extensively discussed in Chapter 6). For a=5 the parameter x_m can be expressed in the following form:

$$x_m = x_0 \left[\sqrt{x_0^4 + 2} - x_0^2 \right]^{1/2} \qquad (3.48)$$

The integral expressing the deflection angle becomes quite complicated and cannot be evaluated in terms of simple functions:

$$\chi(x_0) = \pi - \int_0^1 du \frac{2^{3/2} x_0 \sqrt{\sqrt{x_0^4 + 2} - x_0^2}}{\sqrt{(x_0^4 + 2) - \left\{ x_0^2 + u^2 \left[\sqrt{x_0^4 + 2} - x_0^2 \right] \right\}^2}} \qquad (3.49)$$

The function $\chi(x_0)$ for a=5 is shown in Figure 3.5. Inspection of Figure 3.5 reveals that for $x_0 = 0$ (which corresponds to b=0) the particle is backscattered ($\chi = \pi$), while there is very little deflection when the normalized impact parameter exceeds 2. It can also be seen that for small impact parameters the $\chi(x_0)$ function is almost linear. In this limit the function can be approximated with a very simple expression:

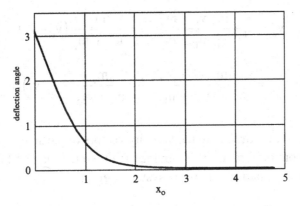

Figure 3.5

$$\lim_{x_0 \to 0} \chi(x_0) = \pi - x_0 \int_0^1 du \, \frac{2^{5/4}}{\sqrt{1-u^4}} = \pi - 3.118 x_0 \qquad (3.50)$$

One could evaluate integral (3.40) for some higher order inverse power interactions, as well. These three examples have given us some insight into the delicate relationship between the impact parameter and the inverse power potential. These insights (and many more) will be needed later in the book when we calculate transport coefficients for different kinds of molecules.

3.1.5 Particle motion in laboratory and center of mass frames

The dynamics of two particle collisions are described in two frames of reference. The reference frame at rest with respect to the container is referred to as the rest frame (or laboratory frame) and our results ultimately have to be expressed in this coordinate system. The center of mass frame (or C system) moves along with the center of mass of the two colliding molecules. This system is particularly simple kinematically and most convenient for formulating conservation laws. It has been shown in subsection 3.1.1 that the two particle collision in the C system reduces to the even simpler case of a single particle deflected in a central field of force. In this subsection we derive the dynamical properties of the colliding particles in the rest frame using the results of the previous subsections of this section.

Let us start by expressing the particle velocities in the center of mass system. Here and throughout this book the subscript 'c' refers to the C system. Before the collision the particle velocities in the C system are

$$v_{1c} = v_1 - v_c = v_1 - \frac{m_1 v_1 + m_2 v_2}{m_1 + m_2} = \frac{m_2}{m_1 + m_2}\left(v_1 - v_2\right) = \frac{m^*}{m_1}g \quad (3.51a)$$

$$v_{2c} = v_2 - v_c = v_2 - \frac{m_1 v_1 + m_2 v_2}{m_1 + m_2} = \frac{m_1}{m_1 + m_2}\left(v_2 - v_1\right) = -\frac{m^*}{m_2}g \quad (3.51b)$$

where $v_1 = dr_1/dt$ and $v_2 = dr_2/dt$ are the velocity vectors of particles 1 and 2, respectively, in the laboratory frame. It is obvious from equations (3.51) that the total momentum of the two particles in the C system is zero:

$$m_1 v_{1c} + m_2 v_{2c} = 0 \quad (3.52)$$

Next we calculate the total kinetic energy in the laboratory frame and express it in terms of the center of mass energy and the energy in the C system:

$$E = \frac{1}{2}m_1 v_1^2 + \frac{1}{2}m_2 v_2^2 = \frac{1}{2}m_1\left(v_c + v_{1c}\right)^2 + \frac{1}{2}m_2\left(v_c + v_{2c}\right)^2$$

$$= \frac{1}{2}m_0 v_c^2 + \frac{1}{2}m_1 v_{1c}^2 + \frac{1}{2}m_2 v_{2c}^2 + \left(m_1 v_{1c} + m_2 v_{2c}\right)v_c \quad (3.53)$$

where m_0 is the total mass of the two particles, $m_0 = m_1 + m_2$. On the other hand we know that $m_1 v_{1c} + m_2 v_{2c} = 0$ because of the conservation of momentum in the C system (see equation (3.52)), therefore we obtain the following expression for the total kinetic energy in the rest frame:

$$E = \frac{1}{2}m_0 v_c^2 + \frac{1}{2}m_1 v_{1c}^2 + \frac{1}{2}m_2 v_{2c}^2 \quad (3.54)$$

Equation (3.54) expresses a well known result: the total kinetic energy in the rest frame is equal to the sum of kinetic energies of the two particles in the C system plus the kinetic energy of the center of mass (the kinetic energy of a particle with mass $m_0 = m_1 + m_2$ moving with the center of mass velocity).

Based on our earlier results one can readily express the total kinetic energy in the C system, E_c, with the help of the relative velocity vector, g:

$$E_c = \frac{1}{2}m_1 v_{1c}^2 + \frac{1}{2}m_2 v_{2c}^2 = \frac{1}{2}\frac{m^{*2}}{m_1}g^2 + \frac{1}{2}\frac{m^{*2}}{m_2}g^2 = \frac{1}{2}m^* g^2 \quad (3.55)$$

which says that the total kinetic energy of the two particles in the C system is equal to the kinetic energy of the relative motion.

Another conserved quantity of our two particle system is the angular momentum vector, **L**. The angular momentum of the two particles in the laboratory frame can be calculated in the following way:

$$
\begin{aligned}
L &= r_1 \times m_1 v_1 + r_2 \times m_2 v_2 = r_1 \times m_1 \left(v_c + v_{1c} \right) + r_2 \times m_2 \left(v_c + v_{2c} \right) \\
&= \left(m_1 r_1 + m_2 r_2 \right) \times v_c + r_1 \times m_1 v_{1c} + r_2 \times m_2 v_{2c} \\
&= r_c \times m_0 v_c + r_1 \times m^* g - r_2 \times m^* g = r_c \times m_0 v_c + r \times m^* g
\end{aligned}
\tag{3.56}
$$

The first term describes the angular momentum of the center of mass, while the second term corresponds to the angular momentum in the C system, L_c. Using the results obtained in the previous subsections of this section it is obvious that the magnitude of the angular momentum in the C system can also be expressed in terms of the impact parameter, b:

$$
L_c = \left| r \times m^* g \right| = m^* g_0 b
\tag{3.57}
$$

where g_0 denotes the magnitude of the relative velocity before the collision.

So far in this subsection we have expressed the fundamental conserved quantities (momentum, energy and angular momentum) in the laboratory and center of mass systems. Next we consider the relations between physical quantities before and after the collision. Unprimed quantities will refer to initial values and primed quantities to final values after the encounter.

We showed at the beginning of this chapter that the velocity of the center of mass remains constant during the two particle collision (we showed in equation (3.4) that the center of mass does not accelerate), therefore $v_c'=v_c$. The conservation of total energy simplifies to the following expression, taking into account the conservation of the center of mass velocity:

$$
\frac{1}{2} m^* g^2 = \frac{1}{2} m^* g'^2
\tag{3.58}
$$

Equation (3.58) means that the magnitude of the relative velocity remains unchanged during the collision, $g'=g$.

Next we consider the consequences of the conservation of momentum. Equations (3.51) express the particle velocities in the C system in terms of the relative velocity vector, **g**. This means that $v_{1c}=m^* g/m_1$ and $v'_{1c}=m^* g'/m_1$ and therefore $v_{1c}=v'_{1c}$ (because $g'=g$). Similarly, $v_{2c}=-m^*g/m_2$ and $v'_{2c}=-m^*g'/m_2$ and therefore $v_{2c}=v'_{2c}$.

These results are schematically shown in Figure 3.6 and they can be summarized as follows. In the C system, the particles approach each other with equal and opposite momenta along parallel paths. The velocity of each particle is changed in direction but not in magnitude by the encounter in the C system.

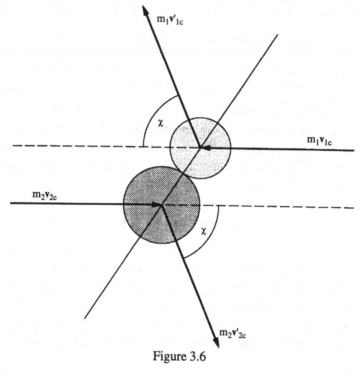

Figure 3.6

We summarize our results by calculating the change of momentum of each particle during the collision in the laboratory frame. The change of momentum of particle 1 during the collision is the following:

$$\delta(m_1v_1)= m_1v'_1 - m_1v_1 = m_1v'_{1c} + m_1v'_c - m_1v_{1c} - m_1v_c \qquad (3.59)$$

On the other hand we know that the center of mass velocity is conserved during the collision, therefore

$$\delta(m_1v_1)= m_1(v'_{1c} - v_{1c}) = m^*(g' - g) \qquad (3.60)$$

Obviously, $\delta(m_1v_1)=-\delta(m_2v_2)$.

Finally, one can calculate the components of the momentum transfer parallel and perpendicular to the relative velocity (these quantities will play very important roles in calculating transport coefficients):

$$\delta\left(m_1 v_1\right)_{\parallel} = m^*(g\cos\chi - g) = -m^* g(1 - \cos\chi) \qquad (3.61a)$$

$$\delta\left(m_1 v_1\right)_{\perp} = m^* g\sin\chi \qquad (3.61b)$$

3.2 Statistical description of collisional effects

In this section we introduce several new statistical quantities such as the mean free time, mean free path, collision frequency, collision rate, collision cross section and differential cross section. We shall also see that these quantities are closely related to each other and to other fundamental molecular quantities. In this section we use very simple physical models to emphasize the basic concepts of these new statistical quantities.

The concept of mean free path was introduced in 1858 by Rudolf Clausius.[10] This revolutionary new idea opened the road to the development of gaskinetic theory.

For the sake of simplicity we consider a single species gas composed of hard sphere molecules. In this simple model all molecules have the same diameter, d. A given molecule will suffer a collision whenever the distance between the centers of this molecule and any other molecule approaches the diameter, d (see Figure 3.4). We may thus imagine each molecule as having a sphere of influence of radius d.

3.2.1 Mean free time

The free time is the travel time of a molecule between two successive collisions. The mean free time is the average time between two successive collisions of a single molecule. Consider a molecule with velocity **v**. Assume that this particle suffers a collision with another molecule at t=0. Let $P_{time}(t)$ denote the probability that this molecule survives a time period t without suffering a second collision. Obviously, $P_{time}(0)=1$, because we have neglected multiple collisions (if $P_{time}(0)\neq 1$, the particle could collide with two separate molecules at the same time: such collisions are not considered in classical gaskinetic theory).

Let vdt denote the probability that a given molecule suffers a collision between time t and t+dt. The quantity ν is now the collision probability per unit time (also called the 'collision frequency') of a particle. It is assumed that the collisional probability density, v, is independent of the past collisional history of the molecule (in other words it does not matter when the molecule suffered its last collision). In general, however, v may depend on the speed of the particle (faster particles travel larger distances and therefore encounter more molecules in the same time interval), so it is written $v(v)$.

With the help of the collision frequency, v, it is possible to derive the functional form of the survival probability, P_{time}. The probability that the particle survives a time interval $t+dt$ without suffering a collision must be equal to the product of the probabilities of no collision in the subsequent time intervals from 0 to t and from t to $t+dt$:

$$P_{time}(t + dt) = P_{time}(t) P_{time}(dt) \qquad (3.62)$$

On the other hand we know that vdt is the probability that a collision occurs during the infinitesimal time interval, dt, therefore the probability that no collision occurs during the same short time period must be $P_{time}(dt) = (1-vdt)$. This means that

$$P_{time}(t + dt) = P_{time}(t)(1 - v\,dt) \qquad (3.63)$$

The left hand side of equation (3.63) can be expanded into a Taylor series and one can stop after the linear term because dt is a very small quantity. This linear expansion yields the following differential equation for P_{time}:

$$\frac{dP_{time}}{dt} = -v P_{time} \qquad (3.64)$$

The solution of this differential equation (taking into account the $P_{time}(0)=1$ condition) is the following:

$$P_{time}(t) = e^{-vt} \qquad (3.65)$$

Equation (3.65) gives the probability that no collision occurs during the time interval t. Based on this probability one can readily derive the probability that the free time between two successive collisions of a given molecule is between t and $t+dt$. Let us denote this probability by $f_{time}(t)dt$. $f_{time}(t)dt$ must be equal to the product of the probability that no collision occurs in the time interval from 0 to t, $P_{time}(t)$, and the probability that a collision occurs between t and $t+dt$, vdt:

$$f_{time}(t)dt = e^{-vt} v\,dt \qquad (3.66)$$

It should be recognized that $f_{time}(t)$ is the distribution of free times. Using this distribution function one can readily derive the mean free time, τ:

$$\tau = \int_0^\infty dt\, t\, f_{time}(t) = \int_0^\infty dt\, t\, \nu e^{-\nu t} = \frac{1}{\nu} \qquad (3.67)$$

The mean free time is the average time between collisions for particles with a given speed (the mean free time is also called the collision time). It should be noted that at this point the collision time (or collision rate) is a formally defined quantity which must be related to known molecular properties (such as size, mass, speed, etc.).

3.2.2 Mean free path

The free path is the distance traveled by a molecule between two successive collisions. The mean free path is the average distance between two successive collisions of a single molecule. The distribution of free paths can be derived in the same way as we derived the distribution of free times.

Consider a molecule with velocity, $\mathbf{v}$. Assume that this particle suffers a collision with another molecule at s=0. Let $P_{path}(s)$ denote the probability that this molecule survives a distance, s, without suffering a second collision. Obviously, $P_{path}(0)=1$, because we neglected multiple collisions. Let αds denote the probability that a given molecule suffers a collision between distance s and s+ds. The quantity α is now the collision probability per unit distance for an individual particle. It is assumed that the collisional probability density, α, is independent of the past collisional history of the molecule. In general, however, α may depend on the speed of the particle (faster particles travel larger distances and therefore encounter more molecules in the same time interval), so it is written $\alpha(v)$.

With the help of the parameter, α, it is possible to derive the functional form of the survival probability, P_{path}. The probability that the particle survives an interval s+ds without suffering a collision must be equal to the product of the probabilities that no collision occurs in the subsequent intervals from 0 to s and from s to s+ds:

$$P_{path}(s + ds) = P_{path}(s)(1 - \alpha\, ds) \qquad (3.68)$$

where we have made use of the fact that $P_{path}(ds)=(1-\alpha ds)$. The left hand side of equation (3.68) can be expanded into a Taylor series and one can stop after the linear term because dt is a very small quantity. This linear expansion yields the following differential equation for P_{path}:

$$\frac{dP_{path}}{ds} = -\alpha P_{path} \tag{3.69}$$

The solution of this differential equation (taking into account the $P_{path}(0)=1$ condition) is the following:

$$P_{path}(s) = e^{-\alpha s} \tag{3.70}$$

Equation (3.70) gives the probability that no collision occurs on the distance interval, s. Based on this probability one can readily derive the probability that the free path between two successive collisions of a given molecule is between s and s+ds. Let us denote this probability by $f_{path}(s)ds$. It is obvious that $f_{path}(s)ds$ must be equal to the product of the probabilities that no collision occurs in the interval from 0 to s, $P_{path}(s)$, and a collision does occur between s to s+ds, αds:

$$f_{path}(s)ds = e^{-\alpha s}\,\alpha ds \tag{3.71}$$

It should be recognized that $f_{path}(s)$ is the distribution of free paths. Using this distribution function one can readily derive the mean free path, λ:

$$\lambda = \int_0^\infty ds\,s\,f_{path}(s) = \int_0^\infty ds\,s\,\alpha e^{-\alpha t} = \frac{1}{\alpha} \tag{3.72}$$

The mean free path is the average distance between collisions for particles (with a given speed). It is obvious that the mean free time, τ, and the mean free path, λ, are not independent quantities. On the average a particle travels a distance of λ between two consecutive collisions in time τ, therefore the ratio of the two, λ/τ, must represent the average speed of the particle:

$$\frac{\lambda}{\tau} = v \tag{3.73}$$

3.2.3 Collision cross section

Collisions between molecules can be described in terms of the scattering cross section, σ, which is determined by the interaction potential between the particles. In the case of a single species gas composed of hard sphere molecules the cross section is simply the projected area of the sphere of influence onto a plane perpendicular to the relative velocity vector of the particles, $\sigma = \pi d_m^2$ (d_m is the diameter of the hard sphere molecule).

A more general definition can be given in the following way. Consider a beam of particles moving with a fixed velocity, v_1, incident upon a single target molecule (moving with velocity v_2) that deflects the beam molecules. In the frame of reference of the target molecule the beam is moving with the relative velocity, $g = v_1 - v_2$. In this frame of reference the beam intensity (number of particles intersecting in unit time a unit area perpendicular to the beam) is denoted by I_0. As a result of collisions with the target molecule a fraction of the beam, dI, will be scattered so that the direction of the scattered molecules is in a small solid angle element, $d\Omega'$ (see Figure 3.7). The number of particles scattered into the solid angle element in unit time is proportional to the intensity of the incident beam, I_0, and to the solid angle element, $d\Omega'$. One can then write

$$d^2N = I_0 \, S \, d\Omega' \qquad (3.74)$$

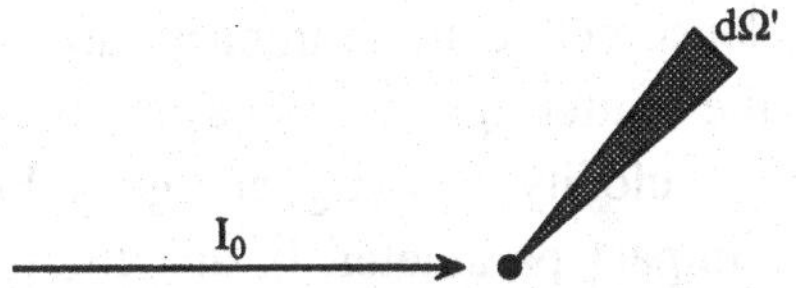

Figure 3.7

where d^2N is the scattering rate into the small solid angle element, and S is a factor of proportionality. This proportionality factor, S, is called differential cross section. In general the differential cross section is a function of the relative velocity vectors before and after the collision, $S = S(g, g')$. If the interaction potential between the molecules is known, the differential cross section can be calculated by using classical or quantum mechanical methods. It can be easily seen from equation (3.74) that the differential cross section has the dimensions of an area. The physical interpretation is that $S(g, g')$ is the perpendicular area of the molecule (in a plane perpendicular to the pre-collision relative velocity vector, g) which scatters incident molecules with relative velocity g to an infinitesimal velocity space solid angle around the relative velocity vector g'.

The total rate of particles scattered in all directions, N, can be obtained by integrating equation (3.74) over all directions of the scattered relative velocity vector, g':

$$N = I_0 \iint_{4\pi} d\Omega' \, S(g, g') = \sigma I_0 \qquad (3.75)$$

where σ is called the total scattering cross section and is defined by the following integral:

$$\sigma(\mathbf{g}, \mathbf{g}') = \iint\limits_{4\pi} d\Omega' \, S(\mathbf{g}, \mathbf{g}') \tag{3.76}$$

In general the total scattering cross section depends on the incident relative velocity vector, **g**, and on the magnitude of the scattered relative velocity, g'.

The intermolecular force field is generally considered to be spherically symmetric. This assumption is approximately valid in many cases. Without this assumption gaskinetic theory becomes almost prohibitively complicated (we have enough difficulties even with this simplifying assumption), therefore spherically symmetric force fields will be assumed throughout this book. At the same time one must realize that this assumption represents a potentially important limitation of gaskinetic theory.

As was discussed in Section 3.1 the relative motion of two molecules interacting with a spherically symmetric central field of force is a planar motion and there is a one-to-one correspondence between the impact parameter, b, and the deflection angle, χ. In this case the scattering process has an axis of symmetry which is parallel to the initial relative velocity vector, g, and goes through the center of the target molecule. Both the impact parameter, b, and the deflection angle, χ, are measured in the plane of motion. This situation is shown schematically in Figure 3.8. The plane of relative motion is defined by the azimuth angle, ε, measured from an arbitrary reference plane in the counter-clockwise direction. Due to the

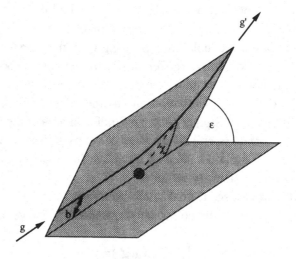

Figure 3.8

spherical symmetry of the central field of force, the differential cross section, S, must be independent of the azimuth angle, ε, and the direction of the initial relative velocity, **g**. This means that $S=S(g,\chi)$.

In the case of a spherically symmetric central field of force the differential cross section can also be expressed in terms of the impact parameter, b, and the azimuth angle, ε. The differential cross section is the infinitesimal area which scatters incident molecules with relative velocity **g** to an infinitesimal velocity space solid angle element, $d\Omega'$, around the scattered relative velocity vector **g'** (see Figure 3.9). We know that there is a one-to-one correspondence between b and χ, therefore particles approaching the center of force along a very thin cylinder with radius b centered at the axis of symmetry will have the same deflection angle, χ. The azimuth angle of the plane of relative motion, ε, can vary from 0 to 2π without having any effect on the deflection angle. The direction of motion of the scattered particle is fully characterized by χ and ε, and the infinitesimal solid angle element around this direction can be written as $d\Omega'=\sin\chi\,d\chi\,d\varepsilon$. The area element of the target molecule which scatters particles with deflection angles between χ and $\chi+d\chi$ can be written as $-2\pi b\,db$, and the area which scatters particles with deflection angles between χ and $\chi+d\chi$ and with azimuth angles between ε and $\varepsilon+d\varepsilon$ is $-b\,db\,d\varepsilon$ (the $-$ sign is needed because χ decreases with increasing values of b, therefore db and $d\chi$ have opposite signs). This area element is exactly the differential cross section, therefore one can write

$$S(g,\chi)\sin\chi\,d\chi\,d\varepsilon = -b\,db\,d\varepsilon \qquad (3.77)$$

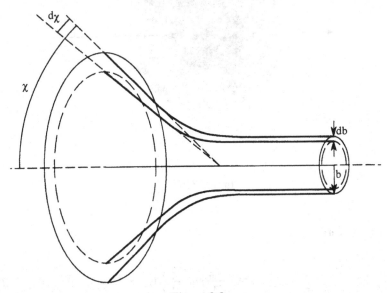

Figure 3.9

This equation can be solved for the differential cross section

$$S(g,\chi) = -\frac{b(\chi)}{\sin\chi}\frac{db}{d\chi} \qquad (3.78)$$

This result can be illustrated by considering the case of hard sphere molecular collisions. Figure 3.10 illustrates the collision of two identical molecules with diameters, d_m, initial relative velocity vector, $\mathbf{g}$, and impact parameter, b. It is obvious from the schematics that the relation between the impact parameter and the deflection angle can be expressed as

$$b = d_m \sin\phi_m = d_m \sin\left(\frac{\pi-\chi}{2}\right) = d_m \cos\left(\frac{\chi}{2}\right) \qquad (3.79)$$

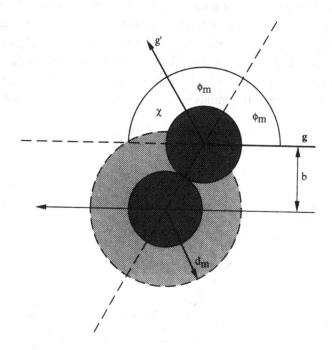

Figure 3.10

or in differential form

$$db = -\frac{d_m}{2}\sin\left(\frac{\chi}{2}\right)d\chi \qquad (3.80)$$

Substituting equations (3.79) and (3.80) into equation (3.78) yields the following expression for the differential cross section of the hard sphere molecule:

$$S(g,\chi) = -\frac{b}{\sin\chi}\frac{db}{d\chi} = \frac{d_m^2}{4} \tag{3.81}$$

The total scattering cross section can be obtained by integrating over the entire solid angle space:

$$\sigma(g) = \int_0^\pi d\chi \sin\chi \int_0^{2\pi} d\varepsilon \, S(g,\chi) = \int_0^\pi d\chi \sin\chi \int_0^{2\pi} d\varepsilon \frac{d_m^2}{4} = \pi d_m^2 \tag{3.82}$$

This result simply tells us that the cross section of a hard sphere molecule is the cross section of the sphere of influence. This result is not surprising and is completely consistent with physical intuition.

Equation (3.78) can be used to obtain the differential cross section for molecular interactions characterized by an arbitrary potential, U(r). In this case the relation between the deflection angle and the impact parameter was found to be the following (see equation (3.33)):

$$\chi = \pi - 2\frac{b}{r_m}\int_1^\infty dx \frac{1}{x\sqrt{x^2 - \frac{b^2}{r_m^2} - \frac{2x^2 U(x)}{m^* g^2}}} \tag{3.83}$$

Here the distance of closest approach, r_m, is given by equation (3.31). In equation (3.83) we have also introduced a new dimensionless integration variable, $x = r/r_m$. Equation (3.83) can be used to prove that $\chi(b)$ is a monotonically decreasing function. This can be seen by taking the derivative of the deflection angle, χ, with respect to the impact parameter, b, and using the definition of the closest approach (equation (3.31)):

$$\frac{d\chi}{db} = -2\frac{d}{db}\left(\frac{b}{r_m}\right)\int_1^\infty dx \, x \frac{1 - \frac{2U(x)}{m^* g^2}}{\left[x^2 - \frac{b^2}{r_m^2} - \frac{2x^2 U(x)}{m^* g^2}\right]^{3/2}} \tag{3.84}$$

The integral in equation (3.84) is positive semidefinite because the expression in the denominator is positive for x>1 and consequently the numerator in the integral is also positive (it should be noted that this integral is improper, because the integrand is singular at x=1). The derivative in equation (3.84) can also be written in the following form:

$$\frac{d}{db}\left(\frac{b}{r_m}\right) = \frac{1}{r_m}\left(1 - \frac{b}{r_m}\frac{dr_m}{db}\right)$$

(3.85)

Equation (3.85) is positive if the increase of the distance of closest approach is slower than the increase of the impact parameter, b. This condition is satisfied by all physically meaningful interparticle potentials. It is interesting to note that if the potential is constant, $U=U_0$, then r_m can be expressed as

$$r_m = \frac{b}{\sqrt{1 - \frac{2U_0}{m^*g^2}}}$$

(3.86)

and consequently

$$\frac{d}{db}\left(\frac{b}{r_m}\right) = 0$$

(3.87)

which means that $d\chi/db=0$.

In addition to the differential and total scattering cross sections several other cross sections appear in non-equilibrium transport theory. These general transport cross sections are defined in the following way:

$$\sigma_n(g) = \int_0^{2\pi} d\varepsilon \int_0^{\pi} d\chi \sin\chi (1 - \cos^n \chi) S(g,\chi)$$

(3.88)

where n is a positive integer (n=1,2,...). For instance σ_1 (called the momentum transfer cross section) plays an important role in the study of diffusion, while viscosity and heat conductivity depend on the next higher order cross section, σ_2. The physical meaning of the transport cross sections will be discussed in detail later (in Chapter 6). Here we only mention that these cross sections are closely related to the transport of various macroscopic gas quantities. For instance, the transfer of the parallel component of the particle momentum is proportional to $1-\cos\chi$ (see equation (3.61a)), therefore σ_1 will be related to the transport of momentum (we will see that the momentum component perpendicular to the ambient relative velocity does not contribute to the macroscopic momentum transport).

For hard sphere molecules the differential cross section is $S(g,\chi)=d_m^2/4$ and the σ_1 and σ_2 quantities can be easily evaluated:

$$\sigma_1(g) = \frac{d_m^2}{4} \int_0^{2\pi} d\varepsilon \int_0^{\pi} d\chi \sin\chi (1-\cos\chi) = \pi d_m^2 \qquad (3.89a)$$

$$\sigma_2(g) = \frac{d_m^2}{4} \int_0^{2\pi} d\varepsilon \int_0^{\pi} d\chi \sin\chi (1-\cos^2\chi) = \frac{2}{3}\pi d_m^2 \qquad (3.89b)$$

3.2.4 Collision cross sections for inverse power interactions

It is of particular interest to examine the differential cross section for inverse power interactions. We start with the $\chi(b)$ function which can be expressed in the following form (see equation (3.40)):

$$\chi(x_0) = \pi - 2\int_0^{x_m} dx \left[1 - x^2 - \frac{2}{(a-1)}\left(\frac{x}{x_0}\right)^{a-1} \right]^{-1/2} \qquad (3.90)$$

where x_0 is the normalized impact parameter:

$$x_0 = b\left(\frac{m^*}{K_a}\right)^{\frac{1}{a-1}} g^{\frac{2}{a-1}} \qquad (3.91)$$

It should be mentioned again that for a given value of the exponent, a, the upper integration boundary, x_m, is only a function of x_0 and is given by equation (3.39).

The differential cross section can be easily calculated for some special inverse power interactions. The lowest order case, a=2, describes electrostatic or gravitational interactions. In the previous subsection we derived the $b(\chi)$ function for this interparticle potential (see equation (3.44)) and one obtains:

$$b = \frac{K_2}{m^*} \frac{1}{g^2} \frac{1}{\tan\left(\frac{\chi}{2}\right)} \qquad (3.92)$$

Using equations (3.78) and (3.92) one can derive the differential cross section for this particular inverse power potential field:

$$S(g,\chi) = \frac{1}{4}\left(\frac{K_2}{m^*}\right)^2 \frac{1}{g^4} \frac{1}{\sin^4\left(\dfrac{\chi}{2}\right)}$$ (3.93)

Equation (3.93) can be used to obtain the well known Rutherford formula describing the differential cross section of electron scattering on atomic nuclei. In this case $K_2 = e^2/4\pi\varepsilon_0$ (where e is the electron charge, $e = 1.602\ 2\times10^{-19}$ coulomb, and ε_0 is the permittivity of vacuum, $\varepsilon_0 = 8.854\ 187\times10^{-12}$ farad m^{-1}). With these values equation (3.93) becomes the Rutherford formula:

$$S(g,\chi) = \left(\frac{e^2}{8\pi\varepsilon_0 m^*}\right)^2 \frac{1}{g^4} \frac{1}{\sin^4\left(\dfrac{\chi}{2}\right)}$$ (3.94)

The differential cross section for the a=3 inverse power potential can be obtained by using equation (3.47) relating the deflection angle and the impact parameter. Using equation (3.44) one can express b in terms of χ:

$$b = b_0 \frac{(\pi-\chi)}{\sqrt{\chi}\sqrt{2\pi-\chi}}$$ (3.95)

where $b_0 = (K_3/m^*g^2)^{1/2}$. Substituting this expression into equation (3.78) yields the differential cross section for the a=3 inverse power potential:

$$S(g,\chi) = -\frac{b}{\sin\chi}\frac{db}{d\chi} = \frac{b_0}{\sin\chi}\frac{\pi^2 b_0(\pi-\chi)}{\chi^2(2\pi-\chi)^2}$$ (3.96)

Next we will examine the total cross section for inverse power interaction potentials. We start this process by investigating the $\chi(b)$ function in the small angle scattering (large impact parameter) limit. For an arbitrary inverse power potential equation (3.90) can be rewritten in the following form:

$$\chi = \pi - 2\int_0^1 dx \frac{1}{\sqrt{\dfrac{1-\delta_a x^{a-1}}{1-\delta_a} - x^2}}$$ (3.97)

where we have introduced a new integration variable and the following dimensionless quantity:

$$\delta_a = \frac{2}{a-1}\left(\frac{x_m}{x_0}\right)^{a-1} \tag{3.98}$$

When deriving equation (3.97) we also used the relationship $x_m^2 = 1 - \delta_a$ (see equation (3.39)). It can be immediately seen that for a given exponent value, a, δ_a is only a function of the dimensionless impact parameter, x_0. For large impact parameters x_0 becomes large and consequently δ_a approaches zero. On the other hand when δ_a approaches zero, x_m goes to 1 (which means that for large impact parameter collisions the closest distance is practically the impact parameter itself).

In the $\delta_a \to 0$ limit one can expand the integrand in equation (3.97) in powers of δ_a and drop the higher order terms (keep only the zeroth and first order terms):

$$\lim_{\delta_a \to 0} \chi = \pi - 2\int_0^1 dx \left[\frac{1}{\sqrt{1-x^2}} - \frac{\delta_a}{2}\frac{1-x^{a-1}}{(1-x^2)^{3/2}}\right] = \delta_a \int_0^1 dx \frac{1-x^{a-1}}{(1-x^2)^{3/2}} \tag{3.99}$$

The integral in equation (3.99) is a pure number which depends only on the inverse power exponent, a. For integer values of a this integral can be evaluated. Let us introduce

$$B_a = \int_0^1 dx \frac{1-x^{a-1}}{(1-x^2)^{3/2}} \tag{3.100}$$

It is easy to see that $B_2 = 1$, while for $a > 2$, B_a assumes values given by

$$B_a = \begin{cases} \dfrac{1\cdot3\cdot5\cdot\ldots\cdot(a-2)}{2\cdot4\cdot\ldots\cdot(a-3)}\dfrac{\pi}{2} & a = 3,5,7,\cdots \\[3mm] \dfrac{2\cdot4\cdot6\cdot\ldots\cdot(a-2)}{1\cdot3\cdot\ldots\cdot(a-3)} & a = 4,6,8,\cdots \end{cases} \tag{3.101}$$

For large impact parameters, x_m can be replaced by 1 and therefore the deflection angle can be expressed in the following form:

$$\lim_{b \to \infty} \chi = \frac{1}{b^{a-1}}\frac{2B_a}{a-1}\left(\frac{K_a}{m^*g^2}\right) \tag{3.102}$$

Substituting this into the expression for the differential cross section yields the following formula:

$$\lim_{b \to \infty} S(g, \chi) = -\frac{b}{\sin \chi} \frac{db}{d\chi} = \chi^{-\frac{2a}{a-1}} \frac{g^{-\frac{4}{a-1}}}{a-1} \left(\frac{2B_a}{a-1} \frac{K_a}{m^*} \right)^{\frac{2}{a-1}} \tag{3.103}$$

where we have used the well known small angle approximation for the sine function, $\sin \chi \approx \chi$.

It can be seen that as the impact parameter goes to infinity ($\chi \to 0$) the differential cross section becomes singular (because $1/\chi^{2a/(a-1)} \to \infty$ as $\chi \to 0$). The physical meaning of this singularity is that collisions with large impact parameters result in highly focused forward scattering (extremely small deflection angles). Inspection of equation (3.103) reveals that the deflection angle dependence of the differential cross section changes from χ^{-4} to χ^{-2} as the inverse power exponent increases from a=2 to a=∞.

The total cross section of molecules interacting via inverse power potentials can be divided into two parts. The first part describes the molecular 'area' resulting in collisions with deflection angles larger than an arbitrary (but small) angle, χ_0 (<<1), while the second integral describes the contribution of small angle collisions ($\chi < \chi_0$):

$$\sigma = \int_0^{2\pi} d\varepsilon \int_{\chi_0}^{\pi} d\chi \sin \chi \, S(g, \chi) + \int_0^{2\pi} d\varepsilon \int_0^{\chi_0} d\chi \sin \chi \, S(g, \chi) \tag{3.104}$$

In the next step we use our earlier results to simplify the first integral and equation (3.103) to evaluate the second integral

$$\sigma = -2\pi \int_{\chi_0}^{\pi} d\chi \, b(\chi) \frac{db}{d\chi} - \pi g^{-\frac{4}{a-1}} \left(\frac{2B_a}{a-1} \frac{K_a}{m^*} \right)^{\frac{2}{a-1}} \left[\chi^{-\frac{2}{a-1}} \right]_0^{\chi_0} \tag{3.105}$$

In equation (3.105) the first integral is finite, but the second term is infinite for any value a≥2. Equation (3.105) indicates that the total scattering cross section of molecules interacting via an r^{-a} potential is infinite, because the integral becomes infinite at the lower boundary ($\chi=0$). The singularity arises because of the large contribution of small deflection angles (large impact parameters). For large impact parameters the differential cross section grows faster than $\sin \chi$ decreases.

The classical cross section for inverse power potentials is infinite indicating the limits of applicability of classical physics. This problem can be resolved by using quantum mechanics[11,12] or by taking into account the shielding of electrical charges in plasma physics[13,14] (for the a=2 case). In plasma physics the impact parameter is limited by the shielding effect of other charged particles, therefore there is a minimum deflection angle for particle collisions, χ_{min}. In this case the total cross section becomes

$$\sigma(g) = -\left(\frac{K_2}{m^*}\right)^2 \frac{2\pi}{g^4}\left[\frac{\cos\chi}{\sin^3\chi}\right]_{\chi_{min}/2}^{\pi/2} = \left(\frac{K_2}{m^*}\right)^2 \frac{2\pi}{g^4}\frac{1}{\sin^3\chi_{min}} \qquad (3.106)$$

Small angle collisions result in only a very small transfer of momentum and energy between the colliding molecules, therefore the total collision cross section is not very important in gaskinetic theory. More important are the cross sections characterizing the transfer of momentum (σ_1) and energy (σ_2) which are weighted by collisions resulting in more significant deflections.

For inverse power interactions the general cross section can be written in the following form:

$$\sigma_n(g) = \int_0^{2\pi} d\varepsilon \int_0^\pi d\chi \sin\chi (1-\cos^n\chi) S(g,\chi) = -2\pi\int_0^\pi d\chi\, b\,(1-\cos^n\chi)\frac{db}{d\chi}$$

$$(3.107)$$

The impact parameter, b, can be replaced by the dimensionless quantity, x_0 (see equation (3.39)). This operation leads to the following expression:

$$\sigma_n(g) = -2\pi\left(\frac{K_a}{m^*}\right)^{\frac{2}{a-1}} g^{-\frac{4}{a-1}}\int_0^\pi d\chi \frac{dx_0}{d\chi} x_0\,(1-\cos^n\chi) \qquad (3.108)$$

It is easy to express the integral in equation (3.108) in terms of x_0 instead of the deflection angle, χ:

$$\sigma_n(g) = 2\pi\left(\frac{K_a}{m^*}\right)^{\frac{2}{a-1}} g^{-\frac{4}{a-1}}\int_0^\infty dx_0\, x_0\left[1-\cos^n\chi(x_0)\right] = 2\pi\left(\frac{K_a}{m^*}\right)^{\frac{2}{a-1}} g^{-\frac{4}{a-1}}A_n(a)$$

$$(3.109)$$

where $A_n(a)$ is a pure number determined only by the exponent of the inverse power potential, a:

$$A_n(a) = \int_0^\infty dx_0 \, x_0 \left[1 - \cos^n \chi(x_0)\right] \tag{3.110}$$

Most of the higher order classical cross sections are finite even for small angle collisions. This can be seen by calculating the contribution of small angle collisions to $\sigma_n(g)$. This can be done by substituting the small angle collision limit of the differential cross section (equation (3.103)) into equation (3.107) and using the series expansion of the $\sin \chi$ and $\cos \chi$ functions:

$$\sigma_n(g)_{\chi < \chi_0} = \int_0^{2\pi} d\varepsilon \int_0^{\chi_0} d\chi \sin \chi (1 - \cos^n \chi) S(g, \chi)$$

$$= 2\pi \frac{g^{-\frac{4}{a-1}}}{a-1} \left(\frac{2B_a}{a-1} \frac{K_a}{m^*}\right)^{\frac{2}{a-1}} \int_0^{\chi_0} d\chi \frac{n}{2} \chi^3 \chi^{-\frac{2a}{a-1}} \tag{3.111}$$

The integral can easily be evaluated to yield the following relation:

$$\sigma_n(g)_{\chi < \chi_0} = \frac{\pi}{2} \frac{n}{a-2} g^{-\frac{4}{a-1}} \left(\frac{2B_a}{a-1} \frac{K_a}{m^*}\right)^{\frac{2}{a-1}} \left[\chi^{\frac{2a-4}{a-1}}\right]_0^{\chi_0} \tag{3.112}$$

It can be seen that the contribution of small angle scattering (and consequently the transport cross section itself) is finite for $a > 2$ values. As previously indicated the still singular $a = 2$ case has to be treated with quantum mechanical or plasmaphysical methods. Tabulated values of $A_n(a)$ are given in Table 3.1 for exponents $a > 2$.

Inverse power interaction potentials are widely used in kinetic theory. On the other hand it was shown above that the classical interpretation of molecular trajectories in inverse power potential fields leads to infinite values of the total cross section, σ. As we shall see in the following sections the total scattering cross section plays an important role in interpreting statistical quantities such as the mean free path or the collision frequency. This problem can be sidestepped by using the momentum transfer cross section, σ_1, in the calculation of transport coefficients and in the classical interpretation of the mean free path and the collision frequency. The use of the momentum transfer cross section is motivated not only by mathematical necessity, but also by the physical argument that in the calculation of mean free path only those collisions should be taken into account which result in appreciable change in the momentum vector.

Table 3.1

a	A_1	A_2	A_3
3	0.794 95	1.055 19	1.424 50
4	0.490 44	0.560 81	0.749 40
5	0.421 94	0.436 19	0.585 20
6	0.395 92	0.384 01	0.518 86
7	0.385 44	0.356 75	0.485 63
8	0.381 54	0.340 66	0.466 98
9	0.380 80	0.330 40	0.455 78
10	0.381 68	0.323 52	0.448 80
11	0.383 43	0.318 73	0.444 37
12	0.385 62	0.315 30	0.441 57
13	0.388 04	0.312 82	0.439 84
14	0.390 54	0.310 99	0.438 84
15	0.393 06	0.309 64	0.438 34
20	0.404 77	0.306 74	0.439 57
25	0.414 40	0.306 55	0.442 87
30	0.422 20	0.307 23	0.446 37
40	0.433 95	0.309 19	0.452 59
50	0.442 35	0.311 13	0.457 56
75	0.455 73	0.314 97	0.466 18
100	0.463 70	0.317 62	0.471 67
∞	0.500 00	0.333 33	0.500 00

3.3 Relations between statistical and molecular quantities

3.3.1 Collision frequency

In this section we shall find quantitative relationships between statistical quantities characterizing molecular collisions (such as mean free path or collision frequency) and basic physical properties of the gas (such as concentration, temperature, etc.). We start with a simplified discussion to introduce the physical concept, and then present a more rigorous method of calculating these relationships.

Consider a gas consisting of only a single kind of molecule. Let us take one single molecule (molecule 1) and calculate the number of its collisions per unit time. We denote the average concentration of molecules (the average number of molecules per unit volume) by n and assume that all other molecules move with the same speed, $\bar{v}$, and their relative speed with respect to our molecule 1 is $\bar{g}$ (where $\bar{v}$ and $\bar{g}$ denote the average speed of the molecules and their mean relative speed, respectively). The fixed molecule represents an effective target area of σ. This cross section is the total scattering cross section of the molecule: in the case of inverse power interaction potentials one can approximate it by the momentum transfer cross section, or by the total cross section obtained by plasmaphysical or quantum mechanical methods. Consider the number of collisions with molecule 1 in time dt. At t=0 those molecules which collide with molecule 1 in time dt have their centers within a circular cylinder of length $\bar{g}\,dt$ and volume $\sigma\,\bar{g}\,dt$. Now the number of collisions of molecule 1 in time dt is the volume of the impact cylinder times the number density of the impinging molecules, $n\sigma\,\bar{g}\,dt$.

Earlier we defined the collision frequency, ν, as the number of collisions of a single molecule per unit time. We conclude that

$$\nu = n\sigma\bar{g} \tag{3.113}$$

Next we calculate the average relative speed, $\bar{g}$. Let us consider the relative velocity between our molecule 1 and another arbitrary molecule, molecule 2. It was assumed that all molecules move with the same speed, $\bar{v}$, therefore the square of the relative speed can be written as

$$g_{12}^2 = v_1^2 + v_2^2 - 2v_1v_2\cos\theta = \bar{v}^2 + \bar{v}^2 - 2\bar{v}^2\cos\theta \tag{3.114}$$

where we have denoted the angle between the two velocity vectors by θ. The average relative speed can be obtained by taking the angular average of equation (3.114):

$$\bar{g}^2 = <g_{12}^2>_\theta = 2\bar{v}^2 - 2\bar{v}^2<\cos\theta>_\theta = 2\bar{v}^2 \tag{3.115}$$

This result can be substituted into equation (3.113) and so we can express the collision frequency (number of collisions of a single molecule in unit time) in terms of well understood fundamental parameters of the gas:

$$\nu = \sqrt{2}n\sigma\bar{v} \tag{3.116}$$

If we assume that the gas is in kinematic equilibrium one can use equation (2.65) to express the average molecular speed in terms of the gas temperature and the molecular mass, m. This substitution results in the following expression for the collision frequency:

$$\nu = 4n\sigma\sqrt{\frac{kT}{\pi m}} \tag{3.117}$$

We were able to express the collision frequency in terms of fundamental gas parameters. This result can be used to obtain specific expressions for the mean free time and the mean free path of the molecules as well. Substituting equation (3.116) into equation (3.67) which directly relates the collision frequency to the mean free time yields the following:

$$\tau = \frac{1}{\sqrt{2}n\sigma\bar{v}} \tag{3.118}$$

We also know that the ratio of the mean free path and mean free time is the average particle speed (see equation (3.73)), therefore one immediately obtains

$$\lambda = \frac{1}{\sqrt{2}n\sigma} \tag{3.119}$$

This is a very important result which tells us that the mean free path of the molecules depends only on the gas density and the scattering cross section of the molecules and is independent of the average molecular speed (or equivalently, the temperature).

3.3.2 Collision rate

Next we calculate the collision rate in a gas mixture. This calculation will be carried out in a rigorous way assuming only that the velocity distribution function is known for all molecular species.

Let us consider two groups of molecules. Molecules in the first group have velocities in the range from $\mathbf{v}_1$ to $\mathbf{v}_1+d^3\mathbf{v}_1$ and the other between $\mathbf{v}_2$ and $\mathbf{v}_2+d^3\mathbf{v}_2$. The number densities of the two groups of molecules are n_1 and n_2, respectively.

The rate at which a single molecule of group 2 encounters molecules of group 1 is the same as if the molecule of group 2 was at rest and the molecules of group 1 were approaching it with velocities between $\mathbf{g}$ and $\mathbf{g}+d^3\mathbf{g}$, where $\mathbf{g}=\mathbf{v}_1-\mathbf{v}_2$ is the relative velocity of the two groups of molecules. The flux of molecules 1 impinging on molecule 2 is

$$d^3I = g\,d^3n_1 = g\,n_1\,f_1(\mathbf{v}_1)d^3\mathbf{v}_1 \tag{3.120}$$

where $f_1(\mathbf{v}_1)$ is the normalized velocity distribution function of molecules 1.

As has been discussed in the previous sections each collision between a molecule 1 and a molecule 2 is characterized by the relative speed, g, and by the direction of the relative velocity vector after the collision, denoted by the angles χ and ε (or equivalently, by the azimuth angle and the impact parameter). The number of encounters per unit time between a single molecule of group 2 and molecules of group 1 resulting in a final relative velocity vector within the solid angle element, $d\Omega$, around direction (χ, ε) is

$$d^5I = S(g, \chi) \sin \chi \, d\chi \, d\varepsilon \, d^3I = g S(g, \chi) n_1 f_1(\mathbf{v}_1) \sin \chi \, d\chi \, d\varepsilon \, d^3v_1 \quad (3.121)$$

Since there are $d^3n_2 = n_2 f_2(\mathbf{v}_2) d^3v_2$ molecules of group 2 per unit volume, the number of encounters per unit time between the two groups which scatter into the solid element $d\Omega$ is

$$d^8I = g S(g, \chi) n_1 n_2 f_1(\mathbf{v}_1) f_2(\mathbf{v}_2) \sin \chi \, d\chi \, d\varepsilon \, d^3v_1 \, d^3v_2 \quad (3.122)$$

At this point we must recognize the potential problem of double counting. If molecules 1 and 2 are identical, equation (3.122) describes the differential collision rate (number of collisions between two groups of molecules with specific velocity vectors and deflection properties in unit time and unit volume) between molecules of the same species. Obviously, by integrating over all eight variables (six velocity components, impact azimuth and deflection angle) one obtains the total collision rate of these molecules (total number of collisions between two groups of molecules in unit time and unit volume). In the case of identical molecules such integration counts each collision twice, because both colliding molecules are denoted once as particle 1 and once as particle 2. In this case both n_1 and n_2 are identical to the number density of all colliding particles.

If molecules 1 and 2 are different, such a double counting cannot occur, because when the integrals are carried out $\mathbf{v}_1$ always denotes molecules 1 and it cannot refer to a molecule 2. In this case the total number density of colliding particles is $n_1 + n_2$.

In order to avoid double counting we introduce a symmetry factor, κ_{12}, which corrects for the potential double counting:

$$\kappa_{12} = \frac{n_{12}}{n_1 + n_2} \quad (3.123)$$

where n_{12} is the total number density of molecules 1 and 2. For identical molecules (also called like molecules) $\kappa_{12} = 1/2$, while for different molecules (unlike molecules) $\kappa_{12} = 1$. With this correction the differential collision rate becomes the following:

$$d^8I = \kappa_{12} g S(g, \chi) n_1 n_2 f_1(\mathbf{v}_1) f_2(\mathbf{v}_2) \sin \chi \, d\chi \, d\varepsilon \, d^3v_1 \, d^3v_2 \quad (3.124)$$

The total collision rate, N_{12}, is the number of collisions between molecules 1 and 2 in unit time and in unit volume. This can be obtained by integrating expression (3.124) over all variables:

$$N_{12} = \kappa_{12} n_1 n_2 \iiint_{\infty} d^3 v_1 \iiint_{\infty} d^3 v_2 \int_0^{2\pi} d\varepsilon \int_0^{\pi} d\chi \sin \chi \, S(g, \chi) g f_1(\mathbf{v}_1) f_2(\mathbf{v}_2) \quad (3.125)$$

The two integrals over the scattering angles define the total scattering cross section (we use a finite scattering cross section obtained by quantum mechanical, plasma-physical or some other approximation), σ. It was shown earlier in this chapter that for spherically symmetric interaction force fields (and our consideration is limited to such forces) the total cross section is only a function of the relative speed, therefore $\sigma = \sigma(g)$. Now equation (3.125) can be written in the following form:

$$N_{12} = \kappa_{12} n_1 n_2 \iiint_{\infty} d^3 v_1 \iiint_{\infty} d^3 v_2 \sigma(g) g f_1(\mathbf{v}_1) f_2(\mathbf{v}_2) \quad (3.126)$$

The integration with respect to the particle velocities is quite difficult, because the relative speed, g, is a transcendental function of the velocity components. To circumvent this difficulty we transform the integration variables from $\mathbf{v}_1$ and $\mathbf{v}_2$ to the center of mass and relative velocities, $\mathbf{v}_c$ and $\mathbf{g}$. One can express $\mathbf{v}_1$ and $\mathbf{v}_2$ in terms of $\mathbf{v}_c$ and $\mathbf{g}$ (see equations (3.51)):

$$\mathbf{v}_1 = \mathbf{v}_c - \frac{m^*}{m_1} \mathbf{g} \qquad \mathbf{v}_2 = \mathbf{v}_c + \frac{m^*}{m_2} \mathbf{g} \quad (3.127)$$

The differential quantities can be transferred in the following way:

$$d^3 v_1 d^3 v_2 = \frac{\partial(\mathbf{v}_1, \mathbf{v}_2)}{\partial(\mathbf{v}_c, \mathbf{g})} d^3 v_c d^3 g \quad (3.128)$$

where the Jacobian of the transformation is given by

$$\frac{\partial(\mathbf{v}_1, \mathbf{v}_2)}{\partial(\mathbf{v}_c, \mathbf{g})} = \begin{vmatrix} \dfrac{\partial \mathbf{v}_1}{\partial \mathbf{v}_c} & \dfrac{\partial \mathbf{v}_2}{\partial \mathbf{v}_c} \\ \dfrac{\partial \mathbf{v}_1}{\partial \mathbf{g}} & \dfrac{\partial \mathbf{v}_2}{\partial \mathbf{g}} \end{vmatrix} = \begin{vmatrix} 1 & 0 & 0 & 1 & 0 & 0 \\ 0 & 1 & 0 & 0 & 1 & 0 \\ 0 & 0 & 1 & 0 & 0 & 1 \\ -m^*/m_1 & 0 & 0 & m^*/m_2 & 0 & 0 \\ 0 & -m^*/m_1 & 0 & 0 & m^*/m_2 & 0 \\ 0 & 0 & -m^*/m_1 & 0 & 0 & m^*/m_2 \end{vmatrix} = 1$$

$$(3.129)$$

Using these results one can express the collision rate in terms of integrals over the center of mass and relative velocities:

$$N_{12} = \kappa_{12} n_1 n_2 \iiint_\infty d^3 v_c \iiint_\infty d^3 g\, \sigma(g)\, g f_1\left(v_c - \frac{m^*}{m_1} g\right) f_2\left(v_c + \frac{m^*}{m_2} g\right) \quad (3.130)$$

Next we evaluate this result for a gas mixture in kinematic equilibrium. Under equilibrium conditions each species individually follows a Maxwell–Boltzmann distribution at the common equilibrium temperature, T. This means that the product of the normalized velocity distribution functions can be written as follows:

$$f_1(v_1) f_2(v_2) = \frac{(m_1 m_2)^{3/2}}{(2\pi kT)^3} e^{-\frac{m_1 v_1^2}{2kT}} e^{-\frac{m_2 v_2^2}{2kT}}$$

$$= \frac{(m_1 m_2)^{3/2}}{(2\pi kT)^3} e^{-\frac{m_1}{2kT}\left(v_c - \frac{m^*}{m_1} g\right)^2} e^{-\frac{m_2}{2kT}\left(v_c + \frac{m^*}{m_2} g\right)^2}$$

$$= \left(\frac{m^*}{2\pi kT}\right)^{3/2} \left(\frac{m_1 + m_2}{2\pi kT}\right)^{3/2} e^{-\frac{(m_1+m_2) v_c^2}{2kT}} e^{-\frac{m^* g^2}{2kT}} \quad (3.131)$$

Substituting this result into equation (3.130) yields the following expression for the collision rate:

$$N_{12} = \kappa_{12} n_1 n_2 \left(\frac{\sqrt{m^*(m_1 + m_2)}}{2\pi kT}\right)^3 \iiint_\infty d^3 v_c\, e^{-\frac{(m_1+m_2) v_c^2}{2kT}} \iiint_\infty d^3 g\, \sigma(g)\, g\, e^{-\frac{m^* g^2}{2kT}}$$

$$(3.132)$$

The three integrals over the center of mass velocity components can be easily carried out. One can use spherical coordinates for the relative velocity space and integrate over the solid angle (the integrand is only a function of the relative speed, g). In this way we can work out all but one of the integrals in equation (3.132) and obtain

$$N_{12} = 4\pi \kappa_{12} n_1 n_2 \left(\frac{m^*}{2\pi kT}\right)^{3/2} \int_0^\infty dg\, g^3\, \sigma(g)\, e^{-\frac{m^* g^2}{2kT}} \quad (3.133)$$

This integral can be evaluated for any specific cross section function. In the simplest case when the cross section can be considered to be constant, $\sigma(g) = \sigma_{12}$, then we can work out the last integral and obtain the following simple expression for the collision rate, N_{12}:

$$N_{12} = \kappa_{12}\sigma_{12}n_1 n_2 \sqrt{\frac{8kT}{\pi m^*}} = \kappa_{12}\sigma_{12}n_1 n_2 \sqrt{\frac{8kT}{\pi m_1} + \frac{8kT}{\pi m_2}} = \kappa_{12}\sigma_{12}n_1 n_2 \sqrt{\bar{v}_1^2 + \bar{v}_2^2}$$

(3.134)

Here we have used the definition of the mean particle speed, $\bar{v}$, which was given by equation (2.65).

The collision rate (collisions per unit volume and unit time) and the collision frequency (number of collisions of a single molecule per unit time) are closely related quantities. The collision rate, N_{12}, is the product of the collision frequency and the number density of target molecules (corrected for potential double counting):

$$N_{12} = \kappa_{12}n_1 v_{12} = \kappa_{12}n_2 v_{21}$$

(3.135)

where v_{12} is the number of collisions per unit time of a single molecule of type 1 with molecules of type 2. Similarly, v_{21} is the number of collisions per unit time of a single molecule of type 2 with molecules of type 1. This means that in a gas under equilibrium conditions the collision frequency can be expressed as

$$v_{12} = \sigma_{12}n_2 \sqrt{\frac{8kT}{\pi m^*}} = \sigma_{12}n_2 \sqrt{\bar{v}_1^2 + \bar{v}_2^2}$$

(3.136)

This result is the rigorous derivation of the collision frequency for a gas mixture in kinematic equilibrium. In the case of a single species gas equation (3.136) simplifies to the equation (3.116).

In a mixture of many gases the collision frequency of a single molecule of species s can be expressed as

$$v_s = v_{s1} + v_{s2} + v_{s3} + \cdots = \sum_{t=1}^{K} v_{st}$$

(3.137)

where K is the total number of species and the summation also includes collisions with molecules of the same species.

The mean free time of a molecule of species s is $\tau_s = 1/v_s$, so the mean free path can be written in the following form:

$$\lambda_s = \frac{\overline{v}_s}{v_s} = \frac{\overline{v}_s}{\sum\limits_{t=1}^{K} \sigma_{st} n_t \sqrt{\overline{v}_s^2 + \overline{v}_t^2}} = \frac{1}{\sum\limits_{t=1}^{K} \sigma_{st} n_t \sqrt{\dfrac{m_s + m_t}{m_t}}} \tag{3.138}$$

This is a very important result which expresses the mean free path of a given molecular species in terms of fundamental gas parameters (assuming equilibrium conditions and a constant scattering cross section). It should be emphasized that in a multicomponent gas there is a mass factor in the expression which is sometimes incorrectly replaced by a factor of $\sqrt{2}$.

3.3.3 Chemical reactions

In the previous subsection we assumed that the gas mixture was composed of chemically inert (non-reacting) molecules, so that the molecules did not suffer any change during collisions. In this subsection we briefly consider a chemically reacting mixture of gases and express the chemical reaction rate constant with the help of fundamental molecular and gas parameters.

We use the simple reaction of hydrogen iodide formation and decomposition to introduce the fundamental concepts of kinetic description of gas phase chemical reactions: $H_2 + I_2 \leftrightarrow 2HI$. A forward reaction takes place when an H_2 and an I_2 molecule encounter each other and the energy of the relative motion satisfies certain conditions. In this case the two molecules briefly unite to form a compound molecule that spontaneously decomposes to two hydrogen iodide molecules. Conversely, the reverse reaction occurs when two HI molecules collide, temporarily form a compound molecule and then decompose to an H_2 and an I_2 molecule.

We use a very simple model to illustrate the formation of the unstable compound molecule. Figure 3.11 schematically shows the intermolecular potential for these collisions. It can be seen that when the centers of the colliding molecules are relatively far from each other the potential is weakly attracting (negative), while it becomes strongly repulsive as the molecules get close to each other. The transient molecular compound is formed when the centers of the colliding molecules approach each other within a distance of d_0 (reaction distance). In order to do so the molecules have to overcome a potential energy barrier, U_0 (activation energy). In this model a chemical reaction is interpreted as a simple collision with the following relative velocity dependent cross section:

$$\sigma(g) = \sigma_0 H(g - g_0) \tag{3.139}$$

where $H(g)$ is the Heaviside step function and the velocity threshold, g_0, is the

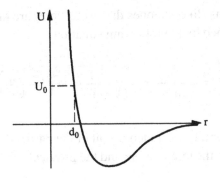

Figure 3.11

velocity of relative motion which is sufficient to overcome the potential energy barrier:

$$g_0 = \sqrt{\frac{2U_0}{m^*}} \tag{3.140}$$

We now calculate the number of reactive collisions per unit volume per unit time, Z_{12}. This can be done by substituting the cross section of the chemical reaction (equation (3.139)) into equation (3.133):

$$Z_{12} = 4\pi\kappa_{12}\sigma_0 n_1 n_2 \left(\frac{m^*}{2\pi kT}\right)^{3/2} \int_{g_0}^{\infty} dg\, g^3\, e^{-\frac{m^* g^2}{2kT}} \tag{3.141}$$

Equation (3.141) is valid for both the forward and the reverse reaction. The main difference is that the forward reaction involves the collision of different (unlike) molecules, therefore $\kappa_f = 1$, while the reverse reaction involves the collision of two like molecules (HI), therefore $\kappa_r = 1/2$. The integral in equation (3.141) can be readily carried out to yield

$$Z_{12} = \kappa_{12}\sigma_0 n_1 n_2 \sqrt{\frac{8kT}{\pi m^*}}\, e^{-\frac{m^* g_0^2}{2kT}} \left(1 + \frac{m^* g_0^2}{2kT}\right) \tag{3.142}$$

The threshold velocity can be expressed in terms of the barrier potential, U_0:

$$Z_{12} = \kappa_{12}\sigma_0 n_1 n_2 \sqrt{\overline{v}_1^2 + \overline{v}_1^2}\, e^{-\frac{U_0}{kT}} \left(1 + \frac{U_0}{kT}\right) \tag{3.143}$$

Bimolecular reactions (like the ones discussed here) are characterized by reaction rate constants defined by the following equation:

$$k_{12} = \frac{Z_{12}}{\kappa_{12} n_1 n_2} = \sigma_0 \sqrt{\frac{8kT}{\pi m^*}} \, e^{-\frac{U_0}{kT}} \left(1 + \frac{U_0}{kT}\right) \tag{3.144}$$

The reaction rate constant, k_{12}, is independent of the particle densities but it is usually a strong function of the temperature and the physical properties of the reacting molecules.

It is clear that the rate constant is a function of the dimensionless quantity, $\alpha = U_0/kT$, which is close to the ratio of the barrier potential and the average molecular kinetic energy. If this ratio is large (the threshold potential is much larger than the average molecular kinetic energy) the reaction rate can be approximated as

$$k_{12} = \sigma_0 \sqrt{\frac{8kT}{\pi m^*}} \, \alpha e^{-\alpha} \tag{3.145}$$

For instance, in the case of our reverse reaction the threshold potential of the $2HI \rightarrow H_2 + I_2$ reaction is $U_0 = 3.0 \times 10^{-19}$ joule (1.9 eV) and therefore $\alpha = 2.2 \times 10^4/T$. It is apparent that at ordinary temperatures very few molecules will have high enough thermal energies and there will be only a small number of reactive collisions. It is also interesting to note that the reaction rate for this collision is extremely sensitive to the gas temperature: a temperature increase from 300 K to 310 K results in a factor of 10 increase in the rate coefficient.

The other extreme case is when the threshold potential is small compared to the average molecular kinetic energy, $\alpha \ll 1$. The reaction rate can be approximated by

$$k_{12} = \sigma_0 \sqrt{\frac{8kT}{\pi m^*}} \tag{3.146}$$

In this case most molecules can penetrate the potential barrier and the reaction rate coefficient only weakly depends on the gas temperature, $k_{12} \propto \sqrt{T}$.

3.4 Problems

3.1 The statement is often made that binary encounters between molecules of masses m_1 and m_2 result in very little relative change in the energy of the heavier molecule if $m_1 \ll m_2$. Verify that this statement is correct (calculate $\Delta E/E$).

3.2 Consider two weakly attracting rigid elastic sphere molecules (Sutherland model) and neglect those collisions when the two molecules do not come into contact. The interparticle potential is shown in Figure 3.12. Both colliding particles have the same mass, m, and diameter, d_0. The relative velocity of the particles before collision is g_0. What is the maximum impact parameter, b_m, resulting in a collision? What is the total cross section of the interaction? (Assume that $2U_0/m^*g_0^2 \ll 1$ at all times.)

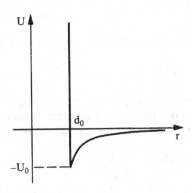

Figure 3.12

3.3 At room temperature (T=300 K) the total scattering cross section for electrons on air molecules is about 10^{-19} m². At what gas pressure will 90 percent of the electrons emitted from a point-like cathode reach a point-like anode 0.2 m away? (Assume that scattered electrons cannot reach the anode.)

3.4 Consider two electrically oppositely charged but otherwise identical Maxwell molecules. The molecular mass is m and the charges of the molecules are ±e (e=1.6022×10⁻¹⁹ coulomb is the electron charge). When the molecules are not charged, they interact via a repulsive potential with a momentum transfer cross section of 10^{-20} m². Assume that the gas temperature is 300 K and all particles move with the average speed. Plot the interparticle potential for these charged Maxwell molecules as a function of intermolecular distance.

3.5 Calculate the total cross section, σ, and the momentum transfer cross section, σ_1, for the differential scattering cross section $S(\chi) = \frac{1}{2}\sigma_0 |3\cos^2\chi - 1|$, where σ_0 is a constant.

3.6 Two different gases (molecules A with mass m_A, temperature T_A and number density n_A, and molecules B with mass m_B, temperature T_B and number density n_B) move through each other with a relative bulk velocity, u_0, where $u_0 \gg (kT_A/m_A)^{1/2}$ and $u_0 \gg (kT_B/m_B)^{1/2}$. Find the (A,B) collision rate (assuming that the collision cross section, σ_{AB}, is independent of the relative velocity). Find the mean free path of the (A,B) collisions. Compare these results to the $u_0=0$ case and explain the difference.

3.5 References

[1] Landau, L.D., and Lifshitz, E.M., *Mechanics*, Pergamon Press, New York, 1976.

[2] Goldstein, H., *Classical mechanics*, Addison-Wesley, Reading, Mass., 1980.

[3] Landau, L.D., and Lifshitz, E.M., *Physical kinetics*, Pergamon Press, New York, 1981.

[4] Present, R.D., *Kinetic theory of gases*, McGraw-Hill, New York, 1958.

[5] Chapman, S., and Cowling, T.G., *The mathematical theory of non-uniform gases*, Cambridge University Press, Cambridge, 1952.

[6] Kennard, E.H., *Kinetic theory of gases*, McGraw-Hill, New York, 1938.

[7] Vincenti, W.G., and Kruger, C.H., *Introduction to physical gas dynamics*, John Wiley and Sons, New York, 1965.

[8] Reif, F., *Fundamentals of statistical and thermal physics*, McGraw-Hill, New York, 1965.

[9] Jeans, J., *An introduction to the kinetic theory of gases*, Cambridge University Press, Cambridge, 1940.

[10] Clausius, R., Über die mittlere Länge der Wege, welche bei der Molecularbewegung gasförmigen Körper von den einzelnen Molecülen zurückgelegt werden, nebst einigen anderen Bemerkungen über die mechanischen Wärmetheorie, *Ann. Phys.* [2], **105**, 239, 1858.

[11] Massey, H.S.W., and Burhop, E.H.S., *Electronic and ionic impact phenomena*, Oxford University Press, Oxford, 1952.

[12] Mott, N.F., and Massey, H.S.W., *The theory of atomic collisions*, Oxford University Press, Oxford, 1949.

[13] Nicholson, D.R., *Introduction to plasma theory*, John Wiley and Sons, New York, 1983.

[14] Bittencourt, J.A., *Fundamentals of plasma physics*, Pergamon Press, Oxford, 1986.

4

Elementary transport theory

So far we have considered gases which were in kinematic equilibrium. It was always assumed that the molecules did not have any organized motion (bulk motion), and that the gas temperature and the number density of the molecules were constant everywhere. Under such conditions there is no net particle transport from one place to another.

In this chapter we relax our earlier assumptions and let the macroscopic quantities (gas bulk velocity, temperature, pressure, concentration, etc.) have slow spatial variations. These configuration space gradients result in macroscopic transport phenomena, such as diffusion or heat flow. Fluid dynamics describes these transport phenomena using empirical transport coefficients, such as viscosity, diffusion coefficient or heat conductivity.[1,2]

In this chapter we shall use the elementary mean free path method to describe macroscopic transport phenomena and to calculate approximate values for the transport coefficients. This simple and remarkably successful method is quite general because it does not depend on the form of the distribution function and, therefore, does not assume equilibrium.[3-7]

The mean free path method is based on the assumption that the gas is not in equilibrium, but the deviation from equilibrium is so small that locally the distribution of molecular velocities can be approximated by Maxwellians. However, the macroscopic parameters of the local Maxwellians slowly vary with configuration space location. Mathematically this assumption can be translated into two basic conditions.

The first condition is that the variation of all macroscopic quantities is small within a mean free path. In other words the mean free path, λ, must be much smaller than the characteristic scale length, L, of the fastest varying macroscopic quantity, G:

$$\lambda \ll L \qquad (4.1)$$

where L is defined by the following equation:

$$\frac{1}{L} \equiv \left|\frac{\nabla G}{G}\right| \qquad\qquad (4.2)$$

The second condition to be satisfied is related to the Taylor series of the fastest varying macroscopic quantity, G. The Taylor series expansion of G about an arbitrary point in the gas, r_0, must rapidly converge for distances comparable to the mean free path, λ. This condition can be written as

$$\left|G(r_0)\right| \gg \lambda \left|\nabla G(r_0)\right| \gg \lambda^2 \left|\nabla\nabla G(r_0)\right| \gg \lambda^3 \left|\nabla\nabla\nabla G(r_0)\right| \gg \dots \qquad (4.3)$$

The mean free path method can be used as long as conditions (4.1) and (4.3) are satisfied. When these conditions are violated different physical processes dominate the gas transport. Some of these processes will be discussed in this chapter in connection to molecular flow in an infinitely long tube. It will be shown that the results of the mean free path method can be used with some modification even when λ and L are comparable (but $\lambda < L$). However, when the mean free path becomes much larger than the smallest characteristic scale size of the problem the mean free path method entirely loses its usefulness and quite different methods have to be applied.

Elementary transport theory is a very powerful way to gain physical insight into the basic processes of molecular transport. In most cases this simple theory leads to the correct dependence of transport coefficients on the gas parameters, and the numerical values are typically within a factor of 2 of the result of more rigorous (and much more complicated) calculations.

4.1 Molecular effusion

First we consider a very simple case of particle transport, where empirical transport coefficients do not play a role. We shall examine the problem of molecular effusion (escape from a container through a small orifice) using continuum gas dynamics and gaskinetic methods. It will be no real surprise to see that the two methods lead to different results.

One of the very important questions that we are interested in is the rate at which gases can leak into vacuum through very small holes. This problem has considerable importance in several applications, such as the design of space vehicles. For other important applications, one must calculate the thrust (net loss of momentum through the hole in unit time) caused by molecular effusion. Among other things this effect can result in spacecraft rotation, which can lead to loss of stabilization.

This problem is very simple, it does not involve macroscopic transport coefficients. We consider two adjacent gas reservoirs with a small orifice connecting them. The thickness of the wall is assumed to be negligible compared to the diameter of the orifice.

4.1.1 Hydrodynamic escape

We consider two reservoirs filled with a frictionless compressible perfect gas characterized by molecular mass, m, and specific heat ratio, γ. The gas pressure and number density in reservoir 1 are p_1 and n_1, and in reservoir 2, p_2 and n_2. The pressure in reservoir 2 is smaller than that in reservoir 1, i.e. $p_1 > p_2$. There is an orifice in the wall between the two reservoirs (this situation is shown schematically in Figure 4.1).

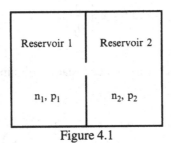

Figure 4.1

If the orifice is sufficiently large (its diameter is much larger than the mean free path of the molecules), the flow problem may be treated by hydrodynamic methods.[1,2] The linear dimensions of the reservoirs are assumed to be very large compared to the size of the orifice and therefore the gas velocity can be taken to be near zero everywhere except in the immediate vicinity of the orifice. We approximate this effusion process by a steady-state adiabatic outflow with no external work. For this flow the specific energy of the fluid is constant along each streamline:[1]

$$\frac{\gamma}{\gamma-1}\frac{p}{mn}+\frac{1}{2}u^2 = \frac{\gamma}{\gamma-1}\frac{p_1}{mn_1} \tag{4.4}$$

where the quantities n, u, and p refer to the gas number density, outflow velocity and pressure at the orifice. The outflow is an adiabatic process, therefore the following relation must hold:

$$\frac{p_1}{n_1^\gamma} = \frac{p}{n^\gamma} \tag{4.5}$$

Combining equations (4.4) and (4.5) we obtain the following expression for the outflow velocity, u:

$$u = \sqrt{\frac{2}{\gamma-1}\frac{\gamma p_1}{mn_1}\left[1-\left(\frac{p}{p_1}\right)^{\frac{\gamma-1}{\gamma}}\right]} = \sqrt{\frac{2a_1^2}{\gamma-1}\left[1-\left(\frac{p}{p_1}\right)^{\frac{\gamma-1}{\gamma}}\right]} \tag{4.6}$$

where $a_1^2 = \gamma p_1/mn_1$ is the speed of sound in reservoir 1. It is interesting to calculate the sound speed at the orifice. This can be directly obtained by using the adiabatic relation, equation (4.5):

$$a^2 = a_1^2\left(\frac{p}{p_1}\right)^{\frac{\gamma-1}{\gamma}} \tag{4.7}$$

Using equations (4.5) and (4.6) one can readily obtain the particle flux through the orifice:

$$j = nu = n_1 a_1\left(\frac{p}{p_1}\right)^{\frac{1}{\gamma}}\sqrt{\frac{2}{\gamma-1}\left[1-\left(\frac{p}{p_1}\right)^{\frac{\gamma-1}{\gamma}}\right]} \tag{4.8}$$

It is easy to see that the particle flux is zero both for $p/p_1=1$ and for $p/p_1=0$ and that it reaches its maximum value at a critical pressure given by

$$p_c = p_1\left(\frac{2}{\gamma+1}\right)^{\frac{\gamma}{\gamma-1}} \tag{4.9}$$

It can be shown that the outflow velocity at this critical pressure is the same as the speed of sound at the orifice:

$$u_c = a_1\sqrt{\frac{2}{\gamma-1}\left[1-\left(\frac{2}{\gamma+1}\right)\right]} = a_1\sqrt{\frac{2}{\gamma+1}} = a_c \tag{4.10}$$

These results have a very important physical meaning. The flux of molecules through the orifice increases with decreasing outside pressure until the pressure in reservoir 2 decreases below the critical value, p_c. At this point the outflow velocity equals the local sonic velocity and the effusing flux reaches its maximum value. The pressure at the orifice cannot decrease any further, because that would violate basic conservation laws (it would involve an 'unprovoked' subsonic to supersonic flow transition). The flow is sonic at the orifice; therefore no information can propagate back to reservoir 1 about the conditions beyond the orifice (the gas in reservoir 1 has no way of 'knowing' how low the pressure is in the other reservoir). The gas pressure eventually decreases from p_c to p_2, but this decrease takes place inside reservoir 2 (near the orifice).

Our results show that for large pressure drops the effusing flux is the maximum hydrodynamic outflow flux, given by

$$ j_c = n_1 a_1 \left(\frac{2}{\gamma + 1} \right)^{\frac{\gamma+1}{2(\gamma-1)}} \tag{4.11} $$

Equation (4.11) is the hydrodynamic escape flux for large pressure differences. This result will be compared to the prediction of the kinetic description, to be obtained in the next section.

4.1.2 Kinetic effusion

Next we consider the same problem with the difference that the orifice is assumed to be much smaller than it was in the hydrodynamic case. It is now assumed that the mean free path of the gas molecules is much larger than the diameter of the orifice, which is still much larger than the molecular size. In the first step we neglect the flow of molecules from reservoir 2 to reservoir 1; these molecules will be taken into account later.

Molecules leave reservoir 1 in the form of a molecular beam. Each molecule leaves the container with the velocity it had as it came up to the hole. The loss of a single molecule now and then through the hole only slightly disturbs their distribution inside the container. A trace of mean motion towards the hole must develop in the reservoir because of the absence of those collisions that the lost molecule would have had with others on its return from the wall. However, this effect will be wiped out promptly by the molecular collisions which always tend to set up and preserve the equilibrium state.

Let us examine the velocity distribution function of escaping molecules. Only those molecules can leave reservoir 1 through the orifice which have an outward pointing velocity component in the direction perpendicular to the orifice. This requirement is schematically illustrated in Figure 4.2. The v_z axis is directed outward (from reservoir 1 to reservoir 2) and therefore all escaping molecules must have positive v_z velocities (in reservoir 1, $v_z < 0$ characterizes molecules coming from the orifice).

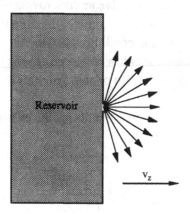

Figure 4.2

In the present approximation all particles which cross the orifice satisfy the condition $v_z > 0$, because we assumed that there is vacuum outside. Consequently, the velocity distribution right outside the orifice can be characterized by a truncated Maxwell–Boltzmann distribution function:

$$F_{esc} = \begin{cases} n\left(\dfrac{m}{2\pi\, kT}\right)^{3/2} e^{-\frac{m}{2\,kT}\left(v_x^2 + v_y^2 + v_z^2\right)} & \text{if } v_z > 0 \\[2ex] 0 & \text{if } v_z \leq 0 \end{cases} \tag{4.12}$$

where n is the number density in reservoir 1. Note that the truncated distribution function, F_{esc}, is normalized to $n/2$.

Using the distribution function of escaping molecules, F_{esc}, we can calculate the escape flux from reservoir 1. Let us denote the orifice area by dS and consider a group of molecules with velocity vectors in an infinitesimal neighborhood of **v** which hit dS during the time interval dt (it is clear that **v** must have a positive v_z component, otherwise these molecules cannot leave reservoir 1). These molecules are all contained at the beginning of the time interval, dt, in an oblique cylinder with base dS and altitude $v_z dt$ (this oblique cylinder is shown schematically in Figure 4.3). The volume of the impact cylinder is

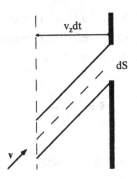

Figure 4.3

$$d\mathcal{V}_{\text{impact}} = v_z \, dt \, dS \qquad (4.13)$$

The name 'impact cylinder' refers to the fact that at the beginning of the time interval, dt, this cylinder contains all particles with velocity **v** which impact the orifice element, dS, during this time interval. The number of particles with velocities between **v** and **v**+d^3v which pass through the orifice element, dS, in time , dt, is given by the following expression:

$$d^6N = F_{\text{esc}} \, d^3v \, d\mathcal{V}_{\text{impact}} = v_z \, F_{\text{esc}} \, d^3v \, dt \, dS \qquad (4.14)$$

The flux escaping from reservoir 1 through the orifice is defined as the number of particles leaving the reservoir per unit orifice area per unit time. The escape flux can be expressed with the help of the distribution function of escaping particles:

$$j_{\text{esc}} = \int\limits_{-\infty}^{\infty} dv_x \int\limits_{-\infty}^{\infty} dv_y \int\limits_{-\infty}^{\infty} dv_z \, v_z \, F_{\text{esc}}(\mathbf{v}) \qquad (4.15)$$

The velocity distribution of escaping particles is given by equation (4.12). Substituting this distribution function into equation (4.15) and recognizing that particles with $v_z < 0$ cannot leave reservoir 1 through the orifice leads to the following expression for the escape flux:

$$j_{\text{esc}} = n \left(\frac{m}{2\pi \, kT} \right)^{3/2} \int\limits_{-\infty}^{\infty} dv_x \int\limits_{-\infty}^{\infty} dv_y \int\limits_{0}^{\infty} dv_z \, v_z \, e^{-\frac{m}{2 \, kT} \left(v_x^2 + v_y^2 + v_z^2 \right)} = n \sqrt{\frac{kT}{2\pi \, m}}$$

$$(4.16)$$

It is important to mention that the boundaries of the v_z integral in equation (4.16) are 0 and ∞ (and not $-\infty$ and $+\infty$), therefore the integral of the odd function is not zero. The mean speed of molecules in reservoir 1 is

$$\bar{v} = \sqrt{\frac{8kT}{\pi m}} \qquad (4.17)$$

(see equation (2.65)). Using this result the escape flux from reservoir 1 through the small orifice can be written as

$$\dot{j}_{esc} = \frac{1}{4}\, n\, \bar{v} \qquad (4.18)$$

It is instructive to compare the escape fluxes derived by hydrodynamic and kinetic methods. The kinetic derivation was carried out with the assumption that reservoir 2 was empty, therefore the maximum hydrodynamic escape flux (given by equation (4.11)) must be compared to the kinetic result given by equation (4.16). The ratio of the two results is the following:

$$\frac{\dot{j}_{hyd}}{\dot{j}_{kin}} = \frac{\left(\dfrac{2}{\gamma+1}\right)^{\frac{\gamma+1}{2(\gamma-1)}} n\, \sqrt{\dfrac{\gamma kT}{m}}}{n\, \sqrt{\dfrac{kT}{2\pi m}}} = \left(\frac{2}{\gamma+1}\right)^{\frac{\gamma+1}{2(\gamma-1)}} \sqrt{2\pi\gamma} \qquad (4.19)$$

It can be shown that in the specific heat ratio interval $1 < \gamma < 2$ (most gases have specific heats in this range), the $\dot{j}_{hyd}/\dot{j}_{kin}$ ratio increases from 1.5 to 1.93. This means that the flux of escaping molecules is about 1.5 to 1.9 times larger under hydrodynamic conditions (when the characteristic size of the orifice is much larger than the mean free path of the molecules) than that under kinetic outflow conditions (when the orifice is small compared to the mean free path).

Our kinetic description of molecular effusion through a small orifice can easily be extended to the situation when the outside reservoir (reservoir 2) is not empty. If the kinetic outflow conditions are satisfied in both reservoirs (i.e. the mean free path of molecular collisions is much larger than the size of the orifice) then the effects of molecular scattering can be neglected in the immediate vicinity of the orifice. This means that molecules leave reservoir 1 through the orifice without noticing the molecules in reservoir 2. By the time a molecule which just escaped from reservoir 1 to reservoir 2 has its next collision (this time inside reservoir 2) it is sufficiently far from the orifice that local effects can be neglected. This molecule will undergo many collisions in reservoir 2 (and accommodate to the gas in reservoir 2) before it reaches the orifice again. The same argument applies to particles leaving reservoir 2. One can conclude that the escape fluxes from reservoirs 1 and 2 do not influence each other and therefore the net particle flux from reservoir 1 to reservoir 2 can be calculated as

$$j = j_1 - j_2 = \frac{1}{4}(n_1 \bar{v}_1 - n_2 \bar{v}_2) \qquad (4.20)$$

4.1.3 Transport of macroscopic quantities through the orifice

Escaping molecules carry a flux of macroscopic quantities through the orifice. One can calculate, for instance, the escaping mass flux, or the momentum flux carried through the orifice by the escaping molecules. In general, a molecular quantity, $Q(\mathbf{v})$, is the microscopic aspect of some macroscopic parameter (note that Q is only a function of the velocity vector and it is independent of time or configuration space location). The average value of this molecular quantity, $\bar{Q}$, multiplied by the average number of molecules in a macroscopically infinitesimal volume element, n, is the density of the macroscopic quantity associated with Q. A simple example is the case Q=m, when the microscopic physical quantity is the mass of the molecules. The related macroscopic quantity is the mass density of the gas, ρ, which can be obtained as $\rho = n\bar{Q} = nm$. Another example is the case $Q=mv^2/2$, when the microscopic physical quantity is the kinetic energy of the molecule. The corresponding macroscopic quantity is the internal energy density of the gas, $h = nm\langle v^2 \rangle/2$. Using our earlier results (see equation (2.70)) we obtain $h = 3nkT/2$, where T is the gas temperature.

The flux of macroscopic quantities through the orifice can be calculated by taking the velocity space averages of the escape fluxes of the corresponding microscopic quantities. The differential escape flux of molecules with velocity vectors in an infinitesimal neighborhood of $\mathbf{v}$ can be obtained from equation (4.14):

$$d^3j = v_z F_{esc}(\mathbf{v}) d^3v \qquad (4.21)$$

The differential flux of the molecular quantity, Q, through the orifice can be written as

$$d^3j_Q = Q(\mathbf{v}) v_z F_{esc}(\mathbf{v}) d^3v \qquad (4.22)$$

Finally, the flux of the macroscopic quantity corresponding to the molecular quantity, Q, is

$$j_Q = \iiint_\infty d^3v \, Q(\mathbf{v}) \, v_z \, F_{esc}(\mathbf{v})$$

$$= n\left(\frac{m}{2\pi kT}\right)^{3/2} \int_{-\infty}^{\infty} dv_x \int_{-\infty}^{\infty} dv_y \int_0^{\infty} dv_z v_z \, Q(\mathbf{v}) \exp\left(-\frac{mv^2}{2kT}\right)$$

$$= n\left(\frac{m}{2\pi kT}\right)^{3/2} \int_0^{\infty} dv \int_0^{2\pi} d\varphi \int_0^{\pi/2} d\theta \, v^3 \sin\theta \cos\theta \, Q(\mathbf{v}) \exp\left(-\frac{mv^2}{2kT}\right) \quad (4.23)$$

It should be noted that the truncated nature of the distribution function of escaping particles is taken into account by modifying the integration boundaries. In Cartesian coordinates this means that the v_z integral goes from 0 to ∞, while in polar coordinates the polar angle varies only from 0 to $\pi/2$, corresponding to positive v_z values only ($v_z = v \cos\theta$).

Next we consider two examples for escape fluxes. The three components of the momentum flux through the orifice can be obtained by substituting the molecular momentum for the Q quantity, $Q = m\mathbf{v}$:

$$j_{M_x} = nm\left(\frac{m}{2\pi kT}\right)^{3/2} \int_{-\infty}^{\infty} dv_x \int_{-\infty}^{\infty} dv_y \int_0^{\infty} dv_z v_z \, v_x \exp\left(-\frac{mv^2}{2kT}\right) = 0 \quad (4.24a)$$

$$j_{M_y} = nm\left(\frac{m}{2\pi kT}\right)^{3/2} \int_{-\infty}^{\infty} dv_x \int_{-\infty}^{\infty} dv_y \int_0^{\infty} dv_z v_z \, v_y \exp\left(-\frac{mv^2}{2kT}\right) = 0 \quad (4.24b)$$

$$j_{M_z} = nm\left(\frac{m}{2\pi kT}\right)^{3/2} \int_{-\infty}^{\infty} dv_x \int_{-\infty}^{\infty} dv_y \int_0^{\infty} dv_z v_z^2 \exp\left(-\frac{mv^2}{2kT}\right) = \frac{1}{2}nkT \quad (4.24c)$$

With the help of this result one can calculate the thrust created by the gas escaping through a microscopic hole with area A_0. Thrust is the net transfer of momentum through the hole in unit time. It is easy to see from equations (4.24) that the net thrust is perpendicular to the hole and its magnitude is $A_0 p/2$, where p is the gas pressure inside the container.

In the second example we calculate the energy flux through the orifice. The corresponding molecular quantity is $Q = mv^2/2$ and the differential energy flux can be written as

$$d^3 j_E = \frac{1}{2}mv^2 \, v_z \, F_{esc}(\mathbf{v}) d^3v \quad (4.25)$$

The energy flux can be obtained by integrating expression (4.25) for all velocities with positive v_z components:

$$j_E = \frac{1}{2} mn \left(\frac{m}{2\pi kT} \right)^{3/2} \int_0^\infty dv \int_0^{2\pi} d\varphi \int_0^{\pi/2} d\Theta \; v^5 \sin\Theta \cos\Theta \exp\left(-\frac{mv^2}{2kT} \right)$$

$$= \frac{1}{2} mn \left(\frac{m}{2\pi kT} \right)^{3/2} \int_0^\infty dv \int_0^{2\pi} d\varphi \int_0^1 d\mu \; \mu \, v^5 \exp\left(-\frac{mv^2}{2kT} \right) = \frac{\pi}{16} mn\bar{v}^3 \quad (4.26)$$

When evaluating the integral we have introduced a new variable, $\mu = \cos\theta$.

It is instructive to calculate the average kinetic energy carried by the escaping particles. A first careless guess would be that the average energy of the particles leaving the reservoir through the small orifice is the same as the average energy of the gas molecules in the container, 3kT/2. One has to think a little more carefully because in this case simple intuition is misleading.

The average energy of the escaping molecules can be obtained by dividing the energy flux (given by equation (4.26)) by the number flux of the outflowing molecules (given by equation (4.18)):

$$< E >_{esc} = \frac{j_E}{j_{esc}} = \frac{\pi}{4} m\bar{v}^2 = 2kT \quad (4.27)$$

This is a very interesting result. The average energy of those particles which leave the container through the orifice is larger than the average energy of all particles in the reservoir. The physical explanation is that the volume of the escape cylinder (which contains all particles with velocity **v** which leave the container through the orifice during a given time interval) is proportional to the v_z velocity of the particles (see equation (4.13)). On the average, faster particles have larger v_z velocity components than slower particles. This means that during the same time interval fast particles can escape from a larger volume than slower particles, consequently the average energy of escaping particles must be higher than the average molecular energy in the container.

4.2 Hydrodynamic transport coefficients

4.2.1 Viscosity

Moving fluids exhibit evidence of internal friction, called viscosity. This point can be illustrated by the following simple example. Consider a moving fluid between two infinite parallel planes (we choose a coordinate system where the (x, y) plane is parallel to the infinite planes). The lower plane (characterized by z=0) is at rest and the upper one (located at z=h) moves in the positive x direction with a constant velocity, u_0. This situation is shown schematically in Figure 4.4.

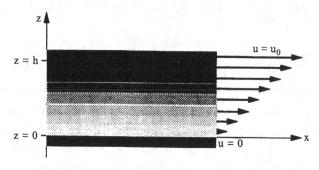

Figure 4.4

Imagine that the moving fluid is subdivided into many thin layers parallel to the two planes. In our case all these layers are parallel to the (x,y) plane and are characterized by their z coordinates. Assume that the top layer adheres to the moving upper plane and the bottom layer to the fixed plane (boundary layers).

The motion of the individual layers is approximated as that of a solid body. This means that the motion of the i-th layer is characterized by a single velocity, $u_i=u(z_i)$, where z_i is the z coordinate of the layer. The individual layers move with different velocities, therefore there is friction between the layers. Each layer will experience a forward drag from the layer above (and will exert a backward drag on the layer above) and a backward drag from below (and will exert a forward drag on the layer below). This process creates a velocity gradient in the fluid. Under steady-state conditions the layer at height z will have a velocity of u(z) with u(0)=0 and $u(h)=u_0$. It is interesting to note that the velocity of the layers points in the x direction, but the velocity gradient is in the z direction.

The frictional forces acting on the two surfaces of a given layer (due to the interaction with the layers above and below) are parallel to the surfaces and they are referred to as shearing forces (these forces are trying to shear the layer). The shearing force per unit surface area of the surface over which the force is acting is called shearing stress. The shearing stress is defined by its magnitude and by two directions: (i) the direction of the force and (ii) the normal vector of the surface. In our case, the shearing stress will be denoted by $-P_{zx}$. The convention is that the first subscript gives the direction of the normal vector and the second refers to the direction of the force. Using this convention P_{zz} measures the normal force per unit area acting on the surface, i.e. P_{zz} is the pressure on the surface, p.

Figure 4.5 illustrates the forces acting along the two interfaces of a given layer. The flow velocity increases upward, therefore the i-th layer exerts a backward shearing stress, $P_{zx(+)}$, on the faster moving fluid layer above it. Conversely, according to Newton's third law the fluid above the i-th layer exerts a stress of $-P_{zx(+)}$ on the layer in the middle. Similarly, the i-th layer exerts a forward stress, $-P_{zx(-)}$, on the layer below, and therefore the stress acting on the lower surface of layer 'i' is $P_{zx(-)}$. In general, the shearing forces generated by the velocity differences between adjoining layers of fluid are trying to reduce the velocity difference. Empirically, the velocity gradient, du/dz, is found to be directly proportional to the decelerating shearing stress, P_{zx}:

$$P_{zx} = -\eta \frac{du}{dz} \qquad (4.28)$$

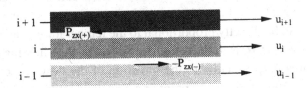

Figure 4.5

Here the proportionality factor, η, is called the coefficient of viscosity of the fluid. The negative sign expresses the fact that the shearing force (generated by the velocity gradient) is trying to reduce the physical effect which generated it. Our example shows that force must be applied to the lower wall to keep it at rest and to the upper wall to maintain its uniform motion. These forces are proportional to the viscosity of the fluid, η.

Kinetic theory is not only able to explain this result qualitatively, but it also gives a theoretical expression for the coefficient of viscosity. This derivation will be carried out later, in connection with the mean free path method. The qualitative explanation of our previous result follows from microscopic considerations of kinetic theory.

In the gas the motion of molecules in the y and z directions is totally random, but in the x direction their average velocity is equal to the velocity of the bulk motion. This average velocity, u_x, slowly varies with the altitude, z. Molecules cross the interface between two adjoining thin gas layers in both directions with equal frequency. Each molecule that crosses from above carries with it momentum and energy which are transferred from the layer above to the layer below via collisions. Molecules that cross from below transfer momentum and energy from the lower to the upper layer. Because of the velocity gradient the molecules above the interface have a somewhat larger average x–momentum than the molecules below. As a result of this microscopic process macroscopic momentum is being slowly transferred from the upper layer to the lower one. The effect is the same as if the upper layer exerted a frictional drag in the x direction on the lower layer.

According to Newton's second law the rate of change in the x–momentum of the lower layer is equal to the force exerted on it. The net transfer of x–momentum per unit time per unit area is equal to the shearing stress. This is the qualitative interpretation of equation (4.28). The coefficient of viscosity will be derived quantitatively in Section 4.3.

4.2.2 *Heat conductivity*

Heat flow takes place in any medium in which a temperature difference exists. In simple cases (like heat flow through a wall) the transport of heat per unit area per unit time is found to be directly proportional to the temperature gradient.

The heat flow vector, **q**, is defined as the quantity of heat flowing per unit time through a unit area perpendicular to the direction of the flow. Experiments show that the direction of heat flow at any point of a heat conducting medium is opposite to the direction of the temperature gradient at that point. Assuming a one dimensional heat flow along the z axis, the empirical heat conduction is

$$\mathbf{q} = -\kappa \nabla T \qquad (4.29)$$

where κ is the coefficient of thermal conduction (sometimes it is also called thermal conductivity) of the medium.

The mechanism of heat conductivity can again be understood in terms of gas-kinetic theory. Consider a gas confined between two infinite parallel planes (a similar situation to the one discussed in connection to viscosity). The lower and upper planes (located at z=0 and z=h) are kept at a temperatures T_0 and $T_h > T_0$, respectively. This situation is shown schematically in Figure 4.6.

Imagine that the gas is again subdivided into many thin layers parallel to the two planes. In our case all these layers are parallel to the (x,y) plane and are characterized by their z coordinates. Assume that the top layer has the same temperature as the upper plane, T_h, and the temperature of the bottom layer is identical to that of the lower plane, T_0. The temperatures of the thin gas layers are increasing with increasing z. The molecules cross the interfaces between the thin gas layers from above and below with equal frequency. In our case dT/dz>0, therefore molecules coming from above have a larger mean kinetic energy than the molecules

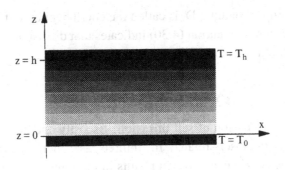

Figure 4.6

coming from below. Thus there is a net flow of energy from the upper layer to the one below.

4.2.3 Diffusion coefficient

Next we consider a very simple example for molecular diffusion, the so-called self diffusion. Consider a gas containing two kinds of molecules with very similar molecular weights and practically identical collision cross sections. For example, consider a situation when a single species gas contains two isotopes of the same molecule. The concentrations of the two isotopes are denoted by n_0 and n_1, where n_0 is the concentration of the regular molecules and $n_1 \ll n_0$ is the concentration of a trace isotope. We assume that n_0 is constant everywhere, as is the temperature, T. This means that the total concentration and pressure are constant and therefore there is no mass motion in the system (the contribution of the tracer molecules is neglected).

Suppose that the concentration of tracer molecules, n_1, is not uniform, and it is a function of the z coordinate, $n_1 = n_1(z)$. In this case the gas can again be subdivided into many thin layers parallel to the (x,y) plane. The tracer molecules cross the interfaces between the thin gas layers from above and below. If $dn_1/dz > 0$, more molecules are crossing the interface from above than from below. Thus there is a net flow from the higher density layer to the lower density one.

The net particle flux in the z direction, j_z, is the difference between the numbers of upward and downward moving tracer molecules crossing unit interface area per unit time. To a good approximation this flux is proportional to the concentration gradient of tracer molecules:

$$j_z = -D \frac{dn_1}{dz} \qquad (4.30)$$

The constant of proportionality, D, is called the coefficient of self diffusion of the gas. The negative sign in equation (4.30) indicates that diffusion is trying to eliminate the concentration gradient of the tracer component.

4.3 Mean free path method

From the previous discussion of self diffusion, viscosity and heat conduction it is clear that the three are closely related. In all three cases there is a gradient of some macroscopic property of the gas which results in the transport of some molecular quantity. This transport is directly proportional to the negative of the gradient of the macroscopic quantity.

In this section we will work out an elementary treatment of these transport phenomena using the mean free path method.

4.3.1 Mean free path treatment of transport phenomena

First let us consider a gas in which the distribution function of molecular velocities varies along the z direction, but does not change in the x or y directions. This one dimensional situation will be used to introduce the fundamental concepts of the mean free path method. At the end the results will be generalized to three dimensional problems.

We assume near-equilibrium conditions, i.e. at every spatial point the velocity distribution function of the molecules is close to the Maxwell-Boltzmann distribution. However, the macroscopic parameters of the distribution function (such as temperature or particle number density) can slowly vary from point to point. The slow variation means that all characteristic scale lengths of these variations are much larger than the mean free path, λ.

The mean free path method is based on the idea that the collisional process can be discretized. It is assumed that all collisions take place simultaneously at discrete instants, $t=0$, τ, 2τ, 3τ, ... (where τ is the mean free time), and at discrete values of the z coordinate, $z=0$, Δz, $2\Delta z$, $3\Delta z$, ... (where Δz is the average distance traveled by a particle in the z direction in a mean free time).

The discretization must be done in such a way that the average properties of the molecular motion be preserved. This can be achieved by the appropriate choice of the discretization distance, Δz.

Consider those molecules which suffer a collision between $t=0$ and $t=\tau$ at a small surface element of the plane $z=0$, dS, and after the collision have their velocity vectors in the infinitesimal vicinity of v which has a positive component in the z direction, $v_z>0$. At $t=\tau$ all these particles will be contained inside an oblique cylinder with base area of dS and altitude $v_z\tau$. The volume of this small cylinder is

$$d\mathcal{V}_{\text{scattered}} = v_z \tau dS \qquad (4.31)$$

The selected particles travel a distance of $v_z\tau$ in the z direction during the time interval, τ. The distance, Δz, can be obtained as the average distance traveled in the z direction during a mean free time, τ:

$$\Delta z = \frac{\int\limits_0^\infty dv \int\limits_0^{2\pi} d\varphi \int\limits_0^{\pi/2} d\theta\, v^2 \sin\theta\, v_z^2 \tau^2\, F(t=0, z=0, v, \theta, \varphi)\, dS}{\int\limits_0^\infty dv \int\limits_0^{2\pi} d\varphi \int\limits_0^{\pi/2} d\theta\, v^2 \sin\theta\, v_z \tau\, F(t=0, z=0, v, \theta, \varphi)\, dS} \qquad (4.32)$$

Here $F(t,z,v,\theta,\varphi)$ is the phase space distribution function of the molecules. Note that the integral in the polar angle, θ, goes only from 0 to $\pi/2$, therefore only positive v_z components are taken into account.

The average distance traveled by the particles between two consecutive collisions is the mean free path, λ. On the average molecules travel a distance of λ in time τ, therefore the average particle speed can be approximated by $\bar{V}=\lambda/\tau$ and the velocity component in the z direction can be written as $v_z=(\lambda/\tau)\cos\theta$. Using this approximation and assuming a locally Maxwellian velocity distribution one can evaluate expression (4.32) to obtain

$$\Delta z = \lambda \frac{\int\limits_0^\infty dv \int\limits_0^{2\pi} d\varphi \int\limits_0^{\pi/2} d\theta\, v^2 \sin\theta \cos^2\theta\, F\!\left(t=0, z=0, v=\frac{\lambda}{\tau}, \theta, \varphi\right) dS}{\int\limits_0^\infty dv \int\limits_0^{2\pi} d\varphi \int\limits_0^{\pi/2} d\theta\, v^2 \sin\theta \cos\theta\, F\!\left(t=0, z=0, v=\frac{\lambda}{\tau}, \theta, \varphi\right) dS} = \frac{2}{3}\lambda$$

$$(4.33)$$

Using this result one can define the mean free path method as follows. The velocity distribution functions at all discrete 'collision planes' are assumed to be locally Maxwellian at each time step (right after the collisions). The macroscopic parameters of these Maxwellian distributions may slowly vary from one collisional plane to another. This means that all particles which at time t (just before collision) are located at the plane $z=z_0$ had their last intermolecular collision at time $t-\tau$ either at the $z=z_0-2\lambda/3$ or at the $z=z_0+2\lambda/3$ planes. This situation is shown schematically in Figure 4.7.

At time t particles arrive from two directions at the plane $z=z_0$. Upward moving particles (those with $v_z>0$) arrive from below, while downward moving molecules (with $v_z<0$) arrive from above. This means that just before the collision the distrib-

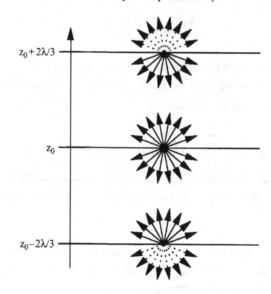

Figure 4.7

ution of molecular velocities is discontinuous at $z=z_0$: it is identical to $F(t-\tau,z_0-2\lambda/3,\mathbf{v})$ for particles with positive v_z, while for downward moving mole-cules it is $F(t-\tau,z_0+2\lambda/3,\mathbf{v})$. Therefore at time t the distribution function at $z=z_0$ can be written as

$$F(t,z_0,\mathbf{v}) = \begin{cases} F\left(t-\tau,z_0-\tfrac{2}{3}\lambda,\mathbf{v}\right) & \text{if } v_z > 0 \\ \\ F\left(t-\tau,z_0+\tfrac{2}{3}\lambda,\mathbf{v}\right) & \text{if } v_z < 0 \end{cases} \qquad (4.34)$$

Using equation (4.34) one can now calculate the net flux of the molecular quan-tity, $Q(\mathbf{v})$, through the plane $z=z_0$. The volume of the infinitesimal 'impact' cylin-der is proportional to the vertical velocity component, v_z (see equation (4.31)); therefore flux of Q is

$$j_Q(t,z_0) = \iiint_{\infty} d^3v\, v_z\, Q(\mathbf{v})\, F(t,z_0,\mathbf{v}) \qquad (4.35)$$

Substituting the mean free path method approximation of the distribution function (given by equation 4.34) into equation (4.35) yields:

$$j_Q\left(t,z_0\right)= \int\limits_{-\infty}^{\infty}dv_x \int\limits_{-\infty}^{\infty}dv_y \int\limits_{-\infty}^{0}dv_z\,v_z\,Q(\mathbf{v})\,F\left(t-\tau,z_0+\tfrac{2}{3}\,\lambda,\mathbf{v}\right)$$

$$+ \int\limits_{-\infty}^{\infty}dv_x \int\limits_{-\infty}^{\infty}dv_y \int\limits_{0}^{\infty}dv_z\,v_z\,Q(\mathbf{v})\,F\left(t-\tau,z_0-\tfrac{2}{3}\,\lambda,\mathbf{v}\right) \qquad (4.36)$$

One of our fundamental assumptions at the beginning of this Chapter was that all macroscopic quantities are slowly varying functions of spatial locations. In other words it is explicitly assumed that the scale length of any macroscopic variation is always much larger than the mean free path. This means that the distribution functions in equation (4.36) can be expanded into Taylor series around the location $z=z_0$ and one can stop after the linear term:

$$F\left(t-\tau,z_0+\tfrac{2}{3}\,\lambda,\mathbf{v}\right)=F\left(t-\tau,z_0,\mathbf{v}\right)+\tfrac{2}{3}\,\lambda\,\frac{\partial F\left(t-\tau,z_0,\mathbf{v}\right)}{\partial z} \qquad (4.37a)$$

$$F\left(t-\tau,z_0-\tfrac{2}{3}\,\lambda,\mathbf{v}\right)=F\left(t-\tau,z_0,\mathbf{v}\right)-\tfrac{2}{3}\,\lambda\,\frac{\partial F\left(t-\tau,z_0,\mathbf{v}\right)}{\partial z} \qquad (4.37b)$$

Substituting expressions (4.37) into equation (4.36) yields the following approximation for the net flux of molecular quantity, $Q(\mathbf{v})$, through the plane: $z=z_0$

$$j_Q\left(t,z_0\right)= \int\limits_{-\infty}^{\infty}dv_x \int\limits_{-\infty}^{\infty}dv_y \int\limits_{-\infty}^{0}dv_z\,v_z\,Q(\mathbf{v})\,F\left(t-\tau,z_0,\mathbf{v}\right)$$

$$+ \int\limits_{-\infty}^{\infty}dv_x \int\limits_{-\infty}^{\infty}dv_y \int\limits_{0}^{\infty}dv_z\,v_z\,Q(\mathbf{v})\,F\left(t-\tau,z_0,\mathbf{v}\right)$$

$$+\tfrac{2}{3}\,\lambda\,\frac{\partial}{\partial z} \int\limits_{-\infty}^{\infty}dv_x \int\limits_{-\infty}^{\infty}dv_y \int\limits_{-\infty}^{0}dv_z\,v_z\,Q(\mathbf{v})\,F\left(t-\tau,z_0,\mathbf{v}\right)$$

$$-\tfrac{2}{3}\,\lambda\,\frac{\partial}{\partial z} \int\limits_{-\infty}^{\infty}dv_x \int\limits_{-\infty}^{\infty}dv_y \int\limits_{0}^{\infty}dv_z\,v_z\,Q(\mathbf{v})\,F\left(t-\tau,z_0,\mathbf{v}\right) \qquad (4.38)$$

The first two integrals add up to an integral over the entire velocity space. The third and fourth integrals can be combined if $Q(\mathbf{v})F(t,z,\mathbf{v})$ is an even function of the velocity coordinate, v_z:

$$j_Q(t,z_0) = \int\limits_{-\infty}^{\infty} dv_x \int\limits_{-\infty}^{\infty} dv_y \int\limits_{-\infty}^{\infty} dv_z v_z\, Q(\mathbf{v})\, F(t-\tau, z_0, \mathbf{v})$$

$$-\tfrac{4}{3}\lambda\frac{\partial}{\partial z}\int\limits_{-\infty}^{\infty} dv_x \int\limits_{-\infty}^{\infty} dv_y \int\limits_{0}^{\infty} dv_z v_z\, Q(\mathbf{v})\, F(t-\tau, z_0, \mathbf{v}) \qquad (4.39)$$

The two terms in equation (4.39) describe two very different physical processes. The first term corresponds to macroscopic, organized transport of physical quantities in the z direction. Such transport can be due to the bulk motion of the gas as a whole. The second term describes the flux of physical quantities in the z direction due to the random (thermal) motion of the gas molecules. This term describes the flux difference carried by particles which cross the plane $z=z_0$ upward and downward due to their thermal motion.

Equation (4.39) is the fundamental equation of the mean free path method. With the help of this equation we are able to obtain specific expressions for the hydrodynamic transport coefficients, η, κ and D. In the following subsections we will consider the physical problems qualitatively discussed in section 4.2 and express the hydrodynamic transport coefficients with the help of the fundamental equation of the mean free path method, equation (4.39).

4.3.2 Diffusion

Let us consider the simplest specific case of equation (4.39), when the situation is steady-state and Q=1. For the flux of molecules through the plane $z=z_0$ the mean free path method yields

$$j_{\text{diff}}(z_0) = \iiint\limits_{\infty} d^3v\, v_z\, F(z_0, \mathbf{v}) - \tfrac{4}{3}\lambda\frac{\partial}{\partial z}\int\limits_{-\infty}^{\infty} dv_x \int\limits_{-\infty}^{\infty} dv_y \int\limits_{0}^{\infty} dv_z v_z\, F(z_0, \mathbf{v}) \quad (4.40)$$

The first integral in equation (4.40) is zero because we integrate the odd function, $v_z F$, from $-\infty$ to $+\infty$. The second integral is the same as the escape flux was through the microscopic orifice (see equation (4.15)) and its value is $n(z_0)\,\bar{v}(z_0)/4$. Using these considerations one can evaluate equation (4.40) and obtain the following result:

$$j_{\text{diff}}(z_0) = -\frac{1}{3}\lambda\frac{\partial}{\partial z}\big[n(z_0)\bar{v}(z_0)\big] \qquad (4.41)$$

In the case of self diffusion the temperature remains constant throughout the container, while the concentration of the tracer component is a slowly varying function of z (see Subsection 4.2.3). Constant temperature corresponds to constant mean particle velocity and therefore equation (4.41) simplifies to

$$j_{\text{diff}}(z_0) = -\frac{1}{3} \lambda \bar{v} \frac{\partial n(z_0)}{\partial z} \qquad (4.42)$$

This result has to be compared to the empirical expression for the density gradient generated self diffusion used in fluid dynamics (this expression was given in equation 4.30). The comparison yields the following expression for the coefficient of self diffusion:

$$D = \frac{1}{3} \lambda \bar{v} \qquad (4.43)$$

Another expression can be obtained for the diffusion coefficient by using our earlier results to express the mean free path and the mean thermal velocity. In this way one can directly relate the diffusion coefficient to fundamental molecular properties:

$$D = \left(\frac{2}{3} \frac{1}{\sigma} \sqrt{\frac{k}{\pi m}} \right) \frac{\sqrt{T}}{n} \qquad (4.44)$$

where σ is the total cross section of molecular collisions. Equation (4.44) predicts that the diffusion coefficient is directly proportional to $T^{1/2}$ and inversely proportional to the particle density, n. The predicted dependence on the temperature and density is well supported by observations. In most cases the numerical factor in the diffusion coefficient is not better than a factor of three or so. However, considering all the simplifications involved in the mean free path method it is quite remarkable that the dependence on the macroscopic gas parameters is correct for perfect gases.

4.3.3 Heat conduction

In the next example we examine the steady-state energy flux through the plane $z=z_0$. In this case the corresponding molecular quantity is the kinetic energy, therefore we choose $Q(\mathbf{v})=mv^2/2$. Substituting this function into the fundamental equation of the mean free path method (equation (4.39)) yields the following:

$$j_E(z_0) = \frac{1}{2} m \iiint_\infty d^3v \, v_z \, v^2 \, F(z_0, \mathbf{v}) - \frac{2}{3} m \lambda \frac{\partial}{\partial z} \left\{ \int_{-\infty}^{\infty} dv_x \int_{-\infty}^{\infty} dv_y \int_0^{\infty} dv_z v_z v^2 \, F(z_0, \mathbf{v}) \right\}$$

$$(4.45)$$

The first integral is again zero, because we integrate an odd function $(v_z v^2)$ from $-\infty$ to $+\infty$. The $\{\}$ term in the second integral is the same as the one evaluated in connection with the escaping energy flux through the microscopic orifice (see equation (4.26)) and its value is $\pi m n \, \bar{v}^3 / 16$, therefore one obtains

$$j_E(z_0) = -\frac{4}{3}\lambda\frac{\partial}{\partial z}\left\{\frac{1}{2}m\int\limits_{-\infty}^{\infty}dv_x\int\limits_{-\infty}^{\infty}dv_y\int\limits_{0}^{\infty}dv_z\,v_z\,v^2\,F(z_0,\mathbf{v})\right\} = -\frac{4}{3}\lambda\frac{\partial}{\partial z}\left\{\frac{\pi}{16}mn\,\bar{v}^3\right\}$$

$$(4.46)$$

In the case of pure heat conduction it is assumed that the particle concentration is constant everywhere, n=constant, and the gas temperature varies with altitude, z. In this case equation (4.46) simplifies to the following:

$$j_E(z_0) = -\frac{\pi}{12}mn\lambda\frac{\partial\bar{v}(z_0)^3}{\partial z} = -n\,\bar{v}\,\lambda\,k\frac{\partial T(z_0)}{\partial z}$$

$$(4.47)$$

Comparing equation (4.47) to the empirical relation between the temperature gradient and the heat flow (given by equation (4.29)) yields the following expression for the heat conductivity, κ:

$$\kappa = n\,\bar{v}\,\lambda\,k$$

$$(4.48)$$

The heat conductivity can also be expressed in terms of molecular parameters:

$$\kappa = \left(\frac{2k}{\sigma}\sqrt{\frac{k}{\pi m}}\right)\sqrt{T}$$

$$(4.49)$$

Equation (4.49) tells us that the heat conductivity is independent of the particle density and it varies with the gas temperature as $T^{1/2}$. This is a quite remarkable result which is well supported by observations. Again, the numerical factor in equation (4.49) cannot be taken too seriously, but the general dependence on the macroscopic gas parameters is true for perfect gases. It should be noted that for some molecules the average cross section is temperature dependent. This fact must be taken into account when calculating transport coefficients.

4.3.4 Viscosity

In this section we apply the mean free path method to calculate the coefficient of viscosity. We consider the same example as was discussed in Subsection 4.2.1. In this example the gas moves between two infinite parallel planes. The lower plane is at rest and the upper one moves in the positive x direction with a constant velocity, u_0. The distance between the two planes is denoted by h. It is assumed that the temperature and concentration are constant everywhere between the two infinite planes, but the bulk flow velocity in the x direction is a slowly varying function of z.

Using the fundamental equation of the mean free path method (equation (4.39)) the steady-state flux of the x component of the momentum through the plane $z=z_0$ can be written as

$$j_{M_{zx}}(z_0) = m \iiint_{\infty} d^3v \, v_x v_z \, F(z_0, \mathbf{v}) - \frac{4}{3} m\lambda \frac{\partial}{\partial z} \int_{-\infty}^{\infty} dv_x \int_{-\infty}^{\infty} dv_y \int_0^{\infty} dv_z v_x v_z F(z_0, \mathbf{v})$$

$$(4.50)$$

The first integral is again zero, because we are integrating an odd function ($v_x v_z$) from $-\infty$ to $+\infty$. The second term can be written in the following form:

$$j_{M_{zx}}(z_0) = -\frac{4}{3} m\lambda \frac{\partial}{\partial z} \left\{ \frac{n\,m}{2\pi kT} \int_{-\infty}^{\infty} dv_x v_x e^{-\frac{m(v_x-u)^2}{2kT}} \int_0^{\infty} dv_z v_z e^{-\frac{mv_z^2}{2kT}} \right\} \quad (4.51)$$

When deriving equation (4.51) we substituted the local Maxwellian for the distribution function and the integration over the variable v_y has already been carried out. The integrals in equation (4.51) can be evaluated to obtain

$$j_{M_{zx}}(z_0) = -\frac{1}{3} m\lambda \frac{\partial}{\partial z} \left\{ n(z_0) u(z_0) \bar{v}(z_0) \right\} \quad (4.52)$$

In this particular case the temperature and the density are constant throughout the gas, while the horizontal velocity, u, is a function of the altitude, z. This means that equation (4.52) can be written as

$$j_{M_{zx}}(z_0) = -\frac{1}{3} n\,m\,\lambda\bar{v} \frac{\partial u(z_0)}{\partial z} \quad (4.53)$$

Equation (4.53) can be directly compared to the empirical formula discussed in Subsection 4.2.1 (see equation (4.28)); one obtains the following expression for the viscosity, η:

$$\eta = \frac{1}{3} n\, m\, \lambda\, \bar{v} \qquad\qquad (4.54)$$

The coefficient of viscosity can also be expressed in terms of molecular parameters:

$$\eta = \left(\frac{2}{3\sigma} \sqrt{\frac{mk}{\pi}} \right) \sqrt{T} \qquad\qquad (4.55)$$

Equation (4.55) represents a very interesting result which indicates that for perfect gases the hydrodynamic viscosity is independent of the density and depends only on the gas temperature. This result contradicts our physical intuition: one would expect the momentum transfer (in this case characterized by the coefficient of viscosity) to be proportional to the number density of molecules. This paradox is resolved by equation (4.53) which shows that the momentum flux is proportional to the product of the density and the mean free path. If the density is increased by a factor of two the mean free path decreases by the same factor and the product of the two remains constant. In other words, when the density is higher more molecules carry the momentum flux but this transport is less efficient because the molecules collide more frequently.

Our result predicts that for perfect gases the viscosity is proportional to $T^{1/2}$. This is quite different than the temperature dependence of the viscosity of liquids, which typically decreases with increasing temperatures. The reason is that in a liquid the molecules are close to each other and the momentum transfer occurs by direct intermolecular forces which become weaker as the temperature increases. On the other hand, in dilute gases the momentum transfer occurs via the translational motion of the particles: this motion increases with increasing gas temperature.

4.3.5 Some general remarks about the mean free path method

In the previous subsections we examined the transport of mass, momentum and energy due to the random motion of the molecules. It was shown that slow variations in the macroscopic parameters of the velocity distribution function give rise to diffusive type transport of molecular quantities. These small fluxes are proportional to the negative gradient of the varying macroscopic parameter of the velocity distribution function. The factors of proportionality are the hydrodynamic transport parameters: diffusion coefficient, heat conductivity and viscosity. With the help of the mean free path method we were able to find specific expressions for the hydrodynamic transport coefficients in terms of fundamental molecular quantities.

All of the specific cases examined so far involved situations where bulk transport of the gas vanished (the first integral in expression (4.39) was always zero). Next we consider a simple example when the bulk transport does not vanish. Let us consider a steady-state gas with constant density and temperature, but with a bulk velocity vector, **u**, which slowly varies in the z direction. Let us consider the transport of the z component of the momentum of the molecules, $Q(\mathbf{v})=mv_z$. In this case $Q(\mathbf{v})F(z,\mathbf{v})$ is not an even function of v_z, therefore one has to start the derivation by using equation (4.38):

$$j_{M_{zz}}(z_0)=m\iiint_{\infty} d^3v\, v_z^2\, F(z_0,\mathbf{v})+\tfrac{2}{3}\, m\lambda\frac{\partial}{\partial z}\int_{-\infty}^{\infty}dv_x\int_{-\infty}^{\infty}dv_y\int_{-\infty}^{0}dv_z v_z^2\, F(z_0,\mathbf{v})$$

$$-\tfrac{2}{3}\, m\lambda\frac{\partial}{\partial z}\int_{-\infty}^{\infty}dv_x\int_{-\infty}^{\infty}dv_y\int_{0}^{\infty}dv_z v_z^2\, F(z_0,\mathbf{v})$$

$$(4.56)$$

The distribution function is locally a drifting Maxwellian with varying bulk velocity components, so

$$j_{M_{zz}}=mn\left(\frac{m}{2\pi kT}\right)^{3/2}\int_{-\infty}^{\infty}dv_x e^{-\frac{mc_x^2}{2kT}}\int_{-\infty}^{\infty}dv_y e^{-\frac{mc_y^2}{2kT}}\int_{-\infty}^{\infty}dv_z v_z^2 e^{-\frac{mc_z^2}{2kT}}$$

$$+\frac{2}{3}\lambda mn\left(\frac{m}{2\pi kT}\right)^{3/2}\frac{\partial}{\partial z}\int_{-\infty}^{\infty}dv_x e^{-\frac{mc_x^2}{2kT}}\int_{-\infty}^{\infty}dv_y e^{-\frac{mc_y^2}{2kT}}\int_{-\infty}^{0}dv_z v_z^2 e^{-\frac{mc_z^2}{2kT}}$$

$$-\frac{2}{3}\lambda mn\left(\frac{m}{2\pi kT}\right)^{3/2}\frac{\partial}{\partial z}\int_{-\infty}^{\infty}dv_x e^{-\frac{mc_x^2}{2kT}}\int_{-\infty}^{\infty}dv_y e^{-\frac{mc_y^2}{2kT}}\int_{0}^{\infty}dv_z v_z^2 e^{-\frac{mc_z^2}{2kT}}\quad(4.57)$$

where we have introduced the random velocity, $\mathbf{c}(z)=\mathbf{v}-\mathbf{u}(z)$. In equation (4.57) all components of u and their derivatives have to be taken at $z=z_0$. Transforming the integration variables from the components of the full velocity, **v**, to the components of the random velocity, **c**, yields the following expression:

$$j_{M_{zz}} = mn\left(\frac{m}{2\pi kT}\right)^{3/2} \int_{-\infty}^{\infty} dc_x e^{-\frac{mc_x^2}{2kT}} \int_{-\infty}^{\infty} dc_y e^{-\frac{mc_y^2}{2kT}} \int_{-\infty}^{\infty} dc_z (c_z + u_z)^2 e^{-\frac{mc_z^2}{2kT}}$$

$$+ \frac{2}{3}\lambda\, mn\left(\frac{m}{2\pi kT}\right)^{3/2} \frac{\partial}{\partial z} \int_{-\infty}^{\infty} dc_x e^{-\frac{mc_x^2}{2kT}} \int_{-\infty}^{\infty} dc_y e^{-\frac{mc_y^2}{2kT}} \int_{-u_z}^{\infty} dc_z (c_z + u_z)^2 e^{-\frac{mc_z^2}{2kT}}$$

$$- \frac{2}{3}\lambda\, mn\left(\frac{m}{2\pi kT}\right)^{3/2} \frac{\partial}{\partial z} \int_{-\infty}^{\infty} dc_x e^{-\frac{mc_x^2}{2kT}} \int_{-\infty}^{\infty} dc_y e^{-\frac{mc_y^2}{2kT}} \int_{-u_z}^{\infty} dc_z (c_z + u_z)^2 e^{-\frac{mc_z^2}{2kT}}$$

$$(4.58)$$

Note that the integration boundaries of the c_x and c_y integrals remain $-\infty$ to $+\infty$, but in the second term the lower integration boundary of the c_z integral is $-u_z$. The integrals from $-\infty$ to $+\infty$ can be readily carried out and the remaining integral can be rewritten in the following form:

$$j_{M_{zz}} = (nkT + nmu_z^2) - \frac{4}{3}\lambda\, mn\left(\frac{m}{2\pi kT}\right)^{1/2} \frac{\partial}{\partial z} \int_{-u_z}^{u_z} dc_z (c_z^2 + u_z^2) e^{-\frac{mc_z^2}{2kT}}$$

$$+ \frac{4}{3}\lambda\, mn\left(\frac{m}{2\pi kT}\right)^{1/2} \frac{\partial}{\partial z}\left[u_z \int_{-\infty}^{-u_z} dc_z\, c_z\, e^{-\frac{mc_z^2}{2kT}} - u_z \int_{-u_z}^{\infty} dc_z\, c_z\, e^{-\frac{mc_z^2}{2kT}} \right] \quad (4.59)$$

After some manipulation equation (4.59) can be written as follows:

$$j_{M_{zz}} = (nkT + nmu_z^2) - \frac{4}{3}\lambda\, nkT \frac{\partial}{\partial z}\left(\frac{2}{\sqrt{\pi}} s_z e^{-s_z^2}\right)$$

$$- \frac{8}{3\sqrt{\pi}}\lambda\, nkT \frac{\partial}{\partial z}\left(\int_0^{s_z} dw_z\, (w_z^2 + s_z^2) e^{-w_z^2}\right) \quad (4.60)$$

where we have introduced a new integration variable, $w_z = c_z/(2kT/m)^{1/2}$. In equation (4.60) we have also introduced the dimensionless bulk flow velocity vector, $\mathbf{s}$, which is defined as $\mathbf{s} = \mathbf{u}/(2kT/m)^{1/2}$. It is interesting to note that the magnitude of $\mathbf{s}$ is very close to the Mach number, M, defined as the ratio of the bulk flow speed and the speed of sound, $s = (\gamma/2)^{1/2} M$.

Now all the integrals can be carried out to yield the following result:

$$j_{M_{zz}} = (p + nmu_z^2) - \frac{2}{3}\lambda\, p \frac{\partial}{\partial z}\left([1 + 2s_z^2]\operatorname{erf}(s_z) + \frac{2}{\sqrt{\pi}} s_z e^{-s_z^2}\right) \quad (4.61)$$

Where erf(x) is the error function (the error function is described in Appendix 9.5.3).

Inspection of equation (4.61) reveals some very interesting results. It can be immediately seen that the slow variation in u_z will not make any significant contribution to the transport of the z momentum in the z direction: as a result of assumption (4.3) the $(2\lambda/3)\partial_z$ operator represents a negligible contribution to the momentum flux. Our result shows that the hydrodynamic momentum flux, $p+mnu_z^2$, is a very good approximation of the momentum transport for slowly varying flows.

Let us return to the transport coefficients derived in the previous subsections. At this point one can ask an interesting question: are the transport coefficients independent? It can be immediately seen that the ratios of the transport coefficients are closely related to fundamental parameters. For instance the ratio of the heat conductivity and viscosity is

$$\frac{\kappa}{\eta} = \frac{n\bar{v}\lambda k}{\frac{1}{3}nm\lambda\bar{v}} = \frac{3k}{m} \tag{4.62}$$

In the case of perfect gases (assumed to be composed of monatomic molecules) the specific heat at constant volume is $C_V = 3k/2m$, therefore equation (4.62) can also be written as

$$\frac{\kappa}{\eta} = 2C_V \tag{4.63}$$

The ratio of the viscosity and the diffusion coefficient can also be calculated:

$$\frac{\eta}{D} = \frac{\frac{1}{3}mn\lambda\bar{v}}{\frac{1}{3}\lambda\bar{v}} = mn \tag{4.64}$$

Measurements of D using tracer molecules combined with viscosity measurements at the same temperature yield values of $(mnD)/\eta$ between 1.3 and 1.5. The measured values of $\kappa/(C_V\eta)$ range from 1.3 to 2.5. This means that even though the mean free path method is not very accurate, it still provides a surprisingly good first guess for the hydrodynamic transport coefficients.

A widely used dimensionless quantity characterizing the relation between various transport coefficients is the so-called Prandtl number. The Prandtl number is defined as

$$Pr = \frac{\eta \, C_p}{\kappa} \tag{4.65}$$

Experiments show that the Prandtl number for a monatomic gas is Pr=2/3. Our elementary mean free path method yields the following value for the Prandtl number:

$$Pr = \frac{\eta \, C_p}{\kappa} = \frac{\frac{1}{3} m n \bar{v} \lambda}{k n \bar{v} \lambda} \frac{5k}{2m} = \frac{5}{6} \tag{4.66}$$

This result shows the limitations of the elementary mean free method: it is not perfect, but a good first approximation. In Chapter 5 we will discuss more sophisticated approximations which yield the correct value for the Prandtl number.

4.3.6 *The mean free path method in three dimensions*

Finally we briefly show how to generalize the mean free path for three dimensional problems.

We again assume near-equilibrium conditions, i.e. at every spatial point the velocity distribution function of the molecules is approximated by a Maxwell–Boltzmann distribution. The macroscopic parameters of the distribution function (such as temperature or particle number density) can slowly vary from point to point. The slow variation means that all characteristic scale lengths of these variations are much larger than the mean free path, λ (see assumption (4.3) at the beginning of this chapter).

The generalized mean free path method is also based on the idea that the collisional process can be discretized. It is assumed that all collisions take place simultaneously at discrete instants, t=0, τ, 2τ, 3τ, ... (where τ is the mean free time), and at discrete values of all three Cartesian coordinates, x=0, Δ, 2Δ, 3Δ, ..., y=0, Δ, 2Δ, 3Δ, ..., and z=0, Δ, 2Δ, 3Δ, ... (where $\Delta=2\lambda/3$ is some kind of an average distance traveled by the particles in the x, y and z directions in a mean free time).

The generalized mean free path method can be described as follows. The velocity distribution functions at all discrete 'collision points' are assumed to be locally Maxwellian at each time step (right after the collisions). The macroscopic parameters of these Maxwellian distributions may slowly vary from point to point. This means that all particles which at time t (just before collision) are located at the point $P_0=(x=x_0, y=y_0, z=z_0)$ had their last intermolecular collision at time t-τ at one of the following eight points: $P_{+,+,+}=(x_0+2\lambda/3, y_0+2\lambda/3, z_0+2\lambda/3)$, $P_{+,+,-}=(x_0+2\lambda/3, y_0+2\lambda/3, z_0-2\lambda/3)$, $P_{+,-,+}=(x_0+2\lambda/3, y_0-2\lambda/3, z_0+2\lambda/3)$, $P_{+,-,-}=(x_0+2\lambda/3, y_0-2\lambda/3,$

$z_0-2\lambda/3$), $P_{-,+,+}=(x_0-2\lambda/3, y_0+2\lambda/3, z_0+2\lambda/3)$, $P_{-,+,-}=(x_0-2\lambda/3, y_0+2\lambda/3, z_0-2\lambda/3)$, $P_{-,-,+}=(x_0-2\lambda/3, y_0-2\lambda/3, z_0+2\lambda/3)$, $P_{-,-,-}=(x_0-2\lambda/3, y_0-2\lambda/3, z_0-2\lambda/3)$.

At time t particles arrive from eight other points at the point P_0. For instance upward moving particles (those with $v_z>0$) arrive from below, while downward moving molecules (with $v_z<0$) arrive from above. At time t the distribution function at P_0 can be written as

$$F(t,x_0,y_0,z_0,\mathbf{v}) = \begin{cases} F(t-\tau,x_0-\Delta,y_0-\Delta,z_0-\Delta,\mathbf{v}) & v_x>0, v_y>0, v_z>0 \\ F(t-\tau,x_0-\Delta,y_0-\Delta,z_0+\Delta,\mathbf{v}) & v_x>0, v_y>0, v_z<0 \\ F(t-\tau,x_0-\Delta,y_0+\Delta,z_0-\Delta,\mathbf{v}) & v_x>0, v_y<0, v_z>0 \\ F(t-\tau,x_0-\Delta,y_0+\Delta,z_0+\Delta,\mathbf{v}) & v_x>0, v_y<0, v_z<0 \\ F(t-\tau,x_0+\Delta,y_0-\Delta,z_0-\Delta,\mathbf{v}) & v_x<0, v_y>0, v_z>0 \\ F(t-\tau,x_0+\Delta,y_0-\Delta,z_0+\Delta,\mathbf{v}) & v_x<0, v_y>0, v_z<0 \\ F(t-\tau,x_0+\Delta,y_0+\Delta,z_0-\Delta,\mathbf{v}) & v_x<0, v_y<0, v_z>0 \\ F(t-\tau,x_0+\Delta,y_0+\Delta,z_0+\Delta,\mathbf{v}) & v_x<0, v_y<0, v_z<0 \end{cases}$$

(4.67)

Equation (4.35) specifies the net flux of an arbitrary molecular quantity, $Q(\mathbf{v})$, in the z direction. Similar results can be obtained for the net flux in the x and y directions as well. The net flux of the molecular quantity $Q(\mathbf{v})$ through the point P_0 can now be written in the following form:

$$\mathbf{j}_Q(t,x_0,y_0,z_0) = \iiint_\infty d^3v\, \mathbf{v}\, Q(\mathbf{v}) F(t,x_0,y_0,z_0,\mathbf{v})$$

(4.68)

Substituting the distribution function given by equation (4.67) we get the following expression for the net flux:

$$j_Q(t,x_0,y_0,z_0) = \int_{-\infty}^{0} dv_x \int_{-\infty}^{0} dv_y \int_{-\infty}^{0} dv_z \, \mathbf{v} Q(\mathbf{v}) F(t-\tau, x_0+\Delta, y_0+\Delta, z_0+\Delta, \mathbf{v})$$

$$+ \int_{-\infty}^{0} dv_x \int_{-\infty}^{0} dv_y \int_{0}^{\infty} dv_z \, \mathbf{v} Q(\mathbf{v}) F(t-\tau, x_0+\Delta, y_0+\Delta, z_0-\Delta, \mathbf{v})$$

$$+ \int_{-\infty}^{0} dv_x \int_{0}^{\infty} dv_y \int_{-\infty}^{0} dv_z \, \mathbf{v} Q(\mathbf{v}) F(t-\tau, x_0+\Delta, y_0-\Delta, z_0+\Delta, \mathbf{v})$$

$$+ \int_{-\infty}^{0} dv_x \int_{0}^{\infty} dv_y \int_{0}^{\infty} dv_z \, \mathbf{v} Q(\mathbf{v}) F(t-\tau, x_0+\Delta, y_0-\Delta, z_0-\Delta, \mathbf{v})$$

$$+ \int_{0}^{\infty} dv_x \int_{-\infty}^{0} dv_y \int_{-\infty}^{0} dv_z \, \mathbf{v} Q(\mathbf{v}) F(t-\tau, x_0-\Delta, y_0+\Delta, z_0+\Delta, \mathbf{v})$$

$$+ \int_{0}^{\infty} dv_x \int_{-\infty}^{0} dv_y \int_{0}^{\infty} dv_z \, \mathbf{v} Q(\mathbf{v}) F(t-\tau, x_0-\Delta, y_0+\Delta, z_0-\Delta, \mathbf{v})$$

$$+ \int_{0}^{\infty} dv_x \int_{0}^{\infty} dv_y \int_{-\infty}^{0} dv_z \, \mathbf{v} Q(\mathbf{v}) F(t-\tau, x_0-\Delta, y_0-\Delta, z_0+\Delta, \mathbf{v})$$

$$+ \int_{0}^{\infty} dv_x \int_{0}^{\infty} dv_y \int_{0}^{\infty} dv_z \, \mathbf{v} Q(\mathbf{v}) F(t-\tau, x_0-\Delta, y_0-\Delta, z_0-\Delta, \mathbf{v})$$

$$(4.69)$$

It is assumed throughout this chapter that the distribution function is locally Maxwellian and that the macroscopic parameters of the distribution function are slowly varying functions of spatial location. This means that the distribution function can be expanded into a Taylor series around the point, P_0, and the series can be truncated after the linear terms:

$$F(t, x_0 \pm \Delta, y_0 \pm \Delta, z_0 \pm \Delta, \mathbf{v}) = F(t, x_0, y_0, z_0, \mathbf{v})$$

$$\pm \frac{2}{3}\lambda \frac{\partial F(t, x_0, y_0, z_0, \mathbf{v})}{\partial x} \pm \frac{2}{3}\lambda \frac{\partial F(t, x_0, y_0, z_0, \mathbf{v})}{\partial y} \pm \frac{2}{3}\lambda \frac{\partial F(t, x_0, y_0, z_0, \mathbf{v})}{\partial z}$$

$$(4.70)$$

Here we have used the numerical value of Δ, $\Delta = 2\lambda/3$. Substituting this expression into equation (4.69) results in the following fundamental equation for the three dimensional mean free path method:

$$\mathbf{j}_Q(t,x_0,y_0,z_0) = \iiint_{\infty} d^3v\,\mathbf{v}\,Q(\mathbf{v})F$$

$$+\tfrac{2}{3}\lambda\frac{\partial}{\partial x}\left(\int_{-\infty}^{0}dv_x\int_{-\infty}^{\infty}dv_y\int_{-\infty}^{\infty}dv_z\,\mathbf{v}\,Q(\mathbf{v})F - \int_{0}^{\infty}dv_x\int_{-\infty}^{\infty}dv_y\int_{-\infty}^{\infty}dv_z\,\mathbf{v}\,Q(\mathbf{v})F\right)$$

$$+\tfrac{2}{3}\lambda\frac{\partial}{\partial y}\left(\int_{-\infty}^{\infty}dv_x\int_{-\infty}^{0}dv_y\int_{-\infty}^{\infty}dv_z\,\mathbf{v}\,Q(\mathbf{v})F - \int_{-\infty}^{\infty}dv_x\int_{0}^{\infty}dv_y\int_{-\infty}^{\infty}dv_z\,\mathbf{v}\,Q(\mathbf{v})F\right)$$

$$+\tfrac{2}{3}\lambda\frac{\partial}{\partial z}\left(\int_{-\infty}^{\infty}dv_x\int_{-\infty}^{\infty}dv_y\int_{-\infty}^{0}dv_z\,\mathbf{v}\,Q(\mathbf{v})F - \int_{-\infty}^{\infty}dv_x\int_{-\infty}^{\infty}dv_y\int_{0}^{\infty}dv_z\,\mathbf{v}\,Q(\mathbf{v})F\right) \quad (4.71)$$

where F denotes $F(t-\tau,x_0,y_0,z_0,\mathbf{v})$.

Let us examine the problem of diffusion as a simple demonstration of the three dimensional mean free path method. We again assume that the gas is in steady state and that only the gas density varies with location (more specifically it is assumed that the bulk velocity is zero everywhere and the temperature is constant). In this case $Q(\mathbf{v})=1$ and equation (4.71) simplifies to the following:

$$\mathbf{j}_{diff}(x_0,y_0,z_0) = -\frac{4}{3}\lambda\frac{\partial}{\partial x}\int_{0}^{\infty}dv_x\int_{-\infty}^{\infty}dv_y\int_{-\infty}^{\infty}dv_z\,\mathbf{v}\,F$$

$$-\frac{4}{3}\lambda\frac{\partial}{\partial y}\int_{-\infty}^{\infty}dv_x\int_{0}^{\infty}dv_y\int_{-\infty}^{\infty}dv_z\,\mathbf{v}\,F - \frac{4}{3}\lambda\frac{\partial}{\partial z}\int_{-\infty}^{\infty}dv_x\int_{-\infty}^{\infty}dv_y\int_{0}^{\infty}dv_z\,\mathbf{v}\,F \quad (4.72)$$

The integrals vanish for all velocity components except the one which has the incomplete integration region (from 0 to ∞). This leads to the following result:

$$\mathbf{j}_{diff}(x_0,y_0,z_0) = -\left(\frac{1}{3}\lambda\,\bar{v}\right)\nabla n(x_0,y_0,z_0) \quad (4.73)$$

Equation (4.73) is the general form of the diffusive particle flux predicted by the mean free path method. Note that the diffusion coefficient is the same as it was in the simple one dimensional case, $D=\lambda\,\bar{v}/3$.

4.4 Flow in a tube

In this section we consider three examples of viscous gas flow at different pressures (densities): collision dominated, collisionless and intermediate flows. The different regimes are characterized by several dimensionless parameters. First we introduce these parameters and then we discuss the various flow regimes.

There are three fundamental dimensionless parameters which determine the flow regime, i.e., the type of applicable approximation. The first of these parameters is the Mach number which was already briefly mentioned above. The Mach number, M, is the ratio of the bulk flow speed, u, and the speed of sound, $a=(\gamma p/mn)^{1/2}$:

$$M = \frac{u}{a} \qquad (4.74)$$

The Knudsen number, Kn, is defined as the ratio of the mean free path of intermolecular collisions (collisions between two molecules), λ, and the characteristic linear size of the problem, L:

$$Kn = \frac{\lambda}{L} \qquad (4.75)$$

The Reynolds number, Re, is the ratio of the bulk momentum flux, nmu^2, and the viscous momentum transfer over the characteristic scale length, $\eta u/L$:

$$Re = \frac{m\,n\,u\,L}{\eta} \qquad (4.76)$$

These three dimensionless parameters are not independent of each other. Using the expression for the coefficient of viscosity obtained by the mean free path method (equation (4.54)) one can write the Reynolds number in the following form:

$$Re = \frac{3u\,L}{\bar{v}\,\lambda} = 3\sqrt{\gamma\frac{\pi}{8}}\,\frac{M}{Kn} = 1.88\,\sqrt{\gamma}\,\frac{M}{Kn} \qquad (4.77)$$

Let us examine the numerical values of these dimensionless quantities under ordinary circumstances. In air at sea level the number density of molecules, n, is about 3×10^{25} molecules/m^3. The cross section of intermolecular collisions, σ, can be well approximated by a typical value of 10^{-19} m^2, therefore one obtains a typical value of about 3×10^{-7} m for the molecular mean free path, λ. This mean free path is so small that for typical size problems L>>λ and therefore Kn<<1. The sound speed in ordinary air, a_s, is about 3×10^2 m/s and most flow problems of compressible fluid dynamics are in the Mach number range of M=0.001–10. These Mach number values are still much larger than the Knudsen number and therefore the Reynolds number is much larger than unity, Re>>1.

The Knudsen number is a very good measure of the collisional regime of a given problem. When Kn<<1 the flow is collision dominated, meaning that the molecules undergo a very large number of collisions before traveling a distance comparable to the scale of the problem. In this case diffusion type processes are responsible for transporting mass, momentum and energy inside the gas (in the absence of macroscopic bulk motion of the gas as a whole). This regime can be very successfully described with the methods of compressible fluid dynamics.

The other extreme case is when the Knudsen number is very large and the effects of intermolecular collisions are negligible. In the case of very large scale problems this regime is called collisionless flow (for instance the motion of the upper atmosphere at very high altitudes is often called collisionless). In the case when we consider the interaction of collisionless gases with finite size objects (such as a long vacuum tube or a spacecraft) this flow is called free molecular flow. In the free molecular regime molecules collide with the wall of the object but their collisions with each other are either neglected or grossly simplified. The description of free molecular flows is conceptually fairly straightforward but mathematically difficult.

The transitional regime, when the Knudsen number is about unity, is the most unexplored parameter regime. There are no well developed standard techniques to handle these problems in general: one has to be innovative to describe flows with Kn~1.

In this section we consider three different regimes of gas flow in an infinite tube. These regimes include a hydrodynamic case (Poiseuille flow), a transition regime situation (slip flow) and finally a free molecular scenario.

4.4.1 Poiseuille flow

Let us first consider the flow of gas through a long straight tube of circular cross section. At room temperature and pressure the molecular mean free path, λ, is negligibly small compared to the radius of the tube, a. Assuming that the flow velocity is not too large and consequently the flow is not turbulent, the steady-state motion of the gas is laminar and is primarily controlled by the viscosity.

This flow problem is axially symmetric around the longitudinal axis of the tube; therefore the bulk velocity of the fluid, u, is a function of the radial distance from the axis of the tube (z–axis). The flow can be subdivided into coaxial cylindrical shells of thickness dr around the z axis (the radial distance in a cross-sectional plane, r, is measured from the axis of symmetry). The velocity, u(r), increases from zero at the wall to a maximum on the axis. The outermost layer of gas adjacent to the wall might be expected to have a negligible flow velocity because of surface irregularities. Translated into a boundary condition, this means u(a)=0. Also, the tube is considered to be long enough so that end effects can be neglected.

Let us consider a short cylinder with radius r and length Δz (see Figure 4.8). In general this cylinder consists of a number of coaxial shells. There are two forces acting on this short cylinder: an accelerating force due to the pressure difference between the two bases of the cylinder, $\Delta p = p(z+\Delta z) - p(z)$, and a retarding viscous force acting on the outside surface of the cylinder. The pressure difference force is the base area times the pressure difference, $(r^2\pi) \times \Delta p$, and the viscous force is the surface area times the viscous stress, $(2\pi r \Delta z) \times (\eta du/dr)$. Under steady-state conditions the pressure gradient and viscous forces balance each other:

$$(2\pi r \, \Delta z) \times \left(\eta \, \frac{du}{dr} \right) - \pi r^2 \Delta p = 0 \qquad (4.78)$$

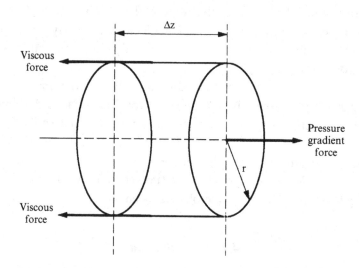

Figure 4.8

It is important to note that the short cylinder is treated as a solid body and therefore the pressure gradient force accelerates all the gas inside the cylinder. Equation (4.78) can be rearranged to express the radial velocity gradient:

$$\frac{du}{dr} = \frac{r}{2\eta} \frac{\Delta p}{\Delta z} \qquad (4.79)$$

This differential equation can be solved for u(r). The boundary condition is u(a)=0, therefore the solution is the following:

$$u(r) = -\frac{a^2 - r^2}{4\eta}\frac{\Delta p}{\Delta z} \qquad (4.80)$$

Equation (4.80) describes the well known parabolic cross-sectional profile of the radial dependence of the flow velocity in a long tube. The negative sign means that the gas will flow toward the lower pressure region (where Δp is negative). Using equation (4.80) one can also calculate the total flux of molecules in the tube. This can be done by integrating the flux in cylindrical shells with radius r over the cross section of the tube:

$$\Phi_{total} = n\int_0^a dr\, 2\pi r\, u(r) = -\frac{2\pi n}{4\eta}\frac{\Delta p}{\Delta z}\int_0^a dr\, r\left(a^2 - r^2\right) = -\frac{\pi a^4}{8\eta}n\frac{\Delta p}{\Delta z} \qquad (4.81)$$

where n is the particle number density inside the short cylinder with length Δz.

4.4.2 Slip flow

Next we consider the case Kn~1, when the shortest characteristic scale of the problem is comparable to the mean free path of intermolecular collisions. In this case the mathematical description of the flow is quite complicated, because the fundamental assumptions of fluid dynamics are no longer valid. Here we discuss a specific case, when the mean free path, λ, is smaller than the radius of the tube, a, but it is not negligibly small. In this particular situation a modified Poiseuille formula can be obtained, corrected for the 'slippage' of the gas at the walls of tube. By slippage we mean that the gas velocity at the wall is different than the wall velocity.

First we introduce the concept of slippage by considering the interaction of gas molecules with two moving infinite parallel planes. The two planes are separated by a distance, h, and they move with velocities u=0 and u=u_0. This situation is illustrated by Figure 4.9. It is assumed that the mean free path of molecular collisions, λ, is somewhat smaller than the distance between the two parallel planes, λ<h. Specifically, it is assumed that after colliding with a wall each molecule undergoes several (but not many) molecular collisions before it collides with the other wall.

First we discuss the collisions with the moving wall at z=h. Let us denote the y components of the average velocity of molecules right before and right after their collision with the upper wall by q_1 and q_2, respectively. Observations indicate that there are two types of collisions between the moving walls and the gas molecules.

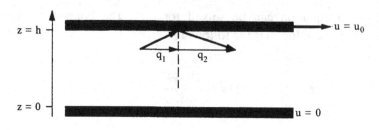

Figure 4.9

The first type is the elastic (or specular) collision, when the tangential velocity of the colliding molecule is conserved, therefore for elastic collisions $q_2=q_1$. The second type of collision is the diffuse reflection, when a molecule is temporarily absorbed and then re-emitted. The average tangential velocity of diffusely reflected particles is the same as the velocity of the wall, therefore for these particles $q_2=u_0$. The relative importance of the two types of reflections is characterized by the accommodation coefficient , σ', which is defined as the fraction of molecules undergoing diffuse reflection.

With the help of the accommodation coefficient we can express the average tangential velocity right after collision. The tangential velocities of diffusely and specularly reflected molecules are u_0 and q_1, respectively, therefore we can write

$$q_2 = \sigma' u_0 + (1-\sigma')q_1 \qquad (4.82)$$

The average gas tangential velocity at the upper wall, u_h, is the mean of the average pre- and post-collision velocities, $u_h=(q_1+q_2)/2$. Substituting equation (4.82) for q_2 yields the following expression for u_h:

$$u_h = \frac{q_1+q_2}{2} = \frac{1}{2}\left(q_1 + \sigma' u_0 + (1-\sigma')q_1\right) = \frac{\sigma'}{2}u_0 + \frac{2-\sigma'}{2} q_1 \qquad (4.83)$$

Equation (4.83) still contains two unknown quantities, q_1 and u_h. One more equation is needed to fully specify these quantities. This equation can be obtained by applying the mean free path method to the slip flow problem. According to the mean free path method, q_1 reflects the average tangential flow velocity at a distance of $2\lambda/3$ from the wall (where the last collision occurred). Expanding the tangential velocity function, $u(z)$, to first order yields the following relation:

$$q_1 = u_h - \frac{2}{3}\lambda \left[\frac{du}{dz}\right]_{z=h} \qquad (4.84)$$

One can combine equations (4.83) and (4.84) to obtain the following equation for u_h:

$$u_h = u_0 - \frac{2-\sigma'}{\sigma'} \frac{2}{3} \lambda \left[\frac{du}{dz} \right]_{z=h}$$

(4.85)

The difference between the velocity of the wall, u_0, and the average gas velocity at the wall, u_h, is called the slip velocity, u_{slip}. This can be expressed as follows:

$$u_{slip} = \frac{2-\sigma'}{\sigma'} \frac{2}{3} \lambda \left[\frac{du}{dz} \right]_{z=h}$$

(4.86)

There are two limiting cases to be considered. In the case when all molecules are reflected diffusely, i.e. if $\sigma'=1$, equation (4.86) simplifies to the following:

$$u_{slip} = \frac{2}{3} \lambda \left[\frac{du}{dz} \right]_{z=h}$$

(4.87)

This expression is well defined mathematically and it describes a finite slip velocity. However, the other limiting case, when all molecules are reflected specularly, is mathematically undefined. In this case $\sigma'=0$ (perfect slip), and therefore $u_{slip} \to \infty$. In most practical cases $\sigma'>0$, therefore the slip velocity remains finite.

Next we apply this result for a low density gas flow in a long circular tube. The physical problem is identical to the one discussed with respect to Poiseuille flow with the exception of 'slippage' at the walls. For the sake of simplicity we assume diffuse reflection ($\sigma'=1$), which is a pretty good approximation in most cases.[6] The radial gradient of the flow is given by equation (4.79) and the slip velocity is

$$u_{slip} = -\frac{1}{3} \lambda \frac{a}{\eta} \frac{\Delta p}{\Delta z}$$

(4.88)

because the gas flows in the direction of smaller pressure ($\Delta p<0$). It should also be mentioned that in equation (4.88) the coordinate z denotes the distance along the longitudinal axis of the tube (and not the coordinate between the two parallel planes). The cross-sectional velocity profile of the flow can be obtained by solving differential equation (4.80) with the slip flow boundary condition, i.e. with $u(a)=u_{slip}$. This solution is the following:

$$u(r) = u_{slip} - \frac{a^2-r^2}{4\eta} \frac{\Delta p}{\Delta z} = -\left(a^2 + \frac{4}{3} \lambda a - r^2 \right) \frac{1}{4\eta} \frac{\Delta p}{\Delta z}$$

(4.89)

It is important to note the correction term, proportional to the mean free path, due to the slippage of the gas flow at the wall. One can use this slip flow cross-sectional velocity profile to calculate a modified Poiseuille formula for the total flux through the tube:

$$\Phi'_{\text{total}} = n \int_0^a dr \, 2\pi r \, u(r) = -\frac{\pi a^4}{8\eta} n \frac{\Delta p}{\Delta z} \left(1 + \frac{8}{3} \frac{\lambda}{a}\right) \qquad (4.90)$$

When compared to the original Poiseuille formula (equation (4.81)) equation (4.90) exhibits a correction factor of $8\lambda/3a$. This means that 'slippage' decreases wall friction and therefore the gas responds more readily to the pressure gradient. The net result is an increase in the gas flux.

4.4.3 Free molecular flow in a tube

Next we consider the flow in a very long tube of circular cross section (with radius a) at a very low pressure. In this case the condition $Kn=\lambda/a >> 1$ is satisfied and the basic assumptions for viscous flow are not valid any more. In this case the gas is described as in the free molecular approximation (this regime is also called Knudsen flow).

In this section we describe the free molecular flow in a long circular cylinder of radius a. The two ends are kept at different pressures, and the temperature is uniform throughout the tube. When $Kn>>1$, a gas molecule collides with the wall many times before it encounters another molecule. The flow of gas is determined almost entirely by wall–molecule collisions and is practically unaffected by intermolecular collisions.

The free molecular approximation assumes that the effects of intermolecular collisions on molecular trajectories are statistically insignificant and, therefore, these collisions are neglected altogether. In this approximation the flow problem is reduced to determination of the effects of molecular collisions with the wall.

We again distinguish two limiting types of collisions between the tube and the gas molecules. The first type is the elastic (or specular) collision, when the tangential velocity of the colliding molecule is conserved. The second type of collision is the diffuse reflection, when a molecule is temporarily absorbed and then re-emitted. Experiments show that the reflection is almost 100% diffusive, as if reflected molecules were first absorbed by the wall and later evaporated.

For the sake of simplicity we adopt the following simple model of diffuse reflection. Imagine that the irregular wall structure is replaced by a smooth mathematical surface. The reflected molecules are envisioned as particles escaping from

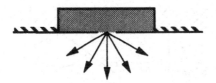

Figure 4.10

a hypothetical equilibrium gas layer on the other side of the surface (assumed to be in thermal equilibrium with the wall) through a very small orifice (see Figure 4.10).

In this model the number of molecules reflected in a particular direction from unit wall area per unit time is equal to the escape flux from the reservoir in the given direction. The differential escape flux is (see equation (4.14)):

$$d^3 j_{ref} = v^3 \cos\theta' F_{esc} \, d\Omega' \, dv \qquad (4.91)$$

where θ' is the angle between the direction of motion of the molecules and normal to the surface element, and $d\Omega'$ is the solid angle element around our given direction. The flux of reflected molecules in a given direction can be obtained by integrating over all speeds, v:

$$d^2 j_{ref} = \cos\theta' \, d\Omega' \, n \left(\frac{m}{2\pi kT}\right)^{3/2} \int_0^\infty dv \, v^3 \, e^{-\frac{mv^2}{2kT}} = n \, \bar{v} \cos\theta' \frac{d\Omega'}{4\pi} \qquad (4.92)$$

where n is the gas number density in the hypothetical gas layer.

Let us denote an area element of the wall and the tube cross section by dS' and dS, respectively. The normal vectors of dS' and dS are perpendicular (see Fig.4.11). Let $d\Omega'$ be the solid angle covered by dS as seen from the center of the surface element, dS'. Similarly, $d\Omega$ is the solid angle covered by the infinitesimal area, dS', as seen from the center of dS. By definition $d\Omega = \sin\theta \, d\theta \, d\varphi$ (and $d\Omega' = \sin\theta' d\theta' d\varphi'$). It should be noted that θ varies from 0 to π (particles can cross the cross-sectional area in either direction) but θ' varies from 0 to $\pi/2$ (reflected particles can only come from the wall). The distance between the centers of dS and dS' is denoted by r.

From these definitions it is obvious that $\theta + \theta' = \pi/2$ (see Figure 4.11), because the normal vectors of dS and dS' are perpendicular to each other (one is a wall surface element and the other is a tube cross-sectional surface element). It can be seen from Figure 4.12 that the following relation holds between the various quantities:

$$r^2 d\Omega = dS' \sin\theta = dS' \sin\left(\frac{\pi}{2} - \theta'\right) = dS' \cos\theta' \qquad (4.93)$$

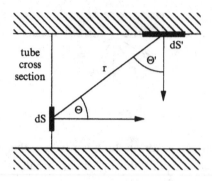

Figure 4.11

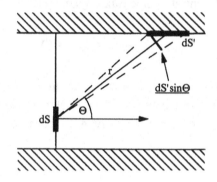

Figure 4.12

and similarly

$$r^2 d\Omega' = dS \sin\theta' = dS \left| \sin\left(\frac{\pi}{2} - \theta\right)\right| = dS \, |\cos\theta| \qquad (4.94)$$

Here the absolute value of $\cos\theta$ appears because θ can vary from 0 to π and the function $\cos\theta$ is negative in the second quadrant.

Consider those molecules which leave the wall element, dS', and pass directly through the cross-sectional element, dS. The number of these molecules per unit time is given by

$$d^4\Phi = d^2 j_{\text{ref}} \, dS' = n \, \overline{v} \cos\theta' \frac{d\Omega'}{4\pi} dS' \qquad (4.95)$$

Equation (4.95) can be rewritten by using equations (4.93) and (4.94) expressing the solid angle elements, $d\Omega$ and $d\Omega'$:

$$d^4\Phi = n\,\bar{v}\cos\theta'\,\frac{d\Omega'}{4\pi}\,dS' = n\,\bar{v}\,\frac{d\Omega'}{4\pi}\,r^2 d\Omega = \frac{n\,\bar{v}}{4\pi}|\cos\theta|\,d\Omega\,dS \qquad (4.96)$$

It should be noted that all variables related to the wall of the tube are eliminated from equation (4.96) and the differential particle flux is expressed in terms of the cross-sectional parameters only.

We choose a coordinate system with the z axis parallel to the longitudinal axis of the tube. Let q denote the projection of r onto the cross-sectional plane containing the surface element dS (this plane is chosen to be the (x,y) plane of our coordinate system): $q=r\sin\theta$ and $z=r\cos\theta$. One can also express the z coordinate in terms of q and θ by combining these two relations:

$$z = q\cot\theta \qquad (4.97)$$

In a very long tube with a finite density difference between the two ends the gas concentration, n(z), is a slowly varying function of the longitudinal coordinate, z (molecular collisions are neglected; therefore the number density is constant in every cross-sectional area). This means that the characteristic scale length of n(z), L, is much larger than the tube radius, a. In other words the following condition must hold:

$$\frac{1}{L} = \left|\frac{1}{n}\frac{dn}{dz}\right| << \frac{1}{a} \qquad (4.98)$$

Those particles which cross the surface element, dS, in the −z direction must be reflected from wall elements with z>0. Assuming a very long tube the total flux of these molecules can be obtained by integrating the differential flux from 0 to $\pi/2$:

$$d^2\Phi_- = \frac{\bar{v}\,dS}{4\pi}\int_0^{2\pi}d\varphi\int_0^{\pi/2}d\theta\,n(z)\cos\theta\sin\theta \qquad (4.99)$$

Similarly, one can calculate the total flux of particles crossing dS in the +z direction:

$$d^2\Phi_+ = -\frac{\bar{v}\,dS}{4\pi}\int_0^{2\pi}d\varphi\int_{\pi/2}^{\pi}d\theta\,n(z)\cos\theta\sin\theta \qquad (4.100)$$

The negative sign appears because $\cos\theta$ is negative between $\pi/2$ and π, and $d^4\Phi$ contains $|\cos\theta|$. The net flux through the cross-sectional surface element, dS, is

$$d^2\Phi = d^2\Phi_+ - d^2\Phi_- = -\frac{\overline{v}\,dS}{4\pi}\int\limits_0^{2\pi}d\varphi\int\limits_0^{\pi}d\theta\,n(z)\cos\theta\sin\theta \qquad (4.101)$$

It should be noted that n(z) cannot be taken out of the integrals, because for a given surface element, dS, z is a function of q (see equation (4.97)). On the other hand n(z) is a slowly varying function of z, so one can expand it into a Taylor series and stop after the first order term:

$$n(z) = n(0) + q\frac{\cos\theta}{\sin\theta}\frac{dn(0)}{dz} + \cdots \qquad (4.102)$$

where we have substituted expression (4.97) for z. Substituting equation (4.102) into (4.101) yields the following equation for the net particle flux going through the cross-sectional surface element, dS:

$$d^2\Phi = -\frac{\overline{v}\,dS}{4\pi}\int\limits_0^{2\pi}d\varphi\int\limits_0^{\pi}d\theta\left[n(0)\sin\theta\cos\theta + q\frac{dn(0)}{dz}\cos^2\theta\right] \qquad (4.103)$$

The first term in the integral makes no contribution, while the integral of $\cos^2\theta$ from 0 to π yields $\pi/2$. This means that equation (4.103) can be written as follows:

$$d^2\Phi = -\frac{\overline{v}}{8}dS\frac{dn(0)}{dz}\int\limits_0^{2\pi}d\varphi\,q \qquad (4.104)$$

It is important to note that the distance from the center of the cross-sectional element to the wall, q, is a function of the azimuth angle, φ, because the surface element is not necessarily at the center of the circular cross section (see Figure 4.13).

Next we want to evaluate the last remaining integral in equation (4.104). In order to do this we introduce a new variable. Let u denote the distance from the axis of the tube to the cross-sectional surface element, dS (see Figure 4.13). Using the schematics shown in Figure 4.13 one can see that the following relation holds between q, u and the angle, φ:

$$a^2 = q^2 + u^2 + 2qu\cos\varphi \qquad (4.105)$$

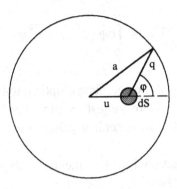

Figure 4.13

One can solve equation (4.105) for q as a function of u and φ:

$$q = \sqrt{a^2 - u^2 \sin^2 \varphi} - u \cos \varphi \qquad (4.106)$$

Here we have kept only the positive root of the quadratic equation, because the distance, q, must be positive. Substituting this expression into equation (4.104) yields the following result:

$$d^2\Phi = -\frac{\overline{v}}{8} dS \frac{dn(0)}{dz} \int_0^{2\pi} d\varphi \left(\sqrt{a^2 - u^2 \sin^2 \varphi} - u \cos \varphi \right) \qquad (4.107)$$

The second term in the integral is zero and the remaining term can be written in the following form:

$$d^2\Phi = -\frac{\overline{v}}{8} a \, dS \frac{dn(0)}{dz} \int_0^{2\pi} d\varphi \sqrt{1 - w^2 \sin^2 \varphi} \qquad (4.108)$$

where w=u/a (it is obvious that $0 \leq w \leq 1$). The remaining integral is periodic and its value can be expressed in terms of a well known special function:

$$\int_0^{2\pi} d\varphi \sqrt{1 - w^2 \sin^2 \varphi} = 4 \int_0^{\pi/2} d\varphi \sqrt{1 - w^2 \sin^2 \varphi} = 4 \, E(w) \qquad (4.109)$$

where the function, E(w), is the complete elliptic integral of the second kind. This function is given by

$$E(w) = \int_0^{\pi/2} d\varphi \sqrt{1 - w^2 \sin^2 \varphi} \tag{4.110}$$

The fundamental properties of the complete elliptic integral of the second kind are discussed in Appendix 9.5.5. Here we just mention that E(w) is defined for values of w between w=0 and w=1. It monotonically decreases from $\pi/2$ at w=0 to a value of 1 at w=1.

The differential cross-sectional area element is a thin ring with a normalized radius of w, $dS = 2\pi a^2 w dw$. Using this value for dS we can express the total net flow through the entire cross section of the tube in the following form:

$$\Phi = -\pi a^3 \, \overline{v} \, \frac{dn(0)}{dz} \int_0^1 dw \int_0^{\pi/2} d\varphi \, w \sqrt{1 - w^2 \sin^2 \varphi} \tag{4.111}$$

First we carry out the integration with respect to w:

$$\Phi = -\frac{\pi}{3} \overline{v} a^3 \frac{dn(0)}{dz} \int_0^{\pi/2} d\varphi \frac{1 - \cos^3 \varphi}{\sin^2 \varphi} \tag{4.112}$$

The remaining integral can also be evaluated and its value turns out to be 2. In this way equation (4.112) can be simplified to the following expression:

$$\Phi = -\frac{2\pi}{3} a^3 \overline{v} \frac{dn(0)}{dz} \tag{4.113}$$

Equation (4.113) is the total net flow through the tube in the free molecular approximation.

Using equation (4.113) one can readily calculate the average flux density (net flow per unit cross-sectional area):

$$j = -\frac{2}{3} a \overline{v} \frac{dn(0)}{dz} \tag{4.114}$$

This result can be compared to the hydrodynamic flux density calculated with the mean free path method (equation (4.42)). This comparison indicates that in the free molecular flow regime the tube diameter behaves as an 'effective' mean free path, λ=2a. This result is quite plausible, because in the free molecular case the molecules interact only with the wall.

In the present model $\bar{v}$ is the average velocity in the hypothetical gas reservoir to be evaluated with the wall temperature. Substituting this expression into the free molecular flow formula one obtains the following expression for the total flux through the tube:

$$\Phi=-\frac{2\pi}{3}a^3\sqrt{\frac{8\,kT}{\pi\,m}}\frac{dn}{dz}=-\frac{4\pi}{3}\frac{a^4}{\lambda}\sqrt{\frac{8\,kT}{\pi\,m}}\frac{dn}{dz} \qquad (4.115)$$

This result can be compared with the Poiseuille formula (using the mean free path approximation to obtain the coefficient of viscosity and assuming constant temperature):

$$\Phi_{\text{Poiseuille}}=-\frac{3\pi}{16}\frac{a^4}{\lambda}\sqrt{\frac{\pi\,kT}{2\,m}}\frac{dn}{dz} \qquad (4.116)$$

The ratio of the fluxes is close to 1/6:

$$\frac{\Phi_{\text{Poiseuille}}}{\Phi_{\text{fm}}}=\frac{9\,\pi}{128\sqrt{2}}=0.156 \qquad (4.117)$$

This result means that the throughput (total flux) carried by the density gradient driven flow decreases with increasing mean free path (assuming the same density gradient). The numerical factor in the expression for the throughput, however, is larger by about a factor of 6.5 in the free molecular regime.

4.5 Problems

4.1 A cylinder contains gas in equilibrium at temperature T. In a reversible, adiabatic expansion of the gas, a piston moves out slowly with velocity u, where u is much smaller than the average thermal speed (see Figure 4.14). What is the average energy loss per molecule rebounding from the piston? Show that the total energy loss of particles rebounding from the piston is equal to the work done by the expanding gas in accordance with the law of energy conservation. (See also Problem 2.12.)

4.2 Consider a very large gas reservoir divided by a thin wall with a tiny hole in the middle. Both sides are filled with the same single species gas and both sides are in equilibrium at different temperatures. The diameter of the hole is much

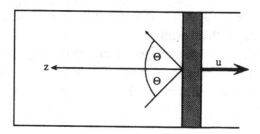

Figure 4.14

smaller than the mean free path of molecular collisions. The number densities are identical in the two parts of the reservoir, $n_1=n_2$, but the temperature in one partition is four times larger than in the other, $T_1=4T_2$. What is the average energy of all the molecules passing through the hole?

4.3 A perfect gas (mass m) fills two compartments of a vessel. The compartments are connected through a very small opening of diameter d, where d is much larger than a molecular diameter, but much smaller than the mean free path of the enclosed gas. The initial temperature in each compartment is T, and the number densities are n_1 and n_2, respectively, with $n_1>n_2$. Calculate the net mass of gas which flows from compartment 1 to 2 in unit time. Calculate the corresponding net rate of energy transport. Calculate the average energy per molecule transported through the hole. Why is the average energy of outflowing molecules different from 3kT/2?

4.4 An apparatus for measuring viscosity consists of two coaxial cylinders, both 0.25 m in length, the outer one rotating at the rate of 1.10 revolutions/min and the inner one attached to a wire suspension whose twist measures the torque on the stationary inner cylinder. The radius of the outer cylinder is 0.06 m and the gap between the cylinders, containing the gas, is 0.002 m wide. If the torque is measured as 4.40×10^{-7} N·m, what is the viscosity of the gas? (Make the approximation that the gap is of negligible width compared to the circumference of the cylinders.)

4.5 An infinitely long tube with radius a=0.01 m is filled with gas (containing only one molecular species) in thermodynamic equilibrium (number density $n=2.6\times10^{25}$ molecules/m³, temperature T=300 K, collision cross section $\sigma=10^{-19}$ m²). Determine the ratio of the number of collisions between molecules in the gas to collisions with the wall.

4.6 Assume that both the particle number density and the temperature are slowly varying functions of the z coordinate and the characteristic scale lengths are much larger than the mean free path, λ. Use the mean free path method and derive the first order approximation of the particle flux and the energy flux.

4.7 Consider an isothermal (T=137 K) pure oxygen ($m=2.67\times10^{-26}$ kg) atmosphere, where the variation of the gas number density is described by the barometric altitude formula:

$$n = n_0 \exp\left[-\frac{mg(z - z_0)}{kT} \right]$$

where k is the Boltzmann constant ($k=1.38\times10^{-23}$ joule K^{-1}), g is the gravitational acceleration ($g=9.81$ m s^{-2}), and the reference density is $n_0 = 2.6\times10^{19}$ m^{-3} at an altitude of $z_0 = 100$ km. Using transport coefficients derived by the mean free path method show that the upward diffusing particle flux (due to the density gradient) is independent of altitude. Show that at z=100 km the magnitude of the diffusive flux is much smaller than n $\bar{v}$, where $\bar{v}$ is the average thermal speed (the collisional cross section is $\sigma=10^{-19}$ m^2).

4.8 Assume that the particle number density is a slowly varying function of the z coordinate (the bulk velocity, **u**=(0,0,u), and the temperature, T, are constant) and the characteristic scale lengths are much larger than the mean free path, λ. Use the mean free path method and derive the first order approximation of the momentum flux in the z direction.

4.6 References

[1] Zucrow, M.J., and Hoffman, J.D., *Gas dynamics*, John Wiley and Sons, New York, 1976.
[2] Landau, L.D., and Lifshitz, E.M., *Fluid mechanics*, 2nd edition, Pergamon Press, New York, 1987.
[3] Landau, L.D., and Lifshitz, E.M., *Physical kinetics*, Pergamon Press, New York, 1981.
[4] Present, R.D., *Kinetic theory of gases*, McGraw-Hill, New York, 1958.
[5] Kennard, E.H., *Kinetic theory of gases*, McGraw-Hill, New York, 1938.
[6] Knudsen, M., *The kinetic theory of gases*, Methuen and Co. Ltd., London, 1934.
[7] Patterson, G.N., *Molecular flow of gases*, John Wiley & Sons, New York, 1956.

5

The Boltzmann equation

In the preceding chapter a highly simplified discussion of molecular transport in non-equilibrium dilute gases was presented. Even though this discussion was highly simplified, it helped us to illuminate the basic physical processes resulting in the transport of macroscopic physical quantities.

In this chapter the theory of macroscopic transport will be treated in a more rigorous way. Our discussion will be based on the Boltzmann equation, a non-linear seven dimensional partial differential equation which describes the evolution of the phase space distribution function in non-equilibrium gases.[1] Conservation equations for macroscopic physical quantities will be obtained in the next chapter as the velocity moments of the Boltzmann equation.

In this chapter our discussions will be based on the following fundamental assumptions:

(i) We assume that the gas density is low enough to ensure that the mean free path is large compared to the effective range of the intermolecular forces. This assumption makes it possible to apply the principle of molecular chaos in the treatment of binary encounters. Molecular chaos means that the velocities of the colliding particles are uncorrelated, i.e. particles which have already collided with each other will have many encounters with other molecules before they meet again. A direct consequence of this assumption is the time irreversibility of the Boltzmann equation.

(ii) The mean free path is short compared to the dimensions of the problem. This assumption eliminates consideration of wall effects and other edge-related problems.

(iii) Time, location and particle velocity (t, **r**, and **v**) are independent variables of the phase space distribution function.

5.1 Derivation of the Boltzmann equation

5.1.1 The evolution of the phase space distribution function

The state of a gas or a gas mixture at a particular instant is completely specified for

the purposes of kinetic theory if the distribution function of the molecular velocities and positions is known throughout the gas. Observable properties of the gas
are obtained as appropriate averages over the phase space distribution function.

The gas may be non-uniform (and consequently non-equilibrium), the distribution function may depend explicitly on time, and external fields of force may also
be present (such as gravity). These externally imposed force fields result in acceleration of the individual molecules. This acceleration, $\mathbf{a}$, is a function of the seven
independent variables, t, $\mathbf{r}$, and $\mathbf{v}$, as well as of some intrinsic properties of the
molecules themselves (such as mass). In gaskinetic theory it is usually assumed
that the acceleration of a molecule, $\mathbf{a}(t,\mathbf{r},\mathbf{v})$, is divergence free in velocity space,
$\nabla_v \cdot \mathbf{a} = 0$. Most external force fields satisfy this condition (all velocity independent
forces, such as gravity, automatically satisfy it, as does the Lorentz force acting on
moving charged particles in electromagnetic fields).

Consider first the case of a gas composed of a single molecular species. Let the
phase space distribution function be denoted by $F(t,\mathbf{r},\mathbf{v})$, so that the number of
molecules in a d^3r volume element around the configuration space location $\mathbf{r}$ with
velocity vectors between $\mathbf{v}$ and $\mathbf{v}+d^3\mathbf{v}$ is given by

$$d^6N = F(t,\mathbf{r},\mathbf{v})d^3r\,d^3v \tag{5.1}$$

At time t these particles are located very close to each other and they move with
nearly identical velocities. If there were no collisions, then time dt later all these
particles would be found in a volume element d^3r' near the spatial location
$\mathbf{r}'=\mathbf{r}+\mathbf{v}dt$ with velocities between $\mathbf{v}'=\mathbf{v}+\mathbf{a}(t,\mathbf{r},\mathbf{v})dt$ and $\mathbf{v}'+d^3\mathbf{v}'$. The number of particles in the modified phase space element around $\mathbf{r}',\mathbf{v}'$ at time t+dt is given by

$$d^6N' = F(t+dt,\mathbf{r}+\mathbf{v}dt,\mathbf{v}+\mathbf{a}dt)d^3r'\,d^3v' \tag{5.2}$$

The relation between the modified phase space volume element, $d^3v'd^3r'$, and
the initial one, $d^3v\,d^3r$, is given by the Jacobian of the transformation:

$$d^3r'd^3v' = |J|d^3rd^3v \tag{5.3}$$

where the Jacobian matrix, J, is given by

$$J = \frac{\partial(\mathbf{r}',\mathbf{v}')}{\partial(\mathbf{r},\mathbf{v})} = \begin{bmatrix} \dfrac{\partial x_i'}{\partial x_j} & \dfrac{\partial v_i'}{\partial x_j} \\[2mm] \dfrac{\partial x_i'}{\partial v_j} & \dfrac{\partial v_i'}{\partial v_j} \end{bmatrix} \tag{5.4}$$

Substituting our expressions for $\mathbf{r}'$ and $\mathbf{v}'$ yields the following 6×6 Jacobian matrix:

$$J = \begin{bmatrix} 1 & 0 & 0 & \frac{\partial a_1}{\partial x_1}dt & \frac{\partial a_1}{\partial x_2}dt & \frac{\partial a_1}{\partial x_3}dt \\ 0 & 1 & 0 & \frac{\partial a_2}{\partial x_1}dt & \frac{\partial a_2}{\partial x_2}dt & \frac{\partial a_2}{\partial x_3}dt \\ 0 & 0 & 1 & \frac{\partial a_3}{\partial x_1}dt & \frac{\partial a_3}{\partial x_2}dt & \frac{\partial a_3}{\partial x_3}dt \\ dt & 0 & 0 & 1+\frac{\partial a_1}{\partial v_1}dt & \frac{\partial a_1}{\partial v_2}dt & \frac{\partial a_1}{\partial v_3}dt \\ 0 & dt & 0 & \frac{\partial a_2}{\partial v_1}dt & 1+\frac{\partial a_2}{\partial v_2}dt & \frac{\partial a_2}{\partial v_3}dt \\ 0 & 0 & dt & \frac{\partial a_3}{\partial v_1}dt & \frac{\partial a_3}{\partial v_2}dt & 1+\frac{\partial a_3}{\partial v_3}dt \end{bmatrix} \tag{5.5}$$

The determinant of the Jacobian can be evaluated as a power series of the infinitesimal time interval, dt. The time interval is very small, therefore one can stop after the linear term:

$$|J| = 1 + \frac{\partial a_i}{\partial v_i}dt + \cdots \tag{5.6}$$

At this point we recall our earlier assumption that the particle acceleration is divergencefree in the velocity coordinates, therefore we conclude that in the first order approximation $|J|=1$. This means that $d^3r'd^3v'=d^3rd^3v$.

If there was no scattering, all particles which were located inside the phase space volume element d^3rd^3v at the initial time, t, would be found inside $d^3r'd^3v'$ at the later time, t+dt. In other words in the absence of particle scattering $d^6N'-d^6N=0$. However, gas molecules do interact with each other. The scattering process changes the velocities of the colliding particles, therefore the total number of molecules inside our infinitesimal phase-space volume element may change during the short time interval, dt:

$$d^6N' - d^6N = d^6N_{coll} \tag{5.7}$$

where d^6N_{coll} denotes the net rate of change (change per unit time) of the number of particles in the phase space volume element due to collisions. Equation (5.7) can be rewritten with the help of equations (5.1) and (5.2):

$$[F(t+dt,\mathbf{r}+\mathbf{v}dt,\mathbf{v}+\mathbf{a}dt) - F(t,\mathbf{r},\mathbf{v})]d^3r\,d^3v = \frac{\delta F(t,\mathbf{r},\mathbf{v})}{\delta t}d^3r\,d^3v\,dt \tag{5.8}$$

where $\delta F/\delta t$ is the rate of change of the phase space distribution function. At this point $\delta F/\delta t$ is only formally introduced, a detailed discussion of this so called 'collision term' will be presented later in this chapter.

In the next step we apply a linear Taylor expansion in the small time interval, dt, to evaluate the change in the total number of particles inside our phase space volume element. The Taylor expansion is centered around time t:

$$\left[\frac{\partial F(t,\mathbf{r},\mathbf{v})}{\partial t} + v_i \frac{\partial F(t,\mathbf{r},\mathbf{v})}{\partial x_i} + a_i \frac{\partial F(t,\mathbf{r},\mathbf{v})}{\partial v_i}\right] d^3r\, d^3v\, dt = \frac{\delta F(t,\mathbf{r},\mathbf{v})}{\delta t} d^3r\, d^3v\, dt$$

(5.9)

The expression in the bracket is the total time derivative of the phase space distribution function, dF/dt. Equation (5.9) is valid for arbitrary phase space volume elements and for arbitrary time intervals, therefore one obtains the following equation:

$$\frac{\partial F(t,\mathbf{r},\mathbf{v})}{\partial t} + v_i \frac{\partial F(t,\mathbf{r},\mathbf{v})}{\partial x_i} + a_i \frac{\partial F(t,\mathbf{r},\mathbf{v})}{\partial v_i} = \frac{\delta F(t,\mathbf{r},\mathbf{v})}{\delta t}$$

(5.10)

This is the Boltzmann equation which describes the evolution of the phase space distribution function, F(t,**r**,**v**). It should be emphasized that in the derivation of the Boltzmann equation we never assumed equilibrium conditions, therefore this equation is equally valid for equilibrium and non-equilibrium gases. It is also important to note again that t, **r**, and **v** are assumed to be independent variables.

The Boltzmann equation can also be expressed in terms of the random velocity (sometimes called peculiar velocity or thermal velocity). The random velocity of a particle is its velocity with respect of the bulk gas flow at time, t, and location, **r**:

$$\mathbf{c}(t,\mathbf{r}) = \mathbf{v} - \mathbf{u}(t,\mathbf{r})$$

(5.11)

Here the average bulk velocity, **u**(t,**r**), is a function of time and configuration space location and is defined as

$$\mathbf{u}(t,\mathbf{r}) = \frac{\iiint_{\infty} d^3v\ \mathbf{v}\ F(t,\mathbf{r},\mathbf{v})}{\iiint_{\infty} d^3v\ F(t,\mathbf{r},\mathbf{v})}$$

(5.12)

It is important to note, that unlike **v**, the random velocity, **c**, is not independent of t and **r**.

In some cases it is more convenient to express the evolution of the phase space distribution, F, in terms of t, $\mathbf{r}$, and $\mathbf{c}$, instead of the independent variables t, $\mathbf{r}$, and $\mathbf{v}$. This means that one has to introduce a 'mixed' phase space, where the configuration space coordinate system is inertial, but the velocity space coordinate system is an accelerating system because it is tied to the instantaneous local bulk velocity of the gas. In spite of the inherent difficulty of dealing with accelerating systems, it is often advantageous to use this mixed phase space.

If we want to transfer the Boltzmann equation into our mixed phase space coordinate system we have to use the following substitutions due to the accelerating nature of the velocity space coordinate system (when replacing the independent variable, $\mathbf{v}$, with the random velocity, $\mathbf{c}$, one has to take into account the t and $\mathbf{r}$ dependence of the new variable):

$$\frac{\partial F(t,\mathbf{r},\mathbf{v})}{\partial t} \Rightarrow \frac{\partial F(t,\mathbf{r},\mathbf{c})}{\partial t} + \frac{\partial c_i(t,\mathbf{r})}{\partial t}\frac{\partial F(t,\mathbf{r},\mathbf{c})}{\partial c_i} = \frac{\partial F(t,\mathbf{r},\mathbf{c})}{\partial t} - \frac{\partial u_i(t,\mathbf{r})}{\partial t}\frac{\partial F(t,\mathbf{r},\mathbf{c})}{\partial c_i}$$

$$(5.13)$$

$$\frac{\partial F(t,\mathbf{r},\mathbf{v})}{\partial x_j} \Rightarrow \frac{\partial F(t,\mathbf{r},\mathbf{c})}{\partial x_j} + \frac{\partial c_i(t,\mathbf{r})}{\partial x_j}\frac{\partial F(t,\mathbf{r},\mathbf{c})}{\partial c_i} = \frac{\partial F(t,\mathbf{r},\mathbf{c})}{\partial x_j} - \frac{\partial u_i(t,\mathbf{r})}{\partial x_j}\frac{\partial F(t,\mathbf{r},\mathbf{c})}{\partial c_i}$$

$$(5.14)$$

$$\frac{\partial F(t,\mathbf{r},\mathbf{v})}{\partial v_j} \Rightarrow \frac{\partial F(t,\mathbf{r},\mathbf{c})}{\partial c_j} \qquad (5.15)$$

Using relations (5.13) through (5.15) one can obtain the Boltzmann equation in the mixed phase space coordinate system:

$$\frac{\partial F}{\partial t} + (u_i + c_i)\frac{\partial F}{\partial x_i} - \left[\frac{\partial u_i}{\partial t} + (u_j + c_j)\frac{\partial u_i}{\partial x_j} - a_i\right]\frac{\partial F}{\partial c_i} = \frac{\delta F}{\delta t} \qquad (5.16)$$

where $F=F(t,\mathbf{r},\mathbf{c})$, $\mathbf{u}=\mathbf{u}(t,\mathbf{r})$, $\mathbf{a}=\mathbf{a}(t,\mathbf{r},\mathbf{v})$.

5.1.2 *The Boltzmann collision integral*

In the previous section the net change per unit time in the number of molecules in the phase space volume element $d^3r\ d^3v$ around the point $\mathbf{r},\mathbf{v}$ as a result of collisions was denoted by $(\delta F/\delta t)d^3r\ d^3v$. In order to calculate the collision term, $\delta F/\delta t$, we shall make the following simplifying assumptions.

(i) The gas is assumed to be sufficiently dilute that only binary collisions are taken into account.

(ii) The velocities of the colliding molecules are assumed to be statistically independent. Any possible correlations between the velocity and position of any single molecule are also neglected. These approximations are also called the assumption of molecular chaos.

(iii) The molecules and their intermolecular forces are assumed to be spherically symmetric. This assumption is highly questionable for most real gases, but it makes the mathematical derivation considerably simpler. It is possible to carry out this derivation with a much relaxed form of this assumption. Such a treatment leads to a significant increase of mathematical complexity and adds very little to our physical insight. In this book we base our derivation on this assumption, with the caveat that it can be relaxed at the expense of increased mathematical difficulty.

(iv) Laws of non-relativistic classical mechanics are applicable (except cross section, which can be taken from quantum physics).

(v) Scale length of the distribution function is much larger than the range of intermolecular forces.

(vi) Effects of the external forces on the magnitude of the collision cross section are neglected.

(vii) The phase space distribution function, $F(t,\mathbf{r},\mathbf{v})$, does not vary appreciably during the brief time interval of a molecular collision, nor does it change significantly over spatial scales comparable to the range of intermolecular forces.

Let us focus our attention on molecules located inside the macroscopically infinitesimal configuration space volume element, d^3r, located at the point $\mathbf{r}$ and consider the effects of binary collisions which take place between times t and t+dt. Our goal is to calculate the net change in the number of molecules inside d^3r with velocity vectors between $\mathbf{v}$ and $\mathbf{v}+d^3v$. The net change is a result of two competing processes. First, the molecules in d^3r can be 'scattered out' of this velocity range as a result of collisions with other molecules. This effect will decrease the number of molecules in the specified velocity range: we denote the rate of this decrease by $(\delta F/\delta t)^{(-)}$. Second, molecules in d^3r whose velocity is originally not between $\mathbf{v}$ and $\mathbf{v}+d^3v$ can be scattered into our specified velocity range: the rate of this increase is denoted by $(\delta F/\delta t)^{(+)}$. Finally, the collision term can be expressed as the difference of these two terms:

$$\frac{\delta F}{\delta t} = \left(\frac{\delta F}{\delta t}\right)^{+} - \left(\frac{\delta F}{\delta t}\right)^{-} \qquad (5.17)$$

Let us start by calculating the loss term. We consider encounters taking place in the configuration space volume element, d^3r, between a molecule in the velocity range of $[\mathbf{v}_1, \mathbf{v}_1+d^3v_1]$ and all other molecules. (Since d^3v_1 is infinitesimally small one can neglect collisions between two molecules in this specified range.) For convenience, we shall use the subscript 1 for molecules in our specific velocity range and the subscript 2 for the velocity of the collision partner.

The number of molecules deflected out of our velocity range is equal to the total number of collisions of molecules 1 in time dt (remember that d^3v_1 is infinitesimally small, therefore practically all collisions will result in a post-collision velocity, v_1', which is outside the range of $[v_1, v_1+d^3v_1]$).

Consider collisions between particles with velocities in the range of v_1 and $v_1+d^3v_1$ (molecule 1) and v_2 and $v_2+d^3v_2$ (molecule 2). The number of encounters between these two groups of molecules taking place for the deflection angle in the range χ to $\chi+d\chi$ and impact azimuths between ε and $\varepsilon+d\varepsilon$ per unit configuration space volume per unit time is given by (see equation (3.122))

$$d^8I = g\,S(g,\chi)\,F(t,\mathbf{r},\mathbf{v}_1)\,F(t,\mathbf{r},\mathbf{v}_2)\sin\chi\,d\chi\,d\varepsilon\,d^3v_1\,d^3v_2 \qquad (5.18)$$

where g is again the magnitude of the relative velocity. The total number of collisions per unit volume per unit time can be obtained by integrating over all possible velocities of the target molecules, all the deflection and azimuth angles:

$$\left(\frac{\delta F}{\delta t}\right)^{(-)} = \iiint_{\infty} d^3v_2 \int_0^{2\pi} d\varepsilon \int_0^{\pi} d\chi \sin\chi\, S(g,\chi)\, g\, F(t,\mathbf{r},\mathbf{v}_1)\,F(t,\mathbf{r},\mathbf{v}_2) \qquad (5.19)$$

Next we evaluate the second part of the collision term. In order to do this we need to know how many encounters terminate with one molecule in the specified infinitesimal range of velocities about v_1.

Let us consider an encounter between two molecules of initial velocities v_1' and v_2' and with deflection angle, χ', and impact azimuth, ε'. Since the conservation equations are invariant against an interchange of initial and final velocities, the final velocities now will be v_1 and v_2. This is an inverse encounter which is also called a replenishing collision (see Figure 5.1).

There is a simple relation between depleting (forward) and replenishing (inverse) collisions. In a depleting collision the initial velocities are v_1 and v_2, and the final velocities are v_1' and v_2'. The conservation of momentum and energy together with the values of the deflection angle, χ, and the impact azimuth, ε, completely determines the trajectory of the particles and their final velocities, v_1' and v_2'.

From the conservation of angular momentum we know that the impact parameters before and after the collision are the same (see equation (3.25)), consequently the deflection angles of the forward and inverse collisions must be identical, $\chi'=\chi$.

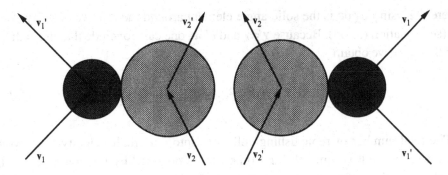

Figure 5.1

It was also shown that the relative motion of the two particles is confined to a plane (see subsection 3.1.2), therefore the impact azimuth angles for the forward and inverse collisions must be the same, $\varepsilon'=\varepsilon$.

It has also been shown that the magnitude of the relative velocity remains constant in the collision process (see equation (3.58)), $g'=g$ and the velocity of center of mass does not change during the collision proc ss (see equation (3.4)), $\mathbf{v}_c'=\mathbf{v}_c$.

In general, it can be shown that for spherically symmetric intermolecular force fields for every depleting collision, $(\mathbf{v}_1,\mathbf{v}_2)\rightarrow(\mathbf{v}_1',\mathbf{v}_2')$, there exists a replenishing encounter, $(\mathbf{v}_1',\mathbf{v}_2')\rightarrow(\mathbf{v}_1,\mathbf{v}_2)$. Similarly, for every inverse collision, $(\mathbf{v}_1',\mathbf{v}_2')\rightarrow(\mathbf{v}_1,\mathbf{v}_2)$, there exists a forward encounter, $(\mathbf{v}_1,\mathbf{v}_2)\rightarrow(\mathbf{v}_1',\mathbf{v}_2')$.

The number of replenishing collisions in volume d^3r during the time interval dt with deflection angle, χ', and impact azimuth, ε', can be expressed as

$$d^8I' = g'S(g',\chi')F(t,\mathbf{r},\mathbf{v}_1')F(t,\mathbf{r},\mathbf{v}_2')\sin\chi'\,d\chi'\,d\varepsilon'\,d^3v_1'\,d^3v_2' \qquad (5.20)$$

On the other hand we just concluded that $\chi'=\chi$, $\varepsilon'=\varepsilon$, $g'=g$, therefore equation (5.20) can be written as

$$d^8I' = gS(g,\chi)F(t,\mathbf{r},\mathbf{v}_1')F(t,\mathbf{r},\mathbf{v}_2')\sin\chi\,d\chi\,d\varepsilon\,d^3v_1'\,d^3v_2' \qquad (5.21)$$

The next question is how can one express $d^3v_1'd^3v_2'$ in terms of unprimed quantities. We know that $d^3v_1'd^3v_2'=d^3v_c'd^3g'$ (see equation (3.129)) and that the center of mass velocity and the magnitude of the relative velocity are conserved, $\mathbf{v}_c'=\mathbf{v}_c$ and $g'=g$. This means that

$$d^3v_1'd^3v_2' = d^3v_c'd^3g' = d^3v_c'g'^2dg'd\Omega_g' = d^3v_c\,g^2dg\,d\Omega_g' \qquad (5.22)$$

where $d\Omega_g = \sin\chi\, d\chi\, d\varepsilon$ is the solid angle element around the relative velocity vector (see equation (3.77)). Because $\chi' = \chi$ and $\varepsilon' = \varepsilon$, one can conclude that $d\Omega_g' = d\Omega_g$, and therefore we obtain

$$d^3v_1'd^3v_2' = d^3v_c\, g^2 dg\, d\Omega_g = d^3v_1\, d^3v_2 \tag{5.23}$$

The total number of replenishing collisions into our small velocity space volume, d^3v_1, per unit volume per unit time can be obtained by integrating over all possible velocities of the target molecules, and over all the deflection and azimuth angles:

$$\left(\frac{\delta F}{\delta t}\right)^{(+)} = \iiint_\infty d^3v_2 \int_0^{2\pi} d\varepsilon \int_0^\pi d\chi \sin\chi\, S(g,\chi)\, g\, F(t,\mathbf{r},\mathbf{v}_1')\, F(t,\mathbf{r},\mathbf{v}_2') \tag{5.24}$$

Finally, one can substitute expressions (5.19) and (5.24) into equation (5.17) to obtain the following form of the Boltzmann collision integral:

$$\frac{\delta F}{\delta t} = \iiint_\infty d^3v_2 \int_0^{2\pi} d\varepsilon \int_0^\pi d\chi \sin{}' S(g,\chi)\, g\left[F'F_2' - FF_2\right] \tag{5.25}$$

where $F = F(t,\mathbf{r},\mathbf{v})$, $F' = F(t,\mathbf{r},\mathbf{v}')$, $F_2 = F(t,\mathbf{r},\mathbf{v}_2)$, and $F_2' = F(t,\mathbf{r},\mathbf{v}_2')$. It should be noted that in equation (5.25) we have changed notation in order to be consistent with the left hand side of the Boltzmann equation: $\mathbf{v}_1$ and $\mathbf{v}_1'$ are now called $\mathbf{v}$ and $\mathbf{v}'$, respectively. This form of the collision term makes the Boltzmann equation a non-linear integro-differential equation with seven independent variables. It should be kept in mind that this integro-differential equation is already a tremendous simplification (we started with 6N equations describing the motion of N particles in the six dimensional phase space). However, the Boltzmann equation is still not a simple equation to solve. With the help of equation (5.25), the Boltzmann equation can be written in the following form:

$$\frac{\partial F}{\partial t} + v_i\frac{\partial F}{\partial x_i} + a_i\frac{\partial F}{\partial v_i} = \iiint_\infty d^3v_2 \int_0^{2\pi} d\varepsilon \int_0^\pi d\chi \sin\chi\, S(g,\chi)\, g\left[F'F_2' - FF_2\right] \tag{5.26}$$

This form of the Boltzmann equation is a non-linear integro-differential equation depending on seven variables: time, t, location, $\mathbf{r}$, and velocity, $\mathbf{v}$. The non-linearity is explicit in the collision integral which explicitly contains products of the distribution function, F. The acceleration, **a**, may also depend on the distribution function and result in additional non-linear effects.

The Boltzmann equation can also be written in terms of the random velocity, **c**:

$$\frac{\partial F}{\partial t} + (u_i + c_i)\frac{\partial F}{\partial x_i} - \left(\frac{\partial u_i}{\partial t} + (u_j + c_j)\frac{\partial u_i}{\partial x_j} - a_i\right)\frac{\partial F}{\partial c_i}$$

$$= \iiint_{\infty} d^3c_2 \int_0^{2\pi} d\varepsilon \int_0^{\pi} d\chi \sin\chi\, S(g,\chi)\, g\left[F' F_2' - F F_2\right] \qquad (5.27)$$

where $F = F(t,\mathbf{r},\mathbf{c})$, $F' = F(t,\mathbf{r},\mathbf{c}')$, $F_2 = F(t,\mathbf{r},\mathbf{c}_2)$, and $F_2' = F(t,\mathbf{r},\mathbf{c}_2')$.

5.1.3 Multispecies gases

For a system composed of different molecular species there is one Boltzmann equation for each molecular species. Let F_s denote the phase space distribution function of particles of type 's'. The velocity vectors referring to particles 's' will be denoted by $\mathbf{v}_s$ (total velocity) or $\mathbf{c}_s$ (random velocity). In this case the Boltzmann collision integral can be written as

$$\left(\frac{\delta F_s}{\delta t}\right) = \sum_t \iiint_{\infty} d^3v_t \int_0^{2\pi} d\varepsilon \int_0^{\pi} d\chi \sin\chi\, S(g,\chi)\, g\left[F_s' F_t' - F_s F_t\right] \qquad (5.28)$$

where $F_s = F_s(t,\mathbf{r},\mathbf{v}_s)$, $F_s' = F_s(t,\mathbf{r},\mathbf{v}_s')$, $F_t = F_t(t,\mathbf{r},\mathbf{v}_t)$, and $F_t' = F_t(t,\mathbf{r},\mathbf{v}_t')$. Each term in the summation refers to encounters between 's' particles and 't' particles. The summation with respect to 't' must run over all types of particles present in the gas, including the t=s case, because 's' molecules also collide with other 's' particles. It should be emphasized again that the velocities, $\mathbf{v}_s'$ and $\mathbf{v}_t'$, are functions of the pre-collision velocities, $\mathbf{v}_s$ and $\mathbf{v}_t$.

With this collision term the Boltzmann equation can be written in the following form:

$$\frac{\partial F_s(t,\mathbf{r},\mathbf{v}_s)}{\partial t} + v_{si}\frac{\partial F_s(t,\mathbf{r},\mathbf{v}_s)}{\partial x_i} + a_{si}(t,\mathbf{r},\mathbf{v}_s)\frac{\partial F_s(t,\mathbf{r},\mathbf{v}_s)}{\partial v_{si}}$$

$$= \sum_t \iiint_{\infty} d^3v_t \int_0^{2\pi} d\varepsilon \int_0^{\pi} d\chi \sin\chi\, S(g,\chi)\, g\left[F_s' F_t' - F_s F_t\right] \qquad (5.29)$$

In terms of the random velocity the Boltzmann equation for composite gases becomes

$$\frac{\partial F_s}{\partial t} + (u_{si} + c_{si}) \frac{\partial F_s}{\partial x_i} - \left(\frac{\partial u_{si}}{\partial t} + (u_{sj} + c_{sj}) \frac{\partial u_{si}}{\partial x_j} - a_{si} \right) \frac{\partial F_s}{\partial c_{si}}$$

$$= \sum_t \iiint_\infty d^3 c_t \int_0^{2\pi} d\varepsilon \int_0^\pi d\chi \sin\chi \, S(g,\chi) \, g \left[F_s' F_t' - F_s F_t \right] \qquad (5.30)$$

where $F_s = F_s(t,\mathbf{r},\mathbf{c}_s)$, $F_s' = F_s(t,\mathbf{r},\mathbf{c}_s')$, $F_t = F_t(t,\mathbf{r},\mathbf{c}_t)$, and $F_t' = F_t(t,\mathbf{r},\mathbf{c}_t')$. The random velocity of species 's', $\mathbf{c}_s$, is defined as $\mathbf{c}_s(t,\mathbf{r}) = \mathbf{v}_s - \mathbf{u}_s(t,\mathbf{r})$.

5.2 The H-theorem and equilibrium distributions

5.2.1 The H-theorem

Next, we explore several important consequences of the Boltzmann integral representation of the collision term. For the sake of simplicity we will consider a gas composed of a single molecular species. In this case the full Boltzmann equation is given by equation (5.26). It will be shown below that an important consequence of this form of the Boltzmann equation is that collisions drive the distribution function towards equilibrium state in an irreversible way. The irreversible character of the Boltzmann collision integral is a consequence of the assumption of molecular chaos (which neglects the correlation effects between colliding particles).

Let us start by introducing the so-called Boltzmann function, H(t):

$$H(t) = \iiint_{\mathcal{V}} d^3 r \iiint_\infty d^3 v \, F(t,\mathbf{r},\mathbf{v}) \, \ln F(t,\mathbf{r},\mathbf{v}) \qquad (5.31)$$

where $\mathcal{V}$ is the volume of the gas. It can be shown that the function H(t) is closely related to the total entropy of the gas[2], S(t):

$$S(t) = -k \, H(t) \qquad (5.32)$$

Boltzmann's H-theorem states that if F satisfies the Boltzmann equation (with the Boltzmann collision integral on the right hand side) then H(t) is a monotonically decreasing function, i.e.

$$\frac{dH(t)}{dt} \leq 0 \qquad (5.33)$$

This theorem is fully consistent with the fundamental laws of thermodynamics, because it means that the total entropy of a closed system is a monotonically increasing function of time. To prove the H-theorem let us consider the time derivative of the H(t) function:

$$\frac{dH(t)}{dt} = \iiint_{\mathcal{V}} d^3r \iiint_{\infty} d^3v \ (1 + \ln F(t,\mathbf{r},\mathbf{v})) \frac{dF(t,\mathbf{r},\mathbf{v})}{dt} \qquad (5.34)$$

The term dF/dt in the integrand is the left hand side of the Boltzmann equation, therefore it can be replaced by the right hand side, which is the Boltzmann collision integral. Now equation (5.34) becomes the following:

$$\frac{dH(t)}{dt} = \iiint_{\mathcal{V}} d^3r \iiint_{\infty} d^3v \iiint_{\infty} d^3v_2 \int_0^{2\pi} d\varepsilon \int_0^{\pi} d\chi \sin\chi \, Sg(1 + \ln F)\left[F'\,F_2' - FF_2\right] \qquad (5.35)$$

where $S=S(g,\chi)$, $F=F(t,\mathbf{r},\mathbf{v})$, $F'=F(t,\mathbf{r},\mathbf{v}')$, $F_2=F(t,\mathbf{r},\mathbf{v}_2)$, and $F_2'=F(t,\mathbf{r},\mathbf{v}_2')$. We integrate over both $\mathbf{v}$ and $\mathbf{v}_2$, therefore the two particles can be interchanged in the integrand (mathematically, both $\mathbf{v}$ and $\mathbf{v}_2$ are 'dummy' variables). The magnitude of the relative velocity, $g = |\mathbf{v} - \mathbf{v}_2|$, is invariant for this interchange, therefore one can write

$$\frac{dH(t)}{dt} = \iiint_{\mathcal{V}} d^3r \iiint_{\infty} d^3v \iiint_{\infty} d^3v_2 \int_0^{2\pi} d\varepsilon \int_0^{\pi} d\chi \sin\chi \, Sg(1 + \ln F_2)\left[F'\,F_2' - FF_2\right] \qquad (5.36)$$

In the next step we add equations (5.35) and (5.36) and divide the result by 2. This operation yields the following equation:

$$\frac{dH(t)}{dt} = \frac{1}{2}\iiint_{\mathcal{V}} d^3r \iiint_{\infty} d^3v \iiint_{\infty} d^3v_2 \int_0^{2\pi} d\varepsilon \int_0^{\pi} d\chi \sin\chi \, Sg(2 + \ln(FF_2))\left[F'\,F_2' - FF_2\right] \qquad (5.37)$$

For every collision there exists an inverse collision with $g'=g$, $\chi'=\chi$, $\varepsilon'=\varepsilon$, and $d^3v'd^3v_2'=d^3vd^3v_2$ (see the derivation of the Boltzmann collision integral in the previous section), therefore in equation (5.37) one can replace $\mathbf{v}$ and $\mathbf{v}_2$ by $\mathbf{v}'$ and $\mathbf{v}_2'$, without altering the value of the integral. Consequently, dH/dt can be rewritten as

$$\frac{dH}{dt} = \frac{1}{2} \iiint_{\mathcal{V}} d^3r \iiint_{\infty} d^3v \iiint_{\infty} d^3v_2 \int_0^{2\pi} d\varepsilon \int_0^{\pi} d\chi \sin\chi \, S\, g \left(2 + \ln\left(F'\, F_2'\right)\right) \left[F\, F_2 - F'\, F_2'\right]$$

$$(5.38)$$

Finally, we add equations (5.37) and (5.38) and divide the result by 2. This operation leads to the following expression:

$$\frac{dH(t)}{dt} = \frac{1}{4} \iiint_{\mathcal{V}} d^3r \iiint_{\infty} d^3v \iiint_{\infty} d^3v_2 \int_0^{2\pi} d\varepsilon \int_0^{\pi} d\chi \sin\chi \, S\, g \ln\left(\frac{F\, F_2}{F'\, F_2'}\right) \left[F'\, F_2' - F\, F_2\right]$$

$$(5.39)$$

Next we examine the sign of the integrand in equation (5.39). The magnitude of the relative velocity, g, and the differential cross section, S, are always positive, therefore they do not need particular attention. We focus on the remaining part of the integrand which can be written in the following form:

$$A = F'\, F_2' \ln(x)(1 - x) \qquad\qquad (5.40)$$

where $x = (F\, F_2 / F'\, F_2')$. It is obvious that the distribution function is positive semi-definite, therefore $0 \le x \le \infty$. The sign of the integrand is determined by the sign of the expression, $\ln(x)(1-x)$. The two elements of this expression are plotted in Figure 5.2. Inspection of Figure 5.2 reveals that for $x \ge 0$ values the signs of $\ln(x)$ and $(1-x)$ are always opposite, therefore their product is always negative, except for $x = 1$, when the product is zero. The conclusion is that the integrand of equation (5.39) is negative for all values $F\, F_2 \ne F'\, F_2'$ and zero only when $F\, F_2 = F'\, F_2'$, therefore the integral itself is negative semidefinite, $dH/dt \le 0$.

This proves the H-theorem and shows that when F satisfies the Boltzmann equation, the Boltzmann function, H(t), decreases monotonically until it reaches a limiting value, which occurs when there is no further change in the system. This limiting value is reached only when $F\, F_2 = F'\, F_2'$, and is a necessary condition for $dH/dt = 0$, and consequently it is also a necessary condition for equilibrium.

Next, we show that the $F\, F_2 = F'\, F_2'$ condition is not only necessary for equilibrium, but also is sufficient. Under equilibrium conditions the total time derivative of the phase space distribution function vanishes, i.e.,

$$\frac{dF(t, r, v)}{dt} = 0 \qquad\qquad (5.41)$$

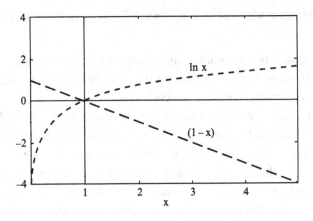

Figure 5.2

Equation (5.41) tells us that under equilibrium conditions the left hand side of the Boltzmann equation vanishes. This naturally means that the Boltzmann collision integral has to vanish as well:

$$\frac{\delta F}{\delta t} = \iiint_{\infty} d^3v_2 \int_0^{2\pi} d\varepsilon \int_0^{\pi} d\chi \sin\chi\, S(g,\chi)\, g\left[F'\, F_2{}' - F F_2\right] = 0 \qquad (5.42)$$

A sufficient condition to satisfy this equation is $F\, F_2 = F'\, F_2{}'$. At the same time we already showed that the same relation is necessary to ensure the equilibrium condition, $dH/dt=0$. The condition $F\, F_2 = F'\, F_2{}'$ is therefore both necessary and sufficient to ensure equilibrium.

A consequence of the H-theorem is that the entropy of the system increases with time until equilibrium is reached. This irreversible behavior is consistent with the laws of thermodynamics, and it is a consequence of the molecular chaos assumption.

5.2.2 Equilibrium distribution

We showed that in equilibrium the conditional probability density of finding two particles with velocities v and v_2, is the same as that of finding two particles with v' and $v_2{}'$:

$$F(t,\mathbf{r},\mathbf{v}')F(t,\mathbf{r},\mathbf{v}_2{}') = F(t,\mathbf{r},\mathbf{v})F(t,\mathbf{r},\mathbf{v}_2) \qquad (5.43)$$

Taking the natural logarithm of both sides of equation (5.43) gives

$$\ln F(t, \mathbf{r}, \mathbf{v}') + \ln F(t, \mathbf{r}, \mathbf{v}_2') = \ln F(t, \mathbf{r}, \mathbf{v}) + \ln F(t, \mathbf{r}, \mathbf{v}_2) \qquad (5.44)$$

which shows that ln F is a summation invariant (or conserved) quantity in the collision process. In other words, the sum of the logarithms of the distribution functions, $\ln F + \ln F_2$, is conserved during the collision.

On the other hand, we know that in the case of molecules without internal degrees of freedom there are four, and only four, independent, velocity dependent, summation invariant, quantities. These quantities are the momentum vector and the kinetic energy of the particles. A fifth summation invariant quantity, the molecular mass, is independent of the molecular velocities. Any other summation invariant, such as lnF, can be expressed as a linear combination of the five independent summation invariants:

$$\ln F(t, \mathbf{r}, \mathbf{v}) = \alpha_0(t, \mathbf{r}) + \alpha_1(t, \mathbf{r}) m v_x + \alpha_2(t, \mathbf{r}) m v_y + \alpha_3(t, \mathbf{r}) m v_z + \alpha_4(t, \mathbf{r}) \frac{1}{2} m v^2$$

$$(5.45)$$

Here α_0 through α_4 are velocity independent coefficients. A more familiar form can be obtained for the distribution function by introducing another set of five independent coefficients:

$$\beta = -\frac{1}{2} m \alpha_4 \qquad u_x = \frac{m \alpha_1}{2\beta} \qquad u_y = \frac{m \alpha_2}{2\beta} \qquad u_z = \frac{m \alpha_3}{2\beta} \qquad \ln A = \alpha_0 + \beta u^2$$

$$(5.46)$$

Using these coefficients the distribution function, F, can be written in the following form:

$$F(t, \mathbf{r}, \mathbf{v}) = A \exp\left\{-\beta\left[(v_x - u_x)^2 + (v_y - u_y)^2 + (v_z - u_z)^2\right]\right\} \qquad (5.47)$$

which is the well known Maxwell–Boltzmann distribution. It should be noted that the parameter, β, is not determined by this derivation. As has been shown earlier, this parameter characterizes the width of the distribution and it is closely related to the average kinetic energy of the molecules in the gas.

This is a tremendously important and far reaching result. It tells us that in equilibrium the distribution of molecular velocities is always Maxwellian. It also tells us that there is a one-to-one correspondence between equilibrium and Maxwellian velocity distributions: if and only if the distribution is Maxwellian, the gas is in equilibrium. This result also gives us a powerful tool to describe near equilibrium conditions, because these distributions must be close to Maxwellian.

5.3 Approximate collision terms

In most cases it is almost impossible to evaluate the Boltzmann collision integral analytically. Several approaches were introduced in order to obtain approximate expressions of varying sophistication (see the detailed monograph of Chapman and Cowling[3]). Here we outline some of the simplest and most widely used approximations.

5.3.1 Relaxation time approximation (BGK approximation)[4-6]

The relaxation time approximation was first suggested by Bhatnagar, Gross, and Krook in 1954 and it is often referred to as the BGK approximation.[4] Boltzmann's H-theorem tells us that collisions randomize the distribution of molecular velocities and bring all gases towards local equilibrium, which is characterized by Maxwellian velocity distributions.

For the sake of mathematical simplicity let us start with a single species gas and evaluate the collision term in the relaxation time approximation. The relaxation time approximation assumes that the velocity distribution of those particles which participate in replenishing collisions is an equilibrium distribution, F_0 (it should be noted that F_0 is always Maxwellian). In this approximation the contribution of replenishing collisions to the collision term can be written in the following form (see equation (5.24)):

$$\left(\frac{\delta F}{\delta t}\right)^{(+)} = \iiint_\infty d^3 v_2 \int_0^{2\pi} d\varepsilon \int_0^\pi d\chi \sin\chi\, S(g,\chi)\, g\, F_0(t,\mathbf{r},\mathbf{v}')\, F_0(t,\mathbf{r},\mathbf{v}_2') \quad (5.48)$$

On the other hand we know that in equilibrium the distribution function satisfies equation (5.43), therefore equation (5.48) can be written as

$$\left(\frac{\delta F}{\delta t}\right)^{(+)} = \iiint_\infty d^3 v_2 \int_0^{2\pi} d\varepsilon \int_0^\pi d\chi \sin\chi\, S(g,\chi)\, g\, F_0(t,\mathbf{r},\mathbf{v})\, F_0(t,\mathbf{r},\mathbf{v}_2) \quad (5.49)$$

Using this approximation the Boltzmann collision integral (equation (5.25)) simplifies to the following:

$$\frac{\delta F}{\delta t} = F_0(t, \mathbf{r}, \mathbf{v}) \iiint_{\infty} d^3 v_2 \int_0^{2\pi} d\epsilon \int_0^{\pi} d\chi \sin \chi \, S(g, \chi) g \, F_0(t, \mathbf{r}, \mathbf{v}_2)$$

$$- F(t, \mathbf{r}, \mathbf{v}) \iiint_{\infty} d^3 v_2 \int_0^{2\pi} d\epsilon \int_0^{\pi} d\chi \sin \chi \, S(g, \chi) g \, F(t, \mathbf{r}, \mathbf{v}_2) \qquad (5.50)$$

It can be seen that the integrals in the first and second terms represent an average collision frequency (the average number of collisions of a single molecule per unit time, see equation (3.113)). The characteristic times can be denoted by τ_0 and τ_F, respectively:

$$\iiint_{\infty} d^3 v_2 \int_0^{2\pi} d\epsilon \int_0^{\pi} d\chi \sin \chi \, S(g, \chi) g \, F_0(t, \mathbf{r}, \mathbf{v}_2) = n_0 \sigma_0 \bar{g} = \frac{1}{\tau_0} \qquad (5.51a)$$

$$\iiint_{\infty} d^3 v_2 \int_0^{2\pi} d\epsilon \int_0^{\pi} d\chi \sin \chi \, S(g, \chi) g \, F(t, \mathbf{r}, \mathbf{v}_2) = n_0 \sigma_0 \bar{g} = \frac{1}{\tau_F} \qquad (5.51b)$$

In the relaxation time approximation the characteristic time scales, τ_0 and τ_F, are treated as free parameters, which can be chosen using considerations outside the scope of gaskinetic theory (such as quantum mechanical effects). In this approximation the collision term can be written in the following form:

$$\frac{\delta F(t, \mathbf{r}, \mathbf{v})}{\delta t} = -\frac{F(t, \mathbf{r}, \mathbf{v})}{\tau_F} + \frac{F_0(t, \mathbf{r}, \mathbf{v})}{\tau_0} \qquad (5.52)$$

The physical meaning of this collision term is the following. Particles obeying the actual distribution function, $F(t, \mathbf{r}, \mathbf{v})$, are removed exponentially (due to collisions) with a time scale of τ_F (the minus sign in the collision term represents loss). The 'removed' particles are 'replaced' with an exponential time scale of τ_0 by another population of particles following the equilibrium phase space distribution function, $F_0(t, \mathbf{r}, \mathbf{v})$.

It is obvious that collisions do not create or destroy particles (our simple collision model does not take into account chemical reactions or other physical processes which alter the identity of the colliding particles, such as ionization or recombination). This means that the normalization of the actual distribution function, F, and that of the equilibrium (Maxwellian) distribution, F_0, must be identical:

$$n = \iiint_{\infty} d^3 v \, F(t, \mathbf{r}, \mathbf{v}) = \iiint_{\infty} d^3 v \, F_0(t, \mathbf{r}, \mathbf{v}) \qquad (5.53)$$

The 'loss' and 'replacement' time scales, τ_F and τ_0, must also be identical, because removed particles must be replaced at the same rate (otherwise one would have a net particle gain or loss as time goes on): $\tau_F = \tau_0$.

We showed in Chapter 3 that elastic binary collisions (the ones which are approximated by the BGK collision term) conserve the total momentum and kinetic energy of the colliding particles. This means that the average momentum and energy described by the instantaneous distribution, F, and the Maxwellian, F_0, must be the same. Therefore the Maxwellian, F_0, can be written in the following form:

$$F_0(t,\mathbf{r},\mathbf{v}) = \frac{n\,m^{3/2}}{(2\pi kT_0)^{3/2}} \exp\left(-\frac{m(\mathbf{v}-\mathbf{u}_0)^2}{2kT_0}\right) \qquad (5.54)$$

where the average velocity, $\mathbf{u}_0$, and temperature, T_0, can be obtained from the conservation of momentum and energy, respectively:

$$m\iiint_\infty d^3v\,\mathbf{v}\,F_0(t,\mathbf{r},\mathbf{v}) = m\,n\,\mathbf{u}_0 = m\iiint_\infty d^3v\,\mathbf{v}\,F(t,\mathbf{r},\mathbf{v}) \qquad (5.55a)$$

$$\frac{1}{2}m\iiint_\infty d^3v\,v^2\,F_0(t,\mathbf{r},\mathbf{v}) = \frac{3}{2}n\,k\,T_0 = \frac{1}{2}m\iiint_\infty d^3v\,v^2\,F(t,\mathbf{r},\mathbf{v}) \qquad (5.55b)$$

In other words, the number density, drift velocity, and temperature of the Maxwellian are identical to the number density, average flow velocity and kinetic temperature of the instantaneous distribution, F. In summary, the relaxation time approximation for a single species gas yields the following expression for the collision term:

$$\frac{\delta F(t,\mathbf{r},\mathbf{v})}{\delta t} = -\frac{1}{\tau_0}\left[\frac{m^{3/2}\,n}{(2\pi kT_0)^{3/2}}\exp\left(-\frac{m(\mathbf{v}-\mathbf{u}_0)^2}{2kT_0}\right) - F(t,\mathbf{r},\mathbf{v})\right] \qquad (5.56)$$

where n is given by the normalization of F, while $\mathbf{u}_0$ and T_0 can be obtained with the help of equations (5.55).

The determination of the Maxwellian parameters is much more complicated in the case of multispecies gases.[7] In this case the rate of change of the distribution function of particles 's' can be written in the following form:

$$\frac{\delta F_s}{\delta t} = -\sum_t \frac{F_s - F_{0s(st)}}{\tau_{st}} \tag{5.57}$$

where τ_{st} is again a formally defined relaxation time, which is treated as a parameter, $F_{0s(st)}$ is an equilibrium (Maxwellian) distribution which would be the time asymptotic distribution function if 's' particles collided only with 't' particles. The notation 's(st)' means that the given quantity would be achieved for particles 's' if there were only s–t collisions in the gas.

The conservation of particles again requires that $F_{0s(st)}$ must be normalized to the number density of particles 's', $n_{0s(st)} = n_s$. The temperature, $T_{s(st)}$, and bulk flow velocity, $\mathbf{u}_{s(st)}$, must be defined in such a way that momentum and energy are conserved in two particle collisions. In the following discussion these conditions will be explored and expressions for $\mathbf{u}_{s(st)}$ and $T_{s(st)}$ will be derived.

In the following treatment we consider the derivation of parameters $\mathbf{u}_{s(st)}$ and $T_{s(st)}$ which enter the expression for $F_{0s(st)}$. These parameters must be defined in such a way that momentum and energy are conserved in binary collisions. Here we outline a derivation which is based on a long series of tedious calculations performed by several authors.[4-7]

Consider purely elastic binary collisions (Chapter 3) of 's' and 't' particles. The conservation of momentum means that the momenta of the particles in the center of mass system can be expressed in terms of the reduced mass, m_{st}, and the relative velocity vector, $\mathbf{g}$ (see equation (3.127)):

$$m_s \mathbf{v}_{sc} = m_{st} \mathbf{g} \qquad m_t \mathbf{v}_{tc} = -m_{st} \mathbf{g} \tag{5.58a}$$

$$m_s \mathbf{v}_{sc}' = m_{st} \mathbf{g}' \qquad m_t \mathbf{v}_{tc}' = -m_{st} \mathbf{g}' \tag{5.58b}$$

where the primed quantities refer to velocities right after the collision, $m_{st} = m_s m_t / (m_s + m_t)$ is the reduced mass, and $\mathbf{g} = \mathbf{v}_s - \mathbf{v}_t$ is the relative velocity vector.

The center of mass velocity vector, $\mathbf{v}_c$, is conserved in the collision (see equation (3.4)), and the magnitude of the relative speed, g, also remains the same during binary collisions (equation (3.58)). Using these conservation relations one can express the velocities of the two particles after the collision in the following form:

$$\mathbf{v}_s' = \mathbf{v}_c + \mathbf{v}_{sc}' = \frac{m_s \mathbf{v}_s + m_t \mathbf{v}_t}{m_s + m_t} + \frac{m_t \mathbf{g}'}{m_s + m_t} \tag{5.59a}$$

$$\mathbf{v}_t' = \mathbf{v}_c + \mathbf{v}_{tc}' = \frac{m_s \mathbf{v}_s + m_t \mathbf{v}_t}{m_s + m_t} - \frac{m_s \mathbf{g}'}{m_s + m_t} \tag{5.59b}$$

It is quite reasonable to assume that while the magnitude of the scattered relative velocity vector, $\mathbf{g}'$, must remain equal to g, it can have any direction in space (because the direction is determined by the impact azimuth and impact parameter, which are independent of the particle velocities), therefore the solid angle average of $\mathbf{g}'$ is zero. This means that the average values of the particle velocities after the collision can be written as follows:

$$\mathbf{u}_{s(st)} = <\mathbf{v}_s'> = \frac{m_s \mathbf{u}_s + m_t \mathbf{u}_t}{m_s + m_t} = \mathbf{u}_{t(ts)} \tag{5.60}$$

In other words the mean flow velocities of the s and t particles after a binary collision are identical and are given by equation (5.60).

Next we calculate the average Maxwellian temperatures, $T_{s(st)}$ and $T_{t(ts)}$. The random velocities after an s–t collision are the following:

$$\mathbf{c}_s' = \mathbf{v}_s' - \mathbf{u}_{s(st)} = \frac{m_s \mathbf{c}_s + m_t \mathbf{c}_t + m_t \mathbf{g}'}{m_s + m_t} \tag{5.61a}$$

$$\mathbf{c}_t' = \mathbf{v}_t' - \mathbf{u}_{t(ts)} = \frac{m_s \mathbf{c}_s + m_t \mathbf{c}_t - m_s \mathbf{g}'}{m_s + m_t} \tag{5.61b}$$

The kinetic temperature of the s particles after the collision is based on random velocity components relative to the Maxwellian flow velocity, $\mathbf{u}_{s(st)}$, therefore one can write

$$T_{s(st)} = \frac{m_s <c_s'^2>}{3k} = \frac{m_s}{3k(m_s + m_t)^2}\left(m_s^2 <c_s^2> + m_t^2 <c_t^2> + m_t^2 <g^2>\right)$$
$$+ \frac{2m_{st}}{3k(m_s + m_t)}\left(m_s <\mathbf{c}_s \cdot \mathbf{c}_t> + m_s <\mathbf{c}_s \cdot \mathbf{g}'> + m_t <\mathbf{c}_t \cdot \mathbf{g}'>\right)$$

$$\tag{5.62}$$

The $<\mathbf{c}_s \cdot \mathbf{c}_t>$ term vanishes, because the colliding particles are assumed to be completely independent of each other. The last two terms are also zero, because the relative velocity vector, $\mathbf{g}'$, can have any direction, and it is independent of $\mathbf{c}_s$ and $\mathbf{c}_t$. The average value of g^2 can be expressed in the following form:

$$< g^2 >=< (\mathbf{v}_s - \mathbf{v}_t)^2 >=< (\mathbf{c}_s - \mathbf{c}_t + \mathbf{u}_s - \mathbf{u}_t)^2 >=< c_s^2 > + < c_t^2 > + (\mathbf{u}_s - \mathbf{u}_t)^2$$
$$-2 < \mathbf{c}_s \cdot \mathbf{c}_t > + < \mathbf{c}_s > \cdot (\mathbf{u}_s - \mathbf{u}_t) - < \mathbf{c}_t > \cdot (\mathbf{u}_s - \mathbf{u}_t) = < c_s^2 > + < c_t^2 > + (\mathbf{u}_s - \mathbf{u}_t)^2$$

$$(5.63)$$

Substituting this into equation (5.62) and using the definition of kinetic temperature we obtain the following expression for $T_{s(st)}$:

$$T_{s(st)} = T_s + \frac{2m_{st}}{(m_s + m_t)}(T_t - T_s) + \frac{m_t}{3k(m_s + m_t)}m_{st}(\mathbf{u}_t - \mathbf{u}_s)^2 \quad (5.64a)$$

A similar expression can be obtained for $T_{t(ts)}$:

$$T_{t(ts)} = T_t + \frac{2m_{st}}{(m_s + m_t)}(T_s - T_t) + \frac{m_s}{3k(m_s + m_t)}m_{st}(\mathbf{u}_s - \mathbf{u}_t)^2 \quad (5.64b)$$

The physical meaning of the individual terms in expressions (5.64) is quite clear. The first term simply represents the present state of the gas: in the absence of a second species the gas would relax to an equilibrium (Maxwellian) distribution with temperature equal to the instantaneous kinetic temperature, T_s. The second term describes the heat exchange between the two species: heat energy is transferred from the 'hotter' gas (higher kinetic temperature) to the colder one. It is interesting to note that the efficiency of the heat transfer depends on the ratio of the reduced mass to the total mass of the colliding molecules. The heat transfer is most efficient between molecules of nearly equal mass, and it becomes very inefficient when the masses of the colliding particles are very different (as in the case of electrons colliding with neutral molecules). The physical interpretation of the third term is very interesting. This term, which is proportional to the square of the mean velocity difference between the two gases, describes friction: kinetic energy of the organized relative (bulk) motion of the gases is transferred into random (thermal) motion. It should be noted that the contribution of this frictional term is positive in both expressions (it increases both $T_{s(st)}$ and $T_{t(ts)}$) and it is distributed according to the mass of the colliding partner. Equations (5.64) tell us that the collisions between the two species gradually reduce the temperature difference and the relative bulk motion between the gases.

The sum of the average thermal energy of the two gases can be expressed the following way:

$$\frac{3}{2}kT_{s(st)} + \frac{3}{2}kT_{t(ts)} = \frac{3}{2}kT_s + \frac{3}{2}kT_t + \frac{1}{2}m_{st}(\mathbf{u}_s - \mathbf{u}_t)^2 \quad (5.65)$$

This is a mathematical formulation of the result we discussed above: the total thermal energy of the system increases due to the frictional transformation of bulk energy to random particle motion.

It is obvious that the two temperatures, $T_{s(st)}$ and $T_{t(ts)}$, are not the same. This means that the temperatures of the two gases approach equilibrium differently: however, eventually the temperatures and flow velocities become identical.

The BGK method is very simple and attractive. First of all the collision term is mathematically simple, which makes it possible to obtain closed analytic expressions for the transport coefficients. Secondly, the approximation does not make any specific assumption about the nature of the intermolecular forces, thus avoiding detailed evaluation of the very complicated Boltzmann collision integral. A third advantage is that the BGK approximation is not limited to two particle collisions only. This is not very important in the case of neutral molecules, but when the Boltzmann equation is used to describe charged particles this question becomes quite relevant.

We have all learned that there are no free lunches in real life, therefore the BGK method must have several disadvantages, as well. First of all the approximation cannot be considered to be a rigorous theory. Secondly the relaxation times, τ_{st}, are treated as given parameters, which depend only on time and location, and their values are determined empirically (and sometimes very inaccurately).

In order to demonstrate the use of the BGK approximation let us consider a simple example. Consider a volume of gas at rest ($\mathbf{u}=0$), with no spatial gradients and no external forces. Assume that the gas is composed of a single molecular species, and that the relaxation time, τ_0, is constant everywhere. In this case the Boltzmann equation simplifies to the following:

$$\frac{\partial F(t, \mathbf{v})}{\partial t} = -\frac{F(t, \mathbf{v})}{\tau_0} + \frac{F_0(t, \mathbf{v})}{\tau_0} \tag{5.66}$$

where F_0 is a drifting Maxwellian with parameters n, $\mathbf{u}_0$, and T_0, where n is the particle concentration, $\mathbf{u}_0$ is the average flow velocity, and T_0 is the kinetic temperature:

$$n = \iiint_\infty d^3v \, F(t, \mathbf{r}, \mathbf{v}) \tag{5.67a}$$

$$\mathbf{u}_0 = \frac{1}{n} \iiint_\infty d^3v \, \mathbf{v} \, F(t, \mathbf{r}, \mathbf{v}) \tag{5.67b}$$

$$T_0 = \frac{m}{3nk} \iiint_\infty d^3v \, v^2 \, F(t, \mathbf{r}, \mathbf{v}) \tag{5.67c}$$

Let us assume that at t=0 we start with an initial distribution function, $F_{in}(v)$. In this case one can solve the simplified Boltzmann equation to obtain

$$F(t,v) = F_0(v) + \left[F_{in}(v) - F_0(v)\right] e^{-\frac{t}{\tau}} \tag{5.68}$$

As a specific example we consider an initial distribution with all particles uniformly distributed inside a velocity space sphere of radius v_0:

$$F_{in}(v) = \begin{cases} \dfrac{3n}{4\pi v_0^3} & v < v_0 \\ 0 & \text{otherwise} \end{cases} \tag{5.69}$$

In this case the concentration is n, the average velocity is zero, and the kinetic temperature is the following:

$$T_0 = \frac{mv_0^2}{5k} \tag{5.70}$$

The solution is

$$F(t,v) = \begin{cases} A\left[1 - \exp\left(-\frac{t}{\tau_0}\right)\right] + \dfrac{3n}{4\pi v_0^3}\exp\left(-\frac{t}{\tau_0}\right) & v < v_0 \\ A\left[1 - \exp\left(-\frac{t}{\tau_0}\right)\right] & v \geq v_0 \end{cases} \tag{5.71}$$

where

$$A = \left(\frac{5}{2\pi}\right)^{3/2} \frac{n}{v_0^3} \exp\left(-\frac{5v^2}{2v_0^2}\right) \tag{5.72}$$

The temporal evolution of this solution is shown in Figure 5.3. In Figure 5.3 velocity and time are measured in dimensionless units, v/v_0 and t/τ_0.

5.3.2 Fokker–Planck approximation

The BGK collision term is a simple approximation to the collision integral which is independent of the actual interparticle potential. A different and more realistic

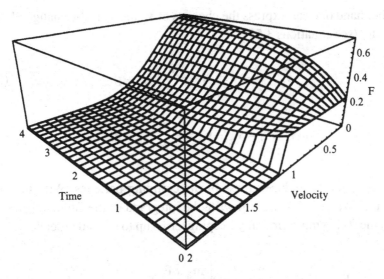

Figure 5.3

approximation can be obtained by realizing that in a broad class of collision models most encounters produce only a small change in the particle velocities (for instance, in the case of inverse power interaction most collisions take place with large impact parameters and therefore result only in small velocity changes). The assumption that collisions result only in small velocity changes leads to another simplified form of the Boltzmann collision integral, called the Fokker–Planck approximation.

Let us start from the Boltzmann collision integral for multispecies gases given by equation (5.28). The distribution function of 's' particles after a collision, $F_s'(t,\mathbf{r},\mathbf{v}_s')$, is only slightly different from $F_s(t,\mathbf{r},\mathbf{v}_s)$, therefore one can use a second order Taylor series expansion to approximate it:

$$F_s(t,\mathbf{r},\mathbf{v}_s') = F_s + (v_{si}'-v_{si})\frac{\partial F_s}{\partial v_{si}} + \frac{1}{2}(v_{si}'-v_{si})(v_{sj}'-v_{sj})\frac{\partial^2 F_s}{\partial v_{si}\,\partial v_{sj}} + \cdots$$

$$(5.73a)$$

where $F_s = F_s(t,\mathbf{r},\mathbf{v}_s)$. A similar approximation can be obtained for the distribution function of scattered 't' particles:

$$F_t(t,\mathbf{r},\mathbf{v}_t') = F_t + (v_{ti}'-v_{ti})\frac{\partial F_t}{\partial v_{ti}} + \frac{1}{2}(v_{ti}'-v_{ti})(v_{tj}'-v_{tj})\frac{\partial^2 F_t}{\partial v_{ti}\,\partial v_{tj}} + \cdots$$

$$(5.73b)$$

On the other hand one can express the changes of $\mathbf{v}_s$ and $\mathbf{v}_t$ as the change of the relative velocity (see equations 3.51):

$$\mathbf{v}_s{}' - \mathbf{v}_s = \mathbf{v}_{sc}{}' - \mathbf{v}_{sc} = \frac{m_{st}}{m_s}(\mathbf{g}' - \mathbf{g}) \tag{5.74a}$$

$$\mathbf{v}_t{}' - \mathbf{v}_t = \mathbf{v}_{tc}{}' - \mathbf{v}_{tc} = -\frac{m_{st}}{m_t}(\mathbf{g}' - \mathbf{g}) \tag{5.74b}$$

where m_{st} is again the reduced mass of the colliding particles. With the help of equations (5.73) and (5.74) we can write the integrand of the Boltzmann collision integral in the following form (only keeping terms up to second order):

$$F_s{}' F_t{}' - F_s F_t = (g_i{}' - g_i) \left[\frac{m_{st}}{m_s} \frac{\partial F_s}{\partial v_{si}} F_t - \frac{m_{st}}{m_t} F_s \frac{\partial F_t}{\partial v_{ti}} \right]$$

$$- (g_i{}' - g_i)(g_j{}' - g_j) \left[\frac{m_{st}^2}{m_s m_t} \frac{\partial F_s}{\partial v_{si}} \frac{\partial F_t}{\partial v_{tj}} - \frac{1}{2} \frac{m_{st}^2}{m_s^2} \frac{\partial^2 F_s}{\partial v_{si} \partial v_{sj}} F_t - \frac{1}{2} \frac{m_{st}^2}{m_t^2} F_s \frac{\partial^2 F_t}{\partial v_{ti} \partial v_{tj}} \right] \tag{5.75}$$

Next, we substitute equation (5.75) into (5.28) to obtain the following expression for the collision integral:

$$\left(\frac{\delta F_s}{\delta t} \right) = \sum_t \iiint_\infty d^3 v_t \int_0^{2\pi} d\varepsilon \int_0^\pi d\chi \sin \chi \, S g \left[(g_i{}' - g_i) A_i^{st} - (g_i{}' - g_i)(g_j{}' - g_j) B_{ij}^{st} \right] \tag{5.76}$$

where

$$A_i^{st} = \left[\frac{m_{st}}{m_s} \frac{\partial F_s}{\partial v_{si}} F_t - \frac{m_{st}}{m_t} F_s \frac{\partial F_t}{\partial v_{ti}} \right] \tag{5.77a}$$

$$B_{ij}^{st} = \left[\frac{m_{st}^2}{m_s m_t} \frac{\partial F_s}{\partial v_{si}} \frac{\partial F_t}{\partial v_{tj}} - \frac{1}{2} \frac{m_{st}^2}{m_s^2} \frac{\partial^2 F_s}{\partial v_{si} \partial v_{sj}} F_t - \frac{1}{2} \frac{m_{st}^2}{m_t^2} F_s \frac{\partial^2 F_t}{\partial v_{ti} \partial v_{tj}} \right] \tag{5.77b}$$

It should be noted that the coefficients, A^{st} and B^{st} are independent of the impact azimuth, ε. The differential cross section, $S(g,\chi)$, is also independent of ε, therefore equation (5.76) can be written in the following form:

$$\left(\frac{\delta F_s}{\delta t}\right) = \sum_t \iiint_\infty d^3 v_t \int_0^\pi d\chi \sin\chi \, S \, g \, A_i^{st} \int_0^{2\pi} d\varepsilon \, (g_i{}' - g_i)$$

$$-\sum_t \iiint_\infty d^3 v_t \int_0^\pi d\chi \sin\chi \, S \, g \, B_{ij}^{st} \int_0^{2\pi} d\varepsilon \left[(g_i{}' g_j{}' - g_i g_j) - g_i (g_j{}' - g_j) - g_j (g_i{}' - g_i) \right]$$

$$(5.78)$$

First we evaluate the integrals over the impact azimuth, ε. Let us choose three orthogonal unit vectors in the following way: $\mathbf{e}_3$ is parallel to the relative velocity vector, $\mathbf{g}$, the impact azimuth, ε, is measured counter-clockwise from the vector $\mathbf{e}_1$, and $\mathbf{e}_1$, $\mathbf{e}_2$, and $\mathbf{e}_3$ form a right handed coordinate system (see Figure 5.4). In this case $\mathbf{g} = g\mathbf{e}_3$, and $\mathbf{g}' = g \sin\chi \cos\varepsilon \, \mathbf{e}_1 + g \sin\chi \sin\varepsilon \, \mathbf{e}_2 + g \cos\chi \, \mathbf{e}_3$ (because the quantities g and ε are conserved during the collision). Using these vectors the integrals over the azimuth angle can be readily evaluated:

$$\int_0^{2\pi} d\varepsilon \, (g_i{}' - g_i) = \mathbf{e}_{1i} \, g \sin\chi \int_0^{2\pi} d\varepsilon \cos\varepsilon + \mathbf{e}_{2i} \, g \sin\chi \int_0^{2\pi} d\varepsilon \sin\varepsilon$$

$$-\mathbf{e}_{3i} \, g (1 - \cos\chi) \int_0^{2\pi} d\varepsilon = -2\pi g_i (1 - \cos\chi) \qquad (5.79)$$

and

$$\int_0^{2\pi} d\varepsilon \, (g_i{}' g_j{}' - g_i g_j) = \pi (1 - \cos^2\chi)(g^2 \delta_{ij} - 3g_i g_j) \qquad (5.80)$$

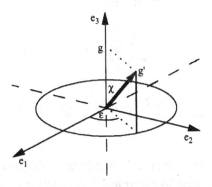

Figure 5.4

Substituting these results into equation (5.78) yields the following expression for the collision integral:

$$\left(\frac{\delta F_s}{\delta t}\right) = -\sum_t \iiint_\infty d^3v_t\, g\left[\sigma_1\, g_i\, A_i^{st} + \tfrac{1}{2}\sigma_2\,(g^2\delta_{ij} - 3g_i g_j)B_{ij}^{st} + 2g_i g_j \sigma_1\, B_{ij}^{st}\right]$$

(5.81)

where $\sigma_1(g)$ and $\sigma_2(g)$ are the higher order transport cross sections introduced in Chapter 3 (see equation (3.88)). Next we substitute equations (5.77a) and (5.77b) into expression (5.81):

$$\left(\frac{\delta F_s}{\delta t}\right) = -\frac{\partial F_s}{\partial v_{si}}G_i^s + \frac{\partial^2 F_s}{\partial v_{si}\partial v_{sj}}K_{ij}^s + \frac{\partial^2 F_s}{\partial v_{si}\partial v_{sj}}L_{ij}^s$$

$$-\frac{\partial F_s}{\partial v_{si}}\sum_t \frac{1}{2}\frac{m_{st}^2}{m_s m_t}\iiint_\infty d^3v_t\,\sigma_2\, g\,(g^2\delta_{ij} - 3g_i g_j)\frac{\partial F_t}{\partial v_{tj}}$$

$$+F_s\sum_t \frac{1}{4}\frac{m_{st}^2}{m_t^2}\iiint_\infty d^3v_t\,\sigma_2\, g\,(g^2\delta_{ij} - 3g_i g_j)\frac{\partial^2 F_t}{\partial v_{ti}\partial v_{tj}} + F_s\sum_t \frac{m_{st}}{m_t}\iiint_\infty d^3v_t\,\sigma_1\, g\, g_i\frac{\partial F_t}{\partial v_{ti}}$$

$$+F_s\sum_t \frac{m_{st}^2}{m_t^2}\iiint_\infty d^3v_t\,\sigma_1\, g\, g_i g_j\frac{\partial^2 F_t}{\partial v_{ti}\partial v_{tj}} - \frac{\partial F_s}{\partial v_{si}}\sum_t 2\frac{m_{st}^2}{m_s m_t}\iiint_\infty d^3v_t\,\sigma_1\, g\, g_i g_j\frac{\partial F_t}{\partial v_{tj}}$$

(5.82)

where we have introduced the following quantities:

$$G_i^s = \sum_t \frac{m_{st}}{m_s}\iiint_\infty d^3v_t\,\sigma_1\, g\, g_i F_t$$

(5.83a)

$$K_{ij}^s = \sum_t \frac{m_{st}^2}{m_s^2}\iiint_\infty d^3v_t\,\sigma_1\, g\, g_i g_j F_t$$

(5.83b)

$$L_{ij}^s = \sum_t \frac{1}{4}\frac{m_{st}^2}{m_s^2}\iiint_\infty d^3v_t\,\sigma_2\, g\,(g^2\delta_{ij} - 3g_i g_j)F_t$$

(5.83c)

The integrals in equation (5.82) still contain derivatives of the velocity distribution function, F_t, with respect to the velocity components, v_{ti}. These integrals can be manipulated so that we end up with derivatives with respect to the velocity components of particles 's'. In order to demonstrate this process we start with one of the integrals in expression (5.82). First we integrate by parts to obtain

$$\iiint_{\infty} d^3v_t \sigma_1 g g_i \frac{\partial F_t}{\partial v_{ti}} = \iiint_{\infty} d^3v_t \frac{\partial(\sigma_1 g g_i F_t)}{\partial v_{ti}} - \iiint_{\infty} d^3v_t F_t \frac{\partial(\sigma_1 g g_i)}{\partial v_{ti}} \quad (5.84)$$

The first integral is zero because the distribution function vanishes for infinite velocity. The second integral can be further manipulated using the following relation:

$$\frac{\partial f(\mathbf{g})}{\partial v_{ti}} = \frac{\partial f(\mathbf{g})}{\partial g_j} \frac{\partial g_j}{\partial v_{ti}} = -\frac{\partial f(\mathbf{g})}{\partial g_j} \frac{\partial g_j}{\partial v_{si}} = -\frac{\partial f(\mathbf{g})}{\partial v_{si}} \quad (5.85)$$

where $f(g_i)$ refers to any arbitrary function of the velocity components. With the help of equation (5.85) one can rewrite equation (5.84) in the following form:

$$\iiint_{\infty} d^3v_t \sigma_1 g g_i \frac{\partial F_t}{\partial v_{ti}} = \frac{\partial}{\partial v_{si}} \iiint_{\infty} d^3v_t \sigma_1 g g_i F_t \quad (5.86)$$

With the help of these relations equation (5.82) can be rewritten in the following form:

$$\left(\frac{\delta F_s}{\delta t}\right) = -\frac{\partial}{\partial v_{si}}\left[(G_i^s + U_i^s)F_s\right] + \frac{\partial^2}{\partial v_{si}\partial v_{sj}}\left[(K_{ij}^s + L_{ij}^s)F_s\right] + W_s F_s \quad (5.87)$$

where

$$W_s = \frac{\partial}{\partial v_{si}} \sum_t \iiint_{\infty} d^3v_t \left\{ \sigma_1 g g_i + \frac{\partial(\sigma_1 g g_i g_j)}{\partial v_{sj}} + \frac{1}{4}\frac{\partial\left[\sigma_2 g(g^2\delta_{ij} - 3g_i g_j)\right]}{\partial v_{sj}} \right\} F_t$$

$$(5.88a)$$

$$U_i^s = \frac{\partial}{\partial v_{sj}} 2\sum_t \frac{m_t}{m_s + m_t} \iiint_{\infty} d^3v_t \left\{ \sigma_1 g g_i g_j + \frac{1}{4}\sigma_2 g(g^2\delta_{ij} - 3g_i g_j) \right\} F_t$$

$$(5.88b)$$

Finally, we evaluate the coefficient of F_s for inverse power interactions. Using equation (3.109) the n-th order transport cross section, σ_n, can be written in the following form:

$$g\sigma_n^{st} = \bar{\sigma}_n^{st} \, g^{\frac{a_{st}-5}{a_{st}-1}} \tag{5.89}$$

where a_{st} is the exponent of the interparticle force between 's' and 't' particles (see Chapter 3) and $\bar{\sigma}_n^{st}$ is a coefficient independent of the relative velocity, $\mathbf{g}$. It can immediately be seen that the ratio of cross sections σ_1 and σ_2 is independent of the relative velocity; therefore one can introduce a new velocity independent parameter:

$$z_{21}^{st} = \frac{\sigma_2^{st}}{\sigma_1^{st}} = \frac{\bar{\sigma}_2^{st}}{\bar{\sigma}_1^{st}} \tag{5.90}$$

Using this form of the cross section, the coefficients in the collision term can be evaluated:

$$W_s = 4 \sum_t \frac{(2 - z_{21}^{st})(a_{st} - 2)(3a_{st} - 5)}{(a_{st} - 1)^2} \, \bar{\sigma}_1^{st} \iiint_\infty d^3 v_t \, g^{\frac{a_{st}-5}{a_{st}-1}} \, F_t \tag{5.91a}$$

$$U_i^s = 2 \sum_t \frac{m_{st}}{m_s} \frac{(5a_{st} - 9) - (3a_{st} - 5)z_{21}^{st}}{a_{st} - 1} \, \bar{\sigma}_1^{st} \iiint_\infty d^3 v_t \, g^{\frac{a_{st}-5}{a_{st}-1}} \, g_i F_t \tag{5.91b}$$

The collision term becomes particularly simple in the case of electrically charged particles, when $a_{st} = 2$. In this case $W_s = 0$ and the collision term can be written in the following form:

$$\left(\frac{\delta F_s}{\delta t} \right) = -\frac{\partial}{\partial v_{si}} \left[H_i^s F_s \right] + \frac{\partial^2}{\partial v_{si} \, \partial v_{sj}} \left[D_{ij}^s F_s \right] \tag{5.92}$$

where

$$H_i^s = G_i^s + U_i^s = \sum_t \frac{m_{st}}{m_s} (3 - 2z_{21}^{st}) \bar{\sigma}_1^{st} \iiint_\infty d^3 v_t \, \frac{g_i}{g^3} F_t \tag{5.93a}$$

$$D_{ij}^s = K_{ij}^s + L_{ij}^s = \sum_t \frac{m_{st}^2}{m_s^2} \overline{\sigma}_1^{st} \iiint_\infty d^3v_t \left[\left(1 - \frac{3}{4} z_{21}^{st}\right) \frac{g_i g_j}{g^2} + \frac{\delta_{ij}}{4} z_{21}^{st} \right] \frac{F_t}{g} \quad (5.93b)$$

Equation (5.92) is the Fokker–Planck approximation of the collision term. The first term describes a phenomenon similar to dynamical friction in which collisions gradually eliminate all velocity gradients in the gas. The second term describes diffusion in velocity space and D_{ij} acts as a diffusion coefficient tensor. This term results in the broadening of the velocity distribution function.

It should be mentioned that the Fokker–Planck collision term is valid for any system in which individual collisions produce only small changes in the velocity of particles, with large changes occurring only as a result of many small changes.

Another important point to be made is that the scattering process is not necessarily constrained to intermolecular collisions. For instance small angle scattering of charged particles by fluctuating electromagnetic fields also results in a Fokker–Planck type collision term.

5.4 Non-equilibrium solutions of the Boltzmann equation[8,9]

In the previous sections we derived a general transport equation for the phase space distribution function, $F(t,\mathbf{r},\mathbf{v})$, the Boltzmann equation. It was also shown that the only equilibrium solution of the Boltzmann equation is the Maxwell–Boltzmann distribution. However, in most practical applications one must find non-equilibrium solutions of the Boltzmann equation. There is no general rule to find a solution and we are usually constrained to solutions which assume that the distribution function is not very far from equilibrium. The near-equilibrium solutions were independently developed in the first decades of the 20th century by Sydney Chapman of Britain and David Enskog of Sweden. It is interesting to mention that Enskog's pioneering work, which represented the first non-equilibrium solutions of the Boltzmann equation, was his Ph.D. thesis defended in 1917 at the University of Uppsala, Sweden.

The Chapman–Enskog solution is based on the recognition that gases in thermodynamic equilibrium can be described by Maxwell–Boltzmann distributions. The Chapman–Enskog solution is based on an equilibrium Maxwellian with number density, n, temperature, T, and flow velocity vector, **u**. Transport phenomena occur only in gases which deviate from equilibrium. In reasonably dense gases (so-called collision dominated gases) the deviation from the equilibrium distribution is small and it can be treated as a perturbation.

Based on this assumption velocity distributions of collision dominated but non-equilibrium gases are approximated in the following form:

$$F(t,\mathbf{r},\mathbf{c}) = F_0(t,\mathbf{r},\mathbf{c})\left\{1 + A_i(t,\mathbf{r})c_i + B_{ij}(t,\mathbf{r})c_i\,c_j + D_{ijk}(t,\mathbf{r})c_i\,c_j\,c_k + ...\right\}$$

(5.94)

where $\mathbf{c}=\mathbf{v}-\mathbf{u}$ is the random velocity, A_i, B_{ij}, and D_{ijk} are coefficient matrices (independent of velocity) and

$$F_0(t,\mathbf{r},\mathbf{c}) = n(t,\mathbf{r})\left(\frac{m}{2\pi\,k\,T(t,\mathbf{r})}\right)^{3/2} \exp\left(-\frac{m(c_1^2 + c_2^2 + c_3^2)}{2k\,T(t,\mathbf{r})}\right) \qquad (5.95)$$

Here $n(t,\mathbf{r})$ is the particle concentration, therefore the normalization of F and F_0 are identical. It is assumed that the higher order terms of the expansion represent a decreasing series of perturbations. It is also clear that the expansion coefficients must be symmetric for the interchange of indices, therefore for instance

$$B_{ij} = B_{ji} \qquad\qquad D_{ijk} = D_{ikj} = D_{kij} = \cdots \qquad (5.96)$$

and so on.

The normalization of F yields the following condition for the expansion coefficients:

$$\iiint_\infty d^3c\,F = \iiint_\infty d^3c\,F_0 + A_i \iiint_\infty d^3c\,c_i F_0$$

$$+ B_{ij} \iiint_\infty d^3c\,c_i c_j F_0 + D_{ijk} \iiint_\infty d^3c\,c_i c_j c_k F_0 + \cdots \qquad (5.97)$$

The normalizations of F and F_0 are the same and the Maxwellian averages of c_i and $c_i c_j c_k$ are zero, therefore equation (5.97) simplifies to the following condition:

$$B_{ii} + (\text{4th order term}) + \cdots = 0 \qquad (5.98)$$

Equation (5.98) means that the normalization condition does not represent any constraint for coefficients of odd order. In lowest order it means that the two index tensor describing second order perturbations must be traceless:

$$B_{ii} = B_{11} + B_{22} + B_{33} = 0 \qquad (5.99)$$

Another constraint can be obtained for the expansion coefficients using the fact that the average random velocity must be identically zero:

$$\iiint_\infty d^3c\, c_h\, F = \iiint_\infty d^3c\, c_h\, F_0 + A_i \iiint_\infty d^3c\, c_h\, c_i\, F_0$$

$$+ B_{ij} \iiint_\infty d^3c\, c_h\, c_i\, c_j\, F_0 + D_{ijk} \iiint_\infty d^3c\, c_h\, c_i\, c_j\, c_k\, F_0 + \cdots = 0 \qquad (5.100)$$

The integrals of odd order again vanish and we obtain the following relation:

$$\iiint_\infty d^3c\, c_h\, F = nA_i <c_h\, c_i>_0 + nD_{ijk} <c_h\, c_i\, c_j\, c_k>_0 + \cdots = 0 \qquad (5.101)$$

where the symbol $<>_0$ denotes averages with respect of the distribution function, F_0. It can be easily verified that

$$<c_i>_0 = 0 \qquad <c_i c_j>_0 = \frac{kT}{m}\delta_{ij}$$

$$<c_i c_j c_k>_0 = 0 \qquad <c_h c_i c_j c_k>_0 = \left(\frac{kT}{m}\right)^2 (\delta_{hi}\delta_{jk} + \delta_{hj}\delta_{ik} + \delta_{hk}\delta_{ij}) \quad (5.102)$$

Using equation (5.102) condition (5.101) can be written in the following form:

$$A_h + 3\frac{kT}{m}D_{iih} + \cdots = 0 \qquad (5.103)$$

Equation (5.103) means that the first and third order perturbations are interrelated and the expansion series can be truncated only at zeroth order (which is a trivial expansion) or at the third order (because the existence of a first order coefficient requires the presence of the third order term as well).

It can be seen that the relative importance of the individual terms in the expansion series can be characterized by the following series of dimensionless quantities:

$$1, \quad v_0 A_i, \quad v_0^2 B_{ij}, \quad v_0^3 D_{ijk}, \quad \cdots \qquad (5.104)$$

where the characteristic velocity, v_0, is given by

$$v_0 = \sqrt{\frac{kT}{m}} \tag{5.105}$$

The convergence of the expansion series requires that the contribution of the higher order terms be rapidly diminishing, therefore coefficients must satisfy the following condition:

$$1 >> \left| v_0 A_i \right| >> \left| v_0^2 B_{ij} \right| >> \left| v_0^3 D_{ijk} \right| >> \cdots \tag{5.106}$$

It is interesting to examine the number of independent parameters characterizing the expansion series of a given order. The expansion coefficients of the n-th order represent a symmetric matrix with n indices. The number of independent components of such a matrix is $(n+1)(n+2)/2$, where n runs from 0 to infinity. This means that the lowest order approximation (when the distribution function is approximated by a Maxwellian) is characterized by the four parameters of the Maxwellian itself (vector velocity and temperature) plus the single independent parameter of the n=0 matrix (normalization). This means that in the lowest order limit the distribution function is characterized by five parameters.

In the most general case the next lowest order approximation must include expansion coefficients up to third order (because the first and third order coefficients are coupled). The number of independent components of the first, second and third order matrices are 3, 6, and 10, respectively, adding 19 additional elements to the five parameters of the lowest order expansion. However, the total number of independent parameters is not 19+5=24, because equations (5.99) and (5.103) must be satisfied by the matrix components. These four equations reduce the number of independent elements by four, therefore the total number of independent components in the third order approximation is 20.

The total number of independent parameters characterizing a given expansion is naturally closely related to the highest order coefficient matrix included in the approximation. If the number of indices of the highest order matrix in a given approximation is k (k>2) then the total number of independent parameters characterizing the expansion is given by $(k+1)(k+2)(k+3)/6$. For the third order expansion this expression gives a value of 20, for fourth order it yields 35, for fifth order it becomes 56, etc.

In the Chapman–Enskog method the solution of the Boltzmann equation is carried out by solving appropriate transport equations for the parameters of the Maxwellian and for the matrix elements of the expansion tensors. These transport equations are obtained by taking the velocity moments of the Boltzmann equation. This technique will be discussed in detail in the next chapter.

5.5 Problems

5.1 Show that the Boltzmann equation is invariant for Galilean transformations.

5.2 Consider a steady-state, isothermal (T= constant) planar atmosphere in thermodynamic equilibrium. Make the assumption that the phase-space distribution function is only a function of the altitude, z, and the particle velocity vector, $\mathbf{v}$. The gravitational acceleration vector is $\mathbf{g} = (0,0,-g)$, where g is constant everywhere (this is true for a thin atmosphere). Derive the phase space distribution function using the Boltzmann equation. The mass of the gas molecules is m, the temperature is T, and the particle number density at the surface is n_0.

5.3 A container of volume $\mathcal{V}$ is uniformly filled with a monatomic gas in thermodynamic equilibrium at temperature T. The total number of molecules in the volume is N. Calculate the total entropy of the gas and show that it satisfies the following thermodynamic relations:

$$\left(\frac{\partial S}{\partial E}\right)_{\mathcal{V},N} = \frac{1}{T} \qquad \left(\frac{\partial S}{\partial \mathcal{V}}\right)_{E,N} = \frac{p}{T}$$

where the total energy of the gas, E, is given as E=3NkT/2.

5.4 Consider a volume of isothermal gas under thermodynamic equilibrium conditions in the presence of a conservative external force. Show that the phase space distribution is the following:

$$F(\mathbf{r},\mathbf{v}) = A\exp\left(-\frac{\frac{1}{2}mv^2 + \Phi(\mathbf{r})}{kT}\right)$$

where A is a normalization constant and Φ is the potential of the external force.

5.5 Maxwell molecules interact via an inverse power potential with a spectral index of a=5. Consider a binary gas mixture of Maxwell molecules composed of 's' and 't' particles. Show that in this case the Fokker–Planck coefficients are proportional to the following quantities:

$$W_s \propto n_t$$

$$H_i^s \propto n_t (v_{si} - u_{ti})$$

$$D_{ij}^s \propto A\, n_t \left[(v_{si} - u_{ti})(v_{sj} - u_{tj}) + \frac{P_{t,ij}}{m_t n_t}\right] + B\delta_{ij}\, n_t \left[(\mathbf{v}_s - \mathbf{u}_t)^2 + \frac{P_{t,kk}}{m_t n_t}\right]$$

where $\mathbf{u}_t$ is the average velocity of 't' particles, while the pressure tensor, $P_{t,ij}$ is defined as

$$P_{t,ij} = m_t \iiint_{\infty} d^3v_t \, c_{ti} c_{tj} \, F_t$$

(here $\mathbf{c}_t$ is the random velocity of 't' particles).

5.6 References

[1]　Boltzmann, L., Über die Natur der Gasmolecüle, *Sitz. Math.-Naturwiss. Cl. Acad. Wiss., Wien II.*, **74**, 55, 1877.

[2]　Reif, F., *Fundamentals of Statistical and thermal physics*, McGraw-Hill, New York, 1965.

[3]　Chapman, S., and Cowling, T.G., *The mathematical theory of non-uniform gases*, Cambridge University Press, Cambridge, 1952.

[4]　Bhatnagar, P.L., Gross, E.P., and Krook, M., A model for collision processes in gases, I., *Phys. Rev.*, **94**, 511, 1954.

[5]　Krook, M., Dynamics of rarefied gases, *Phys. Rev.*, **99**, 1896, 1955.

[6]　Gross, E.P., and Krook, M., A model for collision processes in gases, *Phys. Rev.*, **102**, 593, 1956.

[7]　Burgers, J.M., *Flow equations for composite gases*, Academic Press, New York, 1969.

[8]　Chapman, S., On the kinetic theory of a gas. Part II. – A composite monatomic gas: Diffusion, viscosity, and thermal conduction, *Phil. Trans. Royal Soc. London*, **217A**, 115, 1916.

[9]　Enskog, D., Kinetische Theorie der Vorgänge in massing verdumten Gasen, Ph.D. Thesis, University of Uppsala, Sweden, 1917.

6

Generalized transport equations

In its most general form the Boltzmann equation is a seven dimensional non-linear integro-differential equation (see equation (5.29) or (5.30)). The solutions of the Boltzmann equation provide a full description of the phase space distribution function at all times. In most cases, however, it is next to impossible to solve the full Boltzmann equation and one has to resort to various approximate methods to describe the spatial and temporal evolution of macroscopic quantities characterizing the gas. One successful way to find approximate solutions is the Chapman–Enskog method discussed in the previous Chapter. This method uses a power series expansion around the equilibrium distribution function (Maxwellian) to describe slightly non-equilibrium gases. The method assumes that coefficients of increasing powers of random velocity components are proportional to increasing powers of a smallness parameter, thus ensuring rapid convergence.

An alternative, and mathematically equivalent, method was developed by Grad in the late 1940s.[1] In this method transport equations for macroscopic molecular averages are obtained by taking velocity moments of the Boltzmann equation. This seemingly straightforward technique runs into considerable difficulties because the governing equations for the components of the n-th velocity moment also depend on components of the (n+1)-th moment. In order to get a closed transport equation system one has to use closing relations (expressing a higher order velocity moment of the distribution function in terms of the components of lower moments) and thus make implicit assumptions about the distribution function. The closing relation to be used in this chapter relates the fourth order velocity moments to zeroth and second order ones: this means that one ends up with transport equations for the zeroth, first, second and third velocity moments of the distribution function.

In this chapter we start by examining the physical interpretation of the various velocity moments of the phase space distribution function. Next we derive a general moment equation, the so-called equation of change. Under specific conditions the equation of change yields the simplest set of hydrodynamic equations – the Euler equations. Finally, we derive the 20 moment set of generalized transport equations using Grad's method and examine this equation in various simplifying limits.

It should be emphasized again that the Chapman–Enskog method and the Grad method are both physically and mathematically equivalent. Later in this chapter we will discuss the detailed relation between the Chapman–Enskog coefficients and the velocity moments of the distribution function.

6.1 Moments of the Boltzmann equation

6.1.1 *Velocity moments of the phase space distribution function*

Macroscopic variables, such as number density, average flow velocity, kinetic pressure, and so on, can be considered as average values of molecular properties. These macroscopic variables are related to the velocity moments of the phase space distribution function.

In previous chapters we considered the lowest order velocity moments of the phase space distribution function. Here we repeat some of the earlier results and introduce a consistent hierarchy of velocity moments. This hierarchy is based on the 'order' of a given velocity moment: that is the sum of the powers of velocity components in the moment integral. For instance, the integral

$$A = \iiint_{\infty} d^3v \, v^2 v_1^3 v_2 \, F(t, \mathbf{r}, \mathbf{v}) \tag{6.1}$$

is a sixth order velocity moment of the distribution function, because this is the sum of the velocity powers of the integrand.

Let us start with the lowest velocity moments of the phase space distribution function. We know that the normalization of the distribution function is the particle number density:

$$n(t, \mathbf{r}) = \iiint_{\infty} d^3v \, F(t, \mathbf{r}, \mathbf{v}) \tag{6.2}$$

Obviously the number density is the zeroth velocity moment of the phase space distribution function, because the distribution function is not multiplied by any velocity components.

We know that the components of the average molecular velocity are defined in the following way:

$$\mathbf{u}(t,\mathbf{r}) = \frac{1}{n(t,\mathbf{r})} \iiint_\infty d^3v \, \mathbf{v} \, F(t,\mathbf{r},\mathbf{v}) \tag{6.3}$$

In other words this is the first velocity moment of the distribution function. With the help of the average molecular velocity one can introduce the random velocity vector, $\mathbf{c}$:

$$c_i(t,\mathbf{r}) = v_i - u_i(t,\mathbf{r}) \tag{6.4}$$

Note that while t, $\mathbf{r}$ and $\mathbf{v}$ are independent variables, the new velocity variable, $\mathbf{c}(t,\mathbf{r})$, is not independent of t and $\mathbf{r}$. This fact has to be taken into account when deriving transport equations for the distribution of random velocities.

The zeroth moment of the distribution of random velocities is still the particle number density, because in our non-relativistic approximation the number of particles in a given configuration space volume is independent of the velocity of the coordinate system. On the other hand it follows from the definition of the random velocity that the first moment of this distribution must vanish:

$$\iiint_\infty d^3c \, c_i \, F(t,\mathbf{r},\mathbf{c}) = 0 \tag{6.5}$$

The velocity moments of the distribution function can also be interpreted as net fluxes of molecular quantities. Recall that we showed, in connection with the mean free path method, that the average flux of a molecular quantity, $W(\mathbf{v})$, can be obtained by the following integral (see equations (4.35) and (4.68)):

$$j_i^W(t,\mathbf{r}) = \iiint_\infty d^3v \, v_i \, W(\mathbf{v}) F(t,\mathbf{r},\mathbf{v}) = n(t,\mathbf{r}) \langle v_i \, W(\mathbf{v}) \rangle \tag{6.6}$$

For instance, the first moment of the distribution function can be considered as the net mass flux (W=m) at the configuration space location, $\mathbf{r}$:

$$j_i^m(t,\mathbf{r}) = mn(t,\mathbf{r}) \langle v_i \rangle = mn(t,\mathbf{r}) u_i(t,\mathbf{r}) \tag{6.7}$$

The second moment of the velocity distribution is related to the net flux of particle momentum, therefore one can use $W = mv_j$:

$$j_i^{mv_j} = \iiint_\infty d^3v \, v_i \, mv_j \, F = mn \langle v_i \, v_j \rangle \tag{6.8}$$

This quantity has two indices, i and j, and it describes the net flux of the j-th component of translational momentum in the i-th direction. It should be immediately noticed that equation (6.8) is invariant for the interchange of the two indices; in other words the momentum flux (and consequently the second velocity moment of the distribution function) is a symmetric 3×3 matrix.

The second velocity moment can also be expressed in terms of random velocities:

$$j_i^{mv_j} = mn\langle v_i\, v_j\rangle = mn\langle(u_i + c_i)(u_j + c_j)\rangle$$

$$= mn\left(u_i u_j + u_i\langle c_j\rangle + u_j\langle c_i\rangle + \langle c_i c_j\rangle\right) = mn\, u_i u_j + mn\langle c_i c_j\rangle \qquad (6.9)$$

where we have used the identity $\langle c_i\rangle = 0$. Here the first term describes the momentum flux due to the bulk (organized) motion of the gas as a whole, while the second term corresponds to the momentum flux resulting from the random (thermal) motion of the gas molecules.

It has been discussed earlier (see Chapter 2) how the pressure of a gas is usually defined as the force per unit area exerted by the gas molecules through collisions with the wall. This force is equal to the rate of transfer of molecular momentum to the wall. In connection with our discussions of viscosity (Subsection 4.2.1) we also defined the shearing stress as a tensor quantity: the first index refers to the normal vector of the surface, while the second one represents the direction of the force. Next, we generalize and combine these definitions. Pressure is now defined as the net rate of transport of molecular momentum per unit area, that is, the net flux of momentum across unit area due to the random particle motion. This definition indicates that the pressure in the molecular sense can be represented by the following:

$$P_{ij}(t,\mathbf{r}) = m\iiint_\infty d^3c\, c_i\, c_j\, F = mn\langle c_i c_j\rangle \qquad (6.10)$$

With the help of this definition the total momentum flux can be written as

$$j_i^{mv_j} = mn\, u_i u_j + P_{ij} \qquad (6.11)$$

With the help of the general pressure tensor, P_{ij}, one can define a kinetic 'scalar' pressure (and temperature), as well as a stress tensor. The definition of the scalar pressure is related to the trace of the pressure tensor:

$$p = \frac{1}{3} P_{ii} = \frac{1}{3}(P_{xx} + P_{yy} + P_{zz}) = \frac{1}{3} mn \langle c^2 \rangle \qquad (6.12)$$

The temperature of the particles, T, is a measure of the mean kinetic energy of the random particle motion. According to the thermodynamic definition of the absolute temperature, there is an average molecular thermal energy associated with each translational degree of freedom, so that

$$\frac{1}{2} kT_h = \frac{1}{2} m \langle c_h^2 \rangle = \frac{1}{2} \frac{P_{hh}}{n} \qquad (6.13)$$

Here the repeated index does not refer to summation, P_{hh} is simply a diagonal element of the pressure tensor. It must be recognized that in non-equilibrium gases the translational kinetic temperature might be different in different directions. In the simplest (isotropic) case all diagonal elements of the pressure tensor are identical, and they are the same as the scalar pressure, p=nkT.

There is another frequently used pressure related quantity in addition to the scalar pressure, the stress tensor. Following the definition most frequently used in fluid dynamics we define the stress tensor, τ, in the following way:

$$\tau_{ij} = \delta_{ij} p - P_{ij} \qquad (6.14)$$

In the case of a Maxwellian velocity distribution (with number density, n, and temperature, T) the pressure tensor simply becomes $P_{ij}=nkT\delta_{ij}$. The stress tensor, therefore, expresses the deviation of the pressure tensor from the equilibrium case (characterized by a Maxwellian velocity distribution).

The general form of the third velocity moment of the distribution function is related to the flux of the second moment, the pressure tensor. Mathematically this can be written in the following form:

$$j_i^{mv_jv_k}(t,\mathbf{r}) = \iiint_\infty d^3v \, v_i (m v_j v_k) F(t,\mathbf{r},\mathbf{v}) = mn(t,\mathbf{r}) \langle v_i v_j v_k \rangle \qquad (6.15)$$

This net flux can also be expressed in terms of the random velocities:

$$j_i^{mv_jv_k} = mn \langle (u_i + c_i)(u_j + c_j)(u_k + c_k) \rangle$$

$$= mn \left(u_i u_j u_k + u_i \langle c_j c_k \rangle + u_j \langle c_i c_k \rangle + u_k \langle c_i c_j \rangle + \langle c_i c_j c_k \rangle \right) \qquad (6.16)$$

$$= mn \, u_i u_j u_k + u_i P_{jk} + u_j P_{ik} + u_k P_{ij} + mn \langle c_i c_j c_k \rangle$$

The first term describes the bulk transport of the bulk momentum flux. The second, third and fourth terms describe a combination of bulk and random momentum fluxes, expressed by the presence of the pressure tensor. Finally, the last term describes the net flux of random momentum flux due to the random motion of the particles.

The above description is quite convoluted, but it can be understood by way of a simple example. Let us take the 133 element of the flux and assume that the pressure is isotropic ($P_{ij}=p\delta_{ij}$). In this case equation (6.16) becomes the following:

$$j_1^{mv_3v_3} = 2u_1\left(\frac{1}{2}mn\,u_3^2\right)+u_1p+2\left\langle c_1\left(\frac{1}{2}mn\,c_3^2\right)\right\rangle \qquad (6.17)$$

The interpretation of these three terms is reasonably straightforward. The first term describes the transport of the z component of the organized translational energy in the x direction due to the bulk motion of the gas as a whole. The second term describes the organized (bulk) transport of a translational component of the internal (random) energy of the gas. Finally, the last term describes the transport of the z component of the random energy in the x direction due to the random motion of the molecules. Physically, this last term corresponds to the flow of a random energy component due to the thermal motion of the molecules. In other words, this term describes heat conduction in the gas.

Based on the above interpretation we introduce the generalized heat flow tensor in the following way:

$$Q_{ijk}(t,\mathbf{r})=m\iiint\limits_{\infty} d^3c\,c_i\,c_j\,c_k\,F = mn\left\langle c_i c_j c_k\right\rangle \qquad (6.18)$$

Using this definition the total energy flux can be expressed in the following form:

$$j_i^{mv_jv_k} = mn\,u_i u_j u_k + u_i P_{jk} + u_j P_{ik} + u_k P_{ij} + Q_{ijk} \qquad (6.19)$$

The elements of the general heat flow tensor, Q_{ijk}, describe the transport of random pressure components due to the random motion of the particles. It is quite difficult to visualize the physical meaning of the individual tensor elements. On the other hand, one can introduce the contracted quantity, h_i, which is obviously related to the 'classical' heat flow vector:

$$h_i = n\left\langle c_i\frac{1}{2}mc^2\right\rangle = \frac{1}{2}mn\left\langle c_i c_k c_k\right\rangle = \frac{1}{2}Q_{ikk} \qquad (6.20)$$

This quantity, called the heat flow vector, describes the transport of total translational random energy due to the random motion of the particles.

Finally, we mention that higher order velocity moments can be introduced by analogy to the lower order moments. For instance, the fourth moment is

$$R_{ijkl}(t,\mathbf{r}) = m\iiint_{\infty} d^3c\, c_i\, c_j\, c_k\, c_l\, F = mn\left\langle c_i c_j c_k c_l \right\rangle \tag{6.21}$$

This velocity moment is again a symmetric tensor with four indices. As has been mentioned before the number of independent components of such a tensor is $(n+1)(n+2)/2$, where n is the number of indices (or the order of the velocity moment). The tensor, R, has $5\times6/2=15$ independent components.

6.1.2 Maxwell's equation of change

Generalized transport equations are obtained by taking the velocity moments of the Boltzmann equation. In this subsection we derive a general form of the moment equation, the so-called equation of change. This equation was first derived by Maxwell and it is very useful in describing various transport phenomena. It is also particularly suitable to introduce a hierarchy of moment equations, which can be used as a closed set of generalized transport equations.

Let us consider a $W(\mathbf{c})$ molecular quantity, where $\mathbf{c}$ is the random velocity. It is assumed that $W(\mathbf{c})$ is independent of time and spatial location. Multiply both sides of the single species Boltzmann equation (we use equation (5.27), expressed in terms of the random velocity) by $W(\mathbf{c})$ and integrate over all potential values of the random velocity, $\mathbf{c}$:

$$\iiint_{\infty} d^3c\, W(\mathbf{c})\left\{ \frac{\partial F}{\partial t} + (u_i + c_i)\frac{\partial F}{\partial x_i} - \left(\frac{\partial u_i}{\partial t} + (u_j + c_j)\frac{\partial u_i}{\partial x_j} - a_i \right)\frac{\partial F}{\partial c_i} \right\}$$

$$= \iiint_{\infty} d^3c\, W(\mathbf{c}) \iiint_{\infty} d^3c_2 \int_0^{2\pi} d\varepsilon \int_0^{\pi} d\chi \sin\chi\, S(g,\chi) g\left[F' F_2' - F F_2 \right] \tag{6.22}$$

The average of quantity W over the velocity distribution is again defined in the following way:

$$n<W> = \iiint_{\infty} d^3c\, W(\mathbf{c})F \tag{6.23}$$

because the phase space distribution function, F, is normalized to the particle number density, n.

Using this definition we can rewrite the various terms in equation (6.22). Let us start with the term containing the partial time derivative. We know that variables t, **r**, and **c** can be treated as independent quantities, therefore the sequence of the integral over **c** and the partial time derivative operator can be interchanged (Recall that **c**=**c**(t,**r**) was so defined when the independent variables were t, **r**, and **v**. However, equation (6.22) was derived by changing the independent variables from t,**r**,**v** to t,**r**,**c**. Therefore, when working with equation (6.22) **c** must be seen as an independent variable and **v**=**v**(t,**r**).):

$$\iiint_{\infty} d^3c\, W(\mathbf{c})\frac{\partial F}{\partial t} = \frac{\partial}{\partial t}\iiint_{\infty} d^3c\, W(\mathbf{c})F = \frac{\partial}{\partial t}\left(n\langle W\rangle\right) \qquad (6.24)$$

Similarly, one can evaluate the two terms related to the spatial gradient of the distribution function:

$$\iiint_{\infty} d^3c\, W(\mathbf{c})u_i\frac{\partial F}{\partial x_i} = u_i\frac{\partial}{\partial x_i}\iiint_{\infty} d^3c\, W(\mathbf{c})F = u_i\frac{\partial}{\partial x_i}\left(n\langle W\rangle\right) \qquad (6.25)$$

$$\iiint_{\infty} d^3c\, W(\mathbf{c})c_i\frac{\partial F}{\partial x_i} = \frac{\partial}{\partial x_i}\iiint_{\infty} d^3c\, W(\mathbf{c})c_i F = \frac{\partial}{\partial x_i}\left(n\langle c_i\, W\rangle\right) \qquad (6.26)$$

Terms containing velocity derivatives are somewhat more complicated to evaluate; however, integration by parts usually leads to relatively straightforward results. In particular, we obtain the following:

$$\iiint_{\infty} d^3c\, W(\mathbf{c})\frac{\partial F}{\partial c_i} = \iiint_{\infty} d^3c\, \frac{\partial(W\,F)}{\partial c_i} - \iiint_{\infty} d^3c\, F\frac{\partial W}{\partial c_i} = -n\left\langle\frac{\partial W}{\partial c_i}\right\rangle \qquad (6.27)$$

$$\iiint_{\infty} d^3c\, W(\mathbf{c})c_j\frac{\partial F}{\partial c_i} = \iiint_{\infty} d^3c\, \frac{\partial(W\,c_j\,F)}{\partial c_i} - \iiint_{\infty} d^3c\, F\frac{\partial(c_j\,W)}{\partial c_i}$$

$$= -n\left\langle c_j\frac{\partial W}{\partial c_i}\right\rangle - n\delta_{ij}\langle W\rangle \qquad (6.28)$$

After integration by parts the first integral is an integral of a full differential, therefore its value is given by the integrand evaluated at the integration boundaries. On the other hand we know that the phase space distribution function vanishes for $c_i = \pm\infty$, therefore this term is simply zero. This immediately leads to the simple results given by (6.27) and (6.28).

The last integral on the left hand side contains the acceleration:

$$\iiint_\infty d^3c\, W(\mathbf{c})\, a_i\, \frac{\partial F}{\partial c_i} = \iiint_\infty d^3c\, \frac{\partial(W\, a_i\, F)}{\partial c_i} - \iiint_\infty d^3c\, F\, \frac{\partial(a_i\, W)}{\partial c_i}$$

$$= -\iiint_\infty d^3c\, W\, F\, \frac{\partial a_i}{\partial c_i} - \iiint_\infty d^3c\, a_i\, F\, \frac{\partial W}{\partial c_i} = -n\left\langle a_i\, \frac{\partial W}{\partial c_i}\right\rangle \qquad (6.29)$$

Here we have assumed that the acceleration is divergence free in velocity space: this assumption is justified for most external force fields, including gravitational and electromagnetic forces:

$$\frac{\partial a_i}{\partial v_i} = \frac{\partial a_i}{\partial c_i} = 0 \qquad (6.30)$$

Using equations (6.24) through (6.30) one can rewrite the left hand side of the general moment equation (equation (6.22)) in the following form:

$$\frac{\partial(n\langle W\rangle)}{\partial t} + u_i\, \frac{\partial(n\langle W\rangle)}{\partial x_i} + \frac{\partial(n\langle c_i\, W\rangle)}{\partial x_i} + n\langle W\rangle\, \frac{\partial u_i}{\partial x_i}$$

$$+ n\left\langle \frac{\partial W}{\partial c_i}\right\rangle\left(\frac{\partial u_i}{\partial t} + u_j\, \frac{\partial u_i}{\partial x_j}\right) + n\left\langle c_j\, \frac{\partial W}{\partial c_i}\right\rangle\frac{\partial u_i}{\partial x_j} - n\left\langle a_i\, \frac{\partial W}{\partial c_i}\right\rangle$$

$$= \iiint_\infty d^3c\, W(\mathbf{c}) \iiint_\infty d^3c_2 \int_0^{2\pi} d\varepsilon \int_0^\pi d\chi\, \sin\chi\, S(g,\chi)\, g\left[F'F_2' - FF_2\right] \qquad (6.31)$$

Next we examine the right hand side of this equation. We start by repeating some of the tricks we applied when proving the H-theorem. We integrate over both $\mathbf{c}$ and $\mathbf{c}_2$, therefore the two particles can be interchanged in the integrand without changing the value of the integral (mathematically speaking, both $\mathbf{c}$ and $\mathbf{c}_2$ are 'dummy' variables). We showed earlier that the impact azimuth, ε, the deflection angle, χ, and the magnitude of the relative velocity, $g = |\mathbf{c} - \mathbf{c}_2|$, are invariant for this interchange, therefore the collision term in equation (6.31) can be written in the following form:

$$\frac{\delta W}{\delta t} = \iiint d^3c_2 \iiint_\infty d^3c \int_0^{2\pi} d\varepsilon \int_0^\pi d\chi \sin\chi\, S(g,\chi)\, g\, W_2 \left[F'F_2' - FF_2 \right] \quad (6.32)$$

where $W_2 = W(\mathbf{c}_2)$. Next one can add the two expressions obtained for the same collision term and divide the result by two. This procedure yields the following result:

$$\frac{\delta W}{\delta t} = \frac{1}{2} \iiint_\infty d^3c \iiint_\infty d^3c_2 \int_0^{2\pi} d\varepsilon \int_0^\pi d\chi \sin\chi\, S(g,\chi)\, g\, (W + W_2)\, F'F_2'$$

$$- \frac{1}{2} \iiint_\infty d^3c \iiint_\infty d^3c_2 \int_0^{2\pi} d\varepsilon \int_0^\pi d\chi \sin\chi\, S(g,\chi)\, g\, (W + W_2)\, FF_2 \quad (6.33)$$

In this equation one can replace the velocities before collision, $\mathbf{c}$ and $\mathbf{c}_2$, by the velocities immediately after the collision, $\mathbf{c}'$ and $\mathbf{c}_2'$, without altering the value of the integral. The reason is that for every collision there exists an inverse collision with $\chi'=\chi$, $\varepsilon'=\varepsilon$, $g'=g$ and $d^3c'd^3c_2'=d^3cd^3c_2$ (see the derivation of the Boltzmann collision integral). In the first term of equation (6.33) we interchange the primed and unprimed quantities, while leaving the second term unchanged. This procedure leads to the following form of the collision term:

$$\frac{\delta W}{\delta t} = \frac{1}{2} \iiint_\infty d^3c \iiint_\infty d^3c_2 \int_0^{2\pi} d\varepsilon \int_0^\pi d\chi \sin\chi\, S(g,\chi)\, g\, (W' + W_2' - W - W_2)\, FF_2$$

$$(6.34)$$

It is important to recognize that equation (6.34) represents a weighted average of the change of molecular quantity, W, due to binary collisions. The quantity

$$\Delta W = W' + W_2' - W - W_2 \quad (6.35)$$

is the difference of the total value of molecular quantity, W, before and after a collision. For instance, if W represents the random kinetic energy of the particles, $W = mc^2/2$, the quantity ΔW represents the change of the total kinetic energy during the collision (for elastic collisions this change is naturally zero). One can introduce the following notation for the weighted average of the change in the total value of W:

$$\Delta[W] = \frac{1}{2} \iiint_{\infty} d^3c \iiint_{\infty} d^3c_2 \int_0^{2\pi} d\varepsilon \int_0^{\pi} d\chi \sin\chi \, S(g,\chi) \, g \, (W' + W_2' - W - W_2) F F_2$$

(6.36)

Substituting this expression into the general moment equation we obtain Maxwell's equation of change:

$$\frac{\partial(n\langle W\rangle)}{\partial t} + \frac{\partial(n\langle W\rangle u_i)}{\partial x_i} + \frac{\partial(n\langle c_i W\rangle)}{\partial x_i}$$

$$+ n\left\langle \frac{\partial W}{\partial c_i}\right\rangle \left(\frac{\partial u_i}{\partial t} + u_j \frac{\partial u_i}{\partial x_j}\right) + n\left\langle c_j \frac{\partial W}{\partial c_i}\right\rangle \frac{\partial u_i}{\partial x_j} - n\left\langle a_i \frac{\partial W}{\partial c_i}\right\rangle = \Delta[W] \quad (6.37)$$

It is worth while to note that an alternative form of Maxwell's equation of change can be obtained by using the other form of the Boltzmann equation, when the distribution function is expressed in terms of the total velocity, $\mathbf{v}$. In this case one starts from the alternative form of the Boltzmann equation given by equation (5.26). Taking the general moment of this equation leads to the following:

$$\frac{\partial(n\langle W\rangle)}{\partial t} + \frac{\partial(n\langle v_i W\rangle)}{\partial x_i} - n\left\langle a_i \frac{\partial W}{\partial v_i}\right\rangle = \Delta[W]$$

(6.38)

This is the so-called conservative form of Maxwell's equation of change. The two forms of Maxwell's equation of change are physically identical and one has to decide which form is more convenient to use for a given problem.

6.2 The Euler equations

Maxwell's equation of change can be used to obtain closed transport equation systems of varying levels of sophistication. These equations will be discussed in considerable detail later in this chapter. In this section we consider the simplest set of transport equations resulting from Maxwell's equation of change: the well known set of Euler's equations for perfect gases.

The following derivation is based on the recognition that the equation of change becomes particularly simple for summation invariants: in this case the right hand side of the general moment equation simply vanishes. It has been mentioned earlier (Subsection 5.2.2.) that in the case of molecules without internal degrees of freedom there are four, and only four, independent summations which depend on the particle velocities (the fifth is the particle mass, which is independent of the velocity). This means that five simple transport equations (with vanishing right hand sides) can be obtained by taking W=m, W=mc, and W=mc²/2, respectively.

In order to get a closed set of transport equations, one has to make additional assumptions (the mathematical reason will be discussed later in the chapter). It is assumed that there are enough collisions to insure that the gas is near equilibrium, i.e. at every spatial location the distribution function can be approximated with a Maxwell–Boltzmann distribution function. However, the parameters of the Maxwellian, such as the density or temperature, do vary with time and spatial location. Similar assumptions were used in Chapter 4, when we discussed the mean free path method. This assumption is also called the assumption of local thermodynamic equilibrium (LTE).

Using the assumption of local thermodynamic equilibrium one can evaluate Maxwell's equation of change for the five summation invariant quantities. Let us start with the simplest equation for which W=m. Substituting this function into equation (6.37) yields the transport equation

$$\frac{\partial(mn)}{\partial t} + \frac{\partial(mn\,u_i)}{\partial x_i} = 0 \tag{6.39}$$

because most of the terms fall out since $<c_i>=0$ and $\partial W/\partial c_i=0$. Equation (6.39) is the well known continuity equation which describes the conservation of mass. It is interesting to note that the conservative form of Maxwell's equation of change (equation (6.38)) also leads to equation (6.39).

Next we derive the transport equation obtained by considering the temporal variation of the particle momentum, i.e. $W=mc_k$. Substituting this function into equation (6.37) results in the following transport equation:

$$\frac{\partial\left(mn\langle c_k\rangle\right)}{\partial t} + \frac{\partial\left(mn\langle c_k\rangle u_i\right)}{\partial x_i} + \frac{\partial\left(mn\langle c_i c_k\rangle\right)}{\partial x_i}$$

$$+m\,n\left\langle\frac{\partial c_k}{\partial c_i}\right\rangle\left(\frac{\partial u_i}{\partial t}+u_j\frac{\partial u_i}{\partial x_j}\right)+m\,n\left\langle c_j\frac{\partial c_k}{\partial c_i}\right\rangle\frac{\partial u_i}{\partial x_j}-m\,n\left\langle a_i\frac{\partial c_k}{\partial c_i}\right\rangle=0 \tag{6.40}$$

We know that $<c_k>=0$ and $\partial c_k/\partial c_j=\delta_{kj}$, therefore one obtains the following:

$$mn\frac{\partial u_k}{\partial t} + mnu_j\frac{\partial u_k}{\partial x_j} + \frac{\partial\left(mn\langle c_i c_k\rangle\right)}{\partial x_i} - mn\langle a_k\rangle = 0 \qquad (6.41)$$

This equation expresses the conservation of momentum. It was assumed that the gas is in local thermodynamic equilibrium, therefore one can use a Maxwell–Boltzmann distribution to evaluate the term mn<$c_i c_k$>. It has been shown earlier that for Maxwell–Boltzmann distributions mn<$c_i c_k$>=pδ_{ik} (equation (5.102)), so equation (6.41) becomes

$$mn\frac{\partial u_k}{\partial t} + mnu_j\frac{\partial u_k}{\partial x_j} + \frac{\partial p}{\partial x_k} - mn\langle a_k\rangle = 0 \qquad (6.42)$$

In these equations <a_k> is the macroscopic average of the acceleration caused by external fields (such as gravity, electric and magnetic fields, etc.). Equation (6.42) expresses the conservation of momentum and it is the non-conservative form of the well known momentum equation for perfect fluids. If one uses the alternative form of Maxwell's equation of change (equation (6.38)) and substitutes W=mv_k, the so-called conservative form of the momentum equation is obtained:

$$\frac{\partial\left(mnu_k\right)}{\partial t} + \frac{\partial\left(mnu_i u_k + \delta_{ik}p\right)}{\partial x_i} - mn\langle a_k\rangle = 0 \qquad (6.43)$$

The next equation is obtained by substituting W=$mc^2/2$ into equation (6.37). This results in the following equation:

$$\frac{\partial\left(mn\langle c^2\rangle\right)}{\partial t} + \frac{\partial\left(mn\langle c^2\rangle u_i\right)}{\partial x_i} + \frac{\partial\left(mn\langle c_i c^2\rangle\right)}{\partial x_i}$$

$$+ mn\left\langle\frac{\partial c^2}{\partial c_i}\right\rangle\left(\frac{\partial u_i}{\partial t} + u_j\frac{\partial u_i}{\partial x_j}\right) + mn\left\langle c_j\frac{\partial c^2}{\partial c_i}\right\rangle\frac{\partial u_i}{\partial x_j} - mn\left\langle a_i\frac{\partial c^2}{\partial c_i}\right\rangle = 0 \qquad (6.44)$$

In the present LTE approximation mn<c^2>=3p, <$c_i c^2$>=0, <$\partial c^2/\partial c_i$>=2<$c_i$>=0. Using these relations one can simplify equation (6.44) to the following form:

$$\frac{3}{2}\frac{\partial p}{\partial t} + \frac{3}{2}\frac{\partial\left(pu_i\right)}{\partial x_i} + p\frac{\partial u_i}{\partial x_i} - mn\langle a_i c_i\rangle = 0 \qquad (6.45)$$

For most external forces, such as gravity or electromagnetic fields, the $<a_i c_i>$ term vanishes. In this case equation (6.45) becomes

$$\frac{3}{2}\frac{\partial p}{\partial t} + \frac{3}{2}\frac{\partial (p u_i)}{\partial x_i} + p\frac{\partial u_i}{\partial x_i} = 0 \tag{6.46}$$

This is the well known energy equation for perfect gases. An alternative form of the energy equation can be obtained by starting from equation (6.38) and substituting $W = mv^2/2$. This leads to the conservative form of the energy equation:

$$\frac{\partial \left(\frac{1}{2} mnu^2 + \frac{3}{2} p\right)}{\partial t} + \frac{\partial \left(\frac{1}{2} mnu_i u^2 + \frac{5}{2} u_i p\right)}{\partial x_i} - 2mnu_i\langle a_i\rangle = 0 \tag{6.47}$$

Equations (6.39), (6.42) and (6.46) constitute the Euler equations for a perfect gas. Equation (6.39) expresses the conservation of mass, (6.42) is the equation of motion, and (6.46) describes the evolution of internal energy. Equations (6.39), (6.43) and (6.47) represent the conservative form of the Euler equations, expressing the conservation of total mass, momentum and energy.

The Euler equations represent the simplest form of fluid equations and they are only valid in gases where the assumption of local thermodynamic equilibrium is justified. One has to be careful when applying these equations for low density or non-equilibrium gases and make sure that the fundamental assumptions are satisfied.

6.3 The 20 moment equations

The Euler equations represent a particularly simple set of transport equations which describe the transport of summation invariant molecular quantities under local thermodynamic equilibrium conditions. However, Maxwell's equation of change can also be used to obtain much more general sets of transport equations describing the transport of molecular quantities which are not summation invariants. Furthermore, the assumption of local thermodynamic equilibrium can also be relaxed and one has to assume only that the velocity distribution is not far from local thermodynamic equilibrium.

A major problem with all moment equation systems is that of closure. Inspection of either form of Maxwell's equation of change (equation (6.37) or (6.38)) reveals that the time derivative of an arbitrary macroscopic quantity, $<W>$, depends on the divergence of the next higher velocity moment, $<cW>$. This means that the evolution of the mass density depends on the bulk velocity, the equation of motion contains the pressure tensor, the pressure equation contains the heat flow tensor and so on. Unless we truncate the series somewhere, we will end up with an infinite number of transport equations. In order to close the system of equations, it is necessary to adopt an approximate expression for the velocity distribution function or assume some kind of mathematical symmetry.

There are two ways to truncate the infinite series of transport equations. The first technique is based on the Chapman–Enskog method,[2,3] while the second was introduced by an American, Harold Grad.[1] As we shall show below these two methods are both physically and mathematically equivalent.

6.3.1 The Chapman–Enskog distribution function

We start by deriving moment equations using the Chapman–Enskog method. Let us consider the simplest non-trivial truncation of the Chapman–Enskog expansion series. It has been shown that the first and third order expansion elements are inter-related (equation 5.103), therefore the lowest order non-trivial Chapman–Enskog solution is truncated after the third order term:

$$F(t, \mathbf{r}, \mathbf{c}) = F_0(t, \mathbf{r}, \mathbf{c})\left[1 + A_i(t, \mathbf{r})c_i + B_{ij}(t, \mathbf{r})c_i c_j + D_{ijk}(t, \mathbf{r})c_i c_j c_k\right] \quad (6.48)$$

where F_0 is a Maxwellian (with parameters n_0, $\mathbf{u}_0$, and T_0) and A, B, and D are velocity independent coefficient matrices. As has been discussed in Section 5.4, both the B and D matrices are symmetric and they satisfy the following conditions: $B_{ii}=0$ and $A_i+(3kT/m)D_{ijj}=0$. Using these conditions the lowest moments of the distribution function are the following:

$$m\iiint_\infty d^3c\, F = mn_0 \quad (6.49a)$$

$$m\iiint_\infty d^3c\, c_\alpha\, F = 0 \quad (6.49b)$$

$$m\iiint_\infty d^3c\, c_\alpha c_\beta\, F = p_0\left[\delta_{\alpha\beta} + 2\left(\frac{kT_0}{m}\right)B_{\alpha\beta}\right] \quad (6.49c)$$

$$m\iiint_\infty d^3c\, c_\alpha c_\beta c_\gamma\, F = 6p_0\left(\frac{kT_0}{m}\right)^2 D_{\alpha\beta\gamma} \quad (6.49d)$$

$$m \iiint\limits_{\infty} d^3 c\, c_\alpha c_\beta c_\gamma c_\chi\, F = p_0 \left(\frac{kT_0}{m} \right) \left(\delta_{\alpha\beta} \delta_{\gamma\chi} + \delta_{\alpha\gamma} \delta_{\beta\chi} + \delta_{\alpha\chi} \delta_{\beta\gamma} \right)$$

$$+ 2p_0 \left(\frac{kT_0}{m} \right)^2 \left(\delta_{\alpha\beta} B_{\gamma\chi} + \delta_{\alpha\gamma} B_{\beta\chi} + \delta_{\alpha\chi} B_{\beta\gamma} + \delta_{\beta\gamma} B_{\alpha\chi} + \delta_{\beta\chi} B_{\alpha\gamma} + \delta_{\gamma\chi} B_{\alpha\beta} \right)$$

$$(6.49e)$$

These relations represent explicit one-to-one correspondence between the coefficient matrices and macroscopic physical quantities. For instance, it immediately follows from equations (6.49) and the definitions of the stress and heat flow tensors that

$$\tau_{\alpha\beta} = -2p_0 \left(\frac{kT_0}{m} \right) B_{\alpha\beta} \qquad (6.50a)$$

$$Q_{\alpha\beta\gamma} = 6p_0 \left(\frac{kT_0}{m} \right)^2 D_{\alpha\beta\gamma} \qquad (6.50b)$$

It is important to note that in addition to the parameters of the Maxwellian, n_0, and p_0, the fourth moment of the present distribution function (given by equation (6.48)) contains only elements of the second order tensor, B_{ij} (or the stress tensor, τ_{ij}). This is a direct consequence of our truncation. The distribution function itself does not contain fourth or higher order elements, therefore it is no surprise that the matrix elements of the fourth order velocity moments can be expressed in terms of lower order quantities. The interesting result is that R_{ijkl} is only composed of the Maxwellian parameters and the elements of the stress tensor.

Using these relations the velocity distribution function, F, can also be expressed in terms of the macroscopic quantities:

$$F = F_0 \left[1 - \frac{mn_0}{2p_0^2} \tau_{ij} c_i c_j + \frac{(mn_0)^2}{6p_0^3} \left(Q_{ijk} c_j c_k - \frac{3p_0}{mn_0} Q_{ijj} \right) c_i \right] \qquad (6.51)$$

This expression contains 20 free parameters: 5 parameters of the Maxwellian, F_0 (n_0, $\mathbf{u}_0$, and p_0), 5 elements of the traceless symmetric stress tensor, τ, and 10 elements of the symmetric heat flux matrix, Q. With the help of Maxwell's equation of change one can derive transport equations for these 20 quantities. Not surprisingly, this set of equations is usually referred to as the 20 moment approximation.

It should be emphasized again that the Chapman–Enskog form of the velocity distribution function explicitly assumes that the terms containing elements of the τ and Q matrices represent a rapidly decreasing series of contributions to the distribution function (see discussion in Section 5.4). Higher order terms are neglected, as well as any contribution containing higher order combinations of these coefficient matrices. Specifically, it is assumed that τ is second order and Q is third order in the smallness parameter of the expansion. The distribution function, F, does not contain fourth and higher order terms, and the resulting transport equations also neglect them (such as quadratic terms in the stress tensor, τ).

6.3.2 Transport equations for the Chapman–Enskog coefficients

Next we evaluate Maxwell's equation of change for the lowest velocity moments using the Chapman–Enskog distribution function given by (6.51). Let us start with the zeroth order moment equation obtained using equation (6.37) and setting $W=m$. In this case we obtain

$$\frac{\partial(mn_0)}{\partial t} + \frac{\partial(mn_0 u_{0,i})}{\partial x_i} = \frac{\delta(mn_0)}{\delta t} \tag{6.52}$$

where $\delta(mn_0)/\delta t = \Delta[m]$. This is the well known continuity equation, which expresses the conservation of mass. The three components of the first moment can be obtained by choosing the multiplier $W = mc_k$:

$$mn_0\left(\frac{\partial u_{0,k}}{\partial t} + u_{0,i}\frac{\partial u_{0,k}}{\partial x_i}\right) + \frac{\partial p_0}{\partial x_k} - \frac{\partial \tau_{ik}}{\partial x_i} - mn_0\langle a_k \rangle = mn_0\frac{\delta u_{0,k}}{\delta t} \tag{6.53}$$

where $mn_0\delta(u_{0k})/\delta t = \Delta[mc_k]$. Equation (6.53) is the equation of motion (or the non-conservative form of the momentum equation) for the gas. The second moment of the Boltzmann equation can be obtained by using the multiplier $W=mc_k c_l$. This multiplier yields six independent equations, because $c_k c_l$ is a symmetric tensor. The equations are the following:

$$\frac{\partial(\delta_{kl}p_0 - \tau_{kl})}{\partial t} + \frac{\partial(\delta_{kl}p_0 u_{0,i} - \tau_{kl}u_{0,i})}{\partial x_i} + (\delta_{il}p_0 - \tau_{il})\frac{\partial u_{0,k}}{\partial x_i} + (\delta_{ik}p_0 - \tau_{ik})\frac{\partial u_{0,l}}{\partial x_i}$$

$$+ \frac{\partial Q_{ikl}}{\partial x_i} - mn_0\langle a_k c_l + a_l c_k \rangle = \Delta[mc_k c_l]$$

$$\tag{6.54}$$

The six equations of (6.54) can be separated into an equation for the scalar pressure and five equations for the stress tensor. First we take the trace of equation (6.54) by contracting the indices k and l:

$$\frac{\partial p_0}{\partial t} + \frac{\partial(p_0 u_{0,i})}{\partial x_i} + \frac{2}{3} p_0 \frac{\partial u_{0,i}}{\partial x_i} - \frac{2}{3} \tau_{ih} \frac{\partial u_{0,h}}{\partial x_i} + \frac{2}{3} \frac{\partial h_i}{\partial x_i} = \frac{\delta p_0}{\delta t} \qquad (6.55)$$

where $\delta p_0/\delta t = \Delta[mc^2]/3$, and the heat flow vector, h_i, was defined by equation (6.20). When deriving equation (6.55) we have again applied the condition $\langle a_i c_i \rangle = 0$. The equations for the stress tensor components can be obtained by multiplying equation (6.55) by the unit tensor, δ_{kl}, and subtracting the result from equation (6.54):

$$\frac{\partial \tau_{kl}}{\partial t} + \frac{\partial(\tau_{kl} u_{0,i})}{\partial x_i} - p_0\left(\frac{\partial u_{0,k}}{\partial x_l} + \frac{\partial u_{0,l}}{\partial x_k} - \frac{2}{3}\delta_{kl} \frac{\partial u_{0,i}}{\partial x_i} \right)$$

$$+ \left(\tau_{ik}\frac{\partial u_{0,l}}{\partial x_i} + \tau_{il}\frac{\partial u_{0,k}}{\partial x_i} - \frac{2}{3}\delta_{kl}\tau_{ij}\frac{\partial u_{0,i}}{\partial x_j} \right)$$

$$- \frac{\partial}{\partial x_i}\left(Q_{ikl} - \frac{2}{3}\delta_{kl} h_i \right) + mn_0\langle a_k c_l + a_l c_k \rangle = \frac{\delta \tau_{kl}}{\delta t} \qquad (6.56)$$

where $\delta \tau_{kl}/\delta t = \delta_{kl}\delta p_0/\delta t - \Delta[mc_k c_l]$. The final equation can be obtained by substituting $W = mc_k c_l c_h$ into Maxwell's equation of change:

$$\frac{\partial Q_{klh}}{\partial t} + \frac{\partial(Q_{klh} u_{0,i})}{\partial x_i} + \left(Q_{ilh}\frac{\partial u_{0,k}}{\partial x_i} + Q_{ikh}\frac{\partial u_{0,l}}{\partial x_i} + Q_{ikl}\frac{\partial u_{0,h}}{\partial x_i} \right)$$

$$+ (p_0\delta_{lh} - \tau_{lh})\left(\frac{\partial u_{0,k}}{\partial t} + u_{0,i}\frac{\partial u_{0,k}}{\partial x_i} \right) + (p_0\delta_{kh} - \tau_{kh})\left(\frac{\partial u_{0,l}}{\partial t} + u_{0,i}\frac{\partial u_{0,l}}{\partial x_i} \right)$$

$$+ (p_0\delta_{kl} - \tau_{kl})\left(\frac{\partial u_{0,h}}{\partial t} + u_{0,i}\frac{\partial u_{0,h}}{\partial x_i} \right) + \frac{\partial R_{iklh}}{\partial x_i}$$

$$- mn_0\langle a_k c_l c_h + a_l c_k c_h + a_h c_k c_l \rangle = \frac{\delta Q_{klh}}{\delta t} \qquad (6.57)$$

where $\delta Q_{klh}/\delta t = \Delta[mc_k c_l c_h]$. One can express the convective derivative of the bulk velocity from the equation of motion (equation (6.53)) and substitute it into (6.57):

$$\frac{\partial Q_{klh}}{\partial t} + \frac{\partial\left(Q_{klh}\, u_{0,i}\right)}{\partial x_i} + \left(Q_{ilh}\frac{\partial u_{0,k}}{\partial x_i} + Q_{ikh}\frac{\partial u_{0,l}}{\partial x_i} + Q_{ikl}\frac{\partial u_{0,h}}{\partial x_i}\right)$$

$$-\frac{\left(p_0\delta_{lh}-\tau_{lh}\right)}{mn_0}\frac{\partial\left(p_0\delta_{ik}-\tau_{ik}\right)}{\partial x_i} - \frac{\left(p_0\delta_{kh}-\tau_{kh}\right)}{mn_0}\frac{\partial\left(p_0\delta_{il}-\tau_{il}\right)}{\partial x_i}$$

$$-\frac{\left(p_0\delta_{kl}-\tau_{kl}\right)}{mn_0}\frac{\partial\left(p_0\delta_{ih}-\tau_{ih}\right)}{\partial x_i} + \frac{\partial R_{iklh}}{\partial x_i} + \left\langle a_k\left[(p_0\delta_{lh}-\tau_{lh})-mn_0 c_l c_h\right]\right\rangle$$

$$+\left\langle a_l\left[(p_0\delta_{kh}-\tau_{kh})-mn_0 c_k c_h\right]\right\rangle + \left\langle a_h\left[(p_0\delta_{kl}-\tau_{kl})-mn_0 c_k c_l\right]\right\rangle$$

$$=\frac{\delta Q_{klh}}{\delta t} - (p_0\delta_{lh}-\tau_{lh})\frac{\delta u_{0,k}}{\delta t} - (p_0\delta_{kh}-\tau_{kh})\frac{\delta u_{0,l}}{\delta t} - (p_0\delta_{kl}-\tau_{kl})\frac{\delta u_{0,h}}{\delta t}$$

$$(6.58)$$

Finally, with the help of equation (6.49) one can express the divergence of the tensor R in terms of lower order velocity moments:

$$\frac{\partial R_{iklh}}{\partial x_i} = \frac{\partial}{\partial x_k}\left(\frac{p_0(p_0\delta_{lh}-\tau_{lh})}{mn_0}\right) + \frac{\partial}{\partial x_l}\left(\frac{p_0(p_0\delta_{kh}-\tau_{kh})}{mn_0}\right)$$

$$+\frac{\partial}{\partial x_h}\left(\frac{p_0(p_0\delta_{kl}-\tau_{kl})}{mn_0}\right) - \frac{\partial}{\partial x_i}\left(\frac{p_0}{mn_0}\delta_{kl}\tau_{ih}\right)$$

$$-\frac{\partial}{\partial x_i}\left(\frac{p_0}{mn_0}\delta_{kh}\tau_{il}\right) - \frac{\partial}{\partial x_i}\left(\frac{p_0}{mn_0}\delta_{lh}\tau_{ik}\right)$$

$$(6.59)$$

Equation (6.59) can be substituted into equation (6.58) to get the final form of the generalized heat flow equation:

$$\frac{\partial Q_{klh}}{\partial t} + \frac{\partial \left(Q_{klh} u_{0,i} \right)}{\partial x_i} + \left(Q_{ilh} \frac{\partial u_{0,k}}{\partial x_i} + Q_{ikh} \frac{\partial u_{0,l}}{\partial x_i} + Q_{ikl} \frac{\partial u_{0,h}}{\partial x_i} \right)$$

$$+ p_0 \left(\delta_{ik} \delta_{lh} + \delta_{il} \delta_{kh} + \delta_{ih} \delta_{kl} \right) \frac{\partial}{\partial x_i} \left(\frac{p_0}{mn_0} \right) - p_0 \frac{\partial}{\partial x_i} \left(\frac{\delta_{ik} \tau_{lh} + \delta_{il} \tau_{kh} + \delta_{ih} \tau_{kl}}{mn_0} \right)$$

$$- \left(\delta_{lh} \tau_{ik} + \delta_{kl} \tau_{ih} + \delta_{kh} \tau_{il} \right) \frac{\partial}{\partial x_i} \left(\frac{p_0}{mn_0} \right) + \left\langle a_k \left[(p_0 \delta_{lh} - \tau_{lh}) - mn_0 c_l c_h \right] \right\rangle$$

$$+ \left\langle a_l \left[(p_0 \delta_{kh} - \tau_{kh}) - mn_0 c_k c_h \right] \right\rangle + \left\langle a_h \left[(p_0 \delta_{kl} - \tau_{kl}) - mn_0 c_k c_l \right] \right\rangle$$

$$= \frac{\delta Q_{klh}}{\delta t} - (p_0 \delta_{lh} - \tau_{lh}) \frac{\delta u_{0,k}}{\delta t} - (p_0 \delta_{kh} - \tau_{kh}) \frac{\delta u_{0,l}}{\delta t} - (p_0 \delta_{kl} - \tau_{kl}) \frac{\delta u_{0,h}}{\delta t}$$

$$(6.60)$$

When deriving equation (6.60) the fourth order terms (quadratic in the components of the stress tensor) were neglected.

Equations (6.52), (6.53), and (6.55) represent transport equations for the five parameters of the Maxwellian distribution function, F_0. It should be noted that in addition to the five parameters independent of the smallness parameter, n_0, $\mathbf{u}_0$, and p_0, these equations also depend on the second order quantity, τ, and the third order quantity, $\mathbf{h}$. Equations (6.52), (6.53) and (6.54) define the evolution of the fundamental quantities describing the gas. The evolution of the five components of the second order stress tensor is described by equation (6.56), while the governing equations for the ten components of the third order heat flow tensor are given by equation (6.60). It is important to note that the equations describing the evolution of the components of the stress and heat flow tensors do contain several zeroth order terms in the smallness parameter (terms not containing elements of τ or Q). These zeroth order terms compensate each other up to second (or third) order: this high degree of compensation automatically occurs in analytic solutions, but can result in considerable difficulties in numerical solutions of the generalized transport equations.

6.3.3 Grad's closing relation and the 20 moment approximation

In this subsection we derive the 20 moment approximation using an alternative approach proposed by Grad[1] in the late 1940s. This method does not make a specific assumption for the form of the distribution function, instead it assumes a specific relation between the second and fourth order velocity moments. It will be shown that the Chapman–Enskog and Grad methods are equivalent.

We again start by substituting an increasing series of random velocity moments into Maxwell's equation of change, but this time we do not make any specific assumption about the form of the velocity distribution function. In the lowest order we again use W=m and obtain the continuity equation:

$$\frac{\partial(mn)}{\partial t} + \frac{\partial(mn\,u_i)}{\partial x_i} = \frac{\delta(mn)}{\delta t} \tag{6.61}$$

where n and **u** are related to the zeroth and first moments of the distribution function and are given by equations (6.2) and (6.3). The equation of motion can be obtained by substituting W=mc$_k$ into Maxwell's equation of change (equation (6.37)):

$$mn\left(\frac{\partial u_k}{\partial t} + u_i\frac{\partial u_k}{\partial x_i}\right) + \frac{\partial P_{ik}}{\partial x_i} - mn\langle a_k\rangle = mn\frac{\delta u_k}{\delta t} \tag{6.62}$$

where the pressure tensor, P_{ik}, is defined by (6.10). The second moment of the Boltzmann equation can be obtained by substituting W=mc$_k$c$_l$ into equation (6.37):

$$\frac{\partial P_{kl}}{\partial t} + \frac{\partial(P_{kl}u_i)}{\partial x_i} + P_{il}\frac{\partial u_k}{\partial x_i} + P_{ik}\frac{\partial u_l}{\partial x_i} + \frac{\partial Q_{ikl}}{\partial x_i} - mn\langle a_k c_l + a_l c_k\rangle = \frac{\delta P_{kl}}{\delta t} \tag{6.63}$$

where the heat flow tensor is defined by equation (6.18). The third moment of the Boltzmann equation is obtained by using the multiplier W=mc$_k$c$_l$c$_h$:

$$\frac{\partial Q_{klh}}{\partial t} + \frac{\partial(Q_{klh}u_i)}{\partial x_i} + \left(Q_{ilh}\frac{\partial u_k}{\partial x_i} + Q_{ikh}\frac{\partial u_l}{\partial x_i} + Q_{ikl}\frac{\partial u_h}{\partial x_i}\right) - \frac{P_{lh}}{mn}\frac{\partial P_{ik}}{\partial x_i}$$

$$- \frac{P_{kh}}{mn}\frac{\partial P_{il}}{\partial x_i} - \frac{P_{kl}}{mn}\frac{\partial P_{ih}}{\partial x_i} + \frac{\partial R_{iklh}}{\partial x_i} + \langle a_k[P_{lh} - mnc_l c_h]\rangle$$

$$+ \langle a_l[P_{kh} - mnc_k c_h]\rangle + \langle a_h[P_{kl} - mnc_k c_l]\rangle$$

$$= \frac{\delta Q_{klh}}{\delta t} - P_{lh}\frac{\delta u_k}{\delta t} - P_{kh}\frac{\delta u_l}{\delta t} - P_{kl}\frac{\delta u_h}{\delta t} \tag{6.64}$$

where the fourth velocity moment of the distribution function, R, is defined by equation (6.21). This series could be continued indefinitely, always introducing the divergence of higher velocity moments. The series of equations is truncated by making the following specific assumption about the relation between the components of R and lower order moments of the distribution function:[1]

$$R_{iklh} = \frac{1}{mn}\left(P_{ik}P_{lh} + P_{il}P_{kh} + P_{ih}P_{kl} - \tau_{ik}\tau_{lh} - \tau_{il}\tau_{kh} - \tau_{ih}\tau_{kl}\right) \qquad (6.65)$$

The detailed physical meaning (and all the consequences) of this assumption is not easy to visualize. Originally Grad obtained this relation by requiring that the coefficient of the fourth order term of an Hermite polynomial expansion be identically zero. The simple physical meaning of Grad's relation is that it is satisfied by relatively smooth distribution functions and it does not allow for 'spiky' distribution functions. We will see shortly that this closing relation leads to the same set of equations as the third order Chapman–Enskog distribution function given by equation (6.51).

It is interesting to note that the last three terms in expression (6.65) are of fourth order in the smallness parameter. In his book[4] Burgers neglected these terms and obtained a very simple, symmetric closing relation. This implicit simplification has only recently been called to my attention by Dr A. Barakat (private communication, 1992).

Using Grad's closing relationship one obtains the following transport equation for the heat flow tensor:

$$\frac{\partial Q_{klh}}{\partial t} + \frac{\partial(Q_{klh}u_i)}{\partial x_i} + \left(Q_{ilh}\frac{\partial u_k}{\partial x_i} + Q_{ikh}\frac{\partial u_l}{\partial x_i} + Q_{ikl}\frac{\partial u_h}{\partial x_i}\right)$$

$$+ p\left(\delta_{ik}\delta_{lh} + \delta_{il}\delta_{kh} + \delta_{ih}\delta_{kl}\right)\frac{\partial}{\partial x_i}\left(\frac{p}{mn}\right) - p\frac{\partial}{\partial x_i}\left(\frac{\delta_{ik}\tau_{lh} + \delta_{il}\tau_{kh} + \delta_{ih}\tau_{kl}}{mn}\right)$$

$$- \tau_{lh}\frac{\partial}{\partial x_i}\left(\frac{\tau_{ik}}{mn}\right) - \tau_{kh}\frac{\partial}{\partial x_i}\left(\frac{\tau_{il}}{mn}\right) + \tau_{kl}\frac{\partial}{\partial x_i}\left(\frac{\tau_{ih}}{mn}\right)$$

$$- \left(\delta_{lh}\tau_{ik} + \delta_{kl}\tau_{ih} + \delta_{kh}\tau_{il}\right)\frac{\partial}{\partial x_i}\left(\frac{p}{mn}\right) + \left\langle a_k\left[P_{lh} - mnc_lc_h\right]\right\rangle + \left\langle a_l\left[P_{kh} - mnc_kc_h\right]\right\rangle$$

$$+ \left\langle a_h\left[P_{kl} - mnc_kc_l\right]\right\rangle = \frac{\delta Q_{klh}}{\delta t} - P_{lh}\frac{\delta u_k}{\delta t} - P_{kh}\frac{\delta u_l}{\delta t} - P_{kl}\frac{\delta u_h}{\delta t}$$

$$(6.66)$$

It should be kept in mind that when deriving equations (6.61), (6.62), (6.63) and (6.65) we did not explicitly assume that the stress and heat flow tensors are second and third order quantities in a smallness parameter. In this closed set of transport equations the only assumption is Grad's closing relation, given by equation (6.65).

We have just derived two seemingly different sets of governing equations for the velocity moments of the distribution function. One set of equations ((6.52), (6.53), (6.55), (6.56), and (6.60)) was obtained with the Chapman–Enskog method assuming that the elements of the matrices τ and Q are of second and third order in the smallness parameter of the expansion. The other set of equations ((6.61), (6.62), (6.63), and (6.66)) was obtained using Grad's closure relation and they do not contain explicit assumptions for the form of the velocity distribution function. It should be immediately noted that the continuity equations, (equations (6.52) and (6.61)) are identical, and that the two equations of motion (equations (6.53) and (6.62)) become identical if one uses the decomposition of the pressure tensor given by equation (6.14). The apparent difference is in the energy and heat flow equations.

The two sets of energy and heat flow equations are also identical to third order accuracy (neglecting terms containing τ^2). This can be seen by comparing equations (6.54) and (6.63), and (6.60) and (6.66), respectively. It can easily be seen that these equations become the same if we again use the decomposition of the pressure tensor, P (equation (6.14)). On the other hand we recall that only equation (6.54) was used to obtain the governing equations for the scalar pressure, p, and the stress tensor, τ, therefore we conclude that equations (6.55) and (6.56) are equivalent to equation (6.63).

The conclusion is that the Chapman–Enskog and the Grad forms of the 20 moment equations are identical to third order accuracy. The equivalence of the two approaches was not obvious for quite some time and this fact had resulted in some confusion in generalized transport theory. Today the two approaches are both used: in a particular application it might be more convenient to use one form or the other, but we understand that the result is independent of the method.

6.4 The 13 moment approximation and the Navier–Stokes equations

The 20 moment approximation resulted in a set of 20 partial differential equations for 20 macroscopic quantities, each of which depends on time and spatial location. In other words we traded a single seven dimensional partial differential equation for 20 four dimensional equations. In this trade we also gave up detailed knowledge of the velocity distribution function for 'simplicity'. However, one can hardly call a set of 20 partial differential equations 'simple'. Additional simplification is desirable, and in most physical cases achievable. The most obvious simplification leads to the 13 moment approximation and to the well known Navier–Stokes equation.

6.4.1 The 13 moment equations

In the 20 moment approximation 10 out of the 20 governing equations describe the evolution of various heat flow components. We showed that these quantities represent only small corrections to the velocity distribution function. In most physical problems we do not need the detailed knowledge of all the heat flow components: it is usually sufficient to know the evolution of the heat flow vector, **h**. It is obvious from the definition of the heat flow vector (equation (6.20)) that it describes the transport of the total translational random energy of the particles due to the random motion of the gas molecules. If we are not interested in the transport of the energy components we can reduce the number of independent functions (and therefore the number of governing equations) by seven.

Specifically, it is assumed that the elements of the heat flow tensor, Q, can be approximated by the following expression:

$$Q_{ijk} = \frac{2}{5}\left(\delta_{ij}h_k + \delta_{ik}h_j + \delta_{jk}h_i\right) \tag{6.67}$$

Note that the matrix elements of the tensor, Q, vanish unless at least two indices are the same. This means that 'cross terms' are neglected in this approximation. Equation (6.67) indicated that the heat flow tensor is composed of only three independent physical quantities, therefore the number of unknown functions is reduced by seven (the general form of Q had 10 independent matrix elements). The total number of independent quantities reduces to 13: density, bulk velocity vector, scalar pressure, stress tensor (which is symmetric and traceless, therefore has five independent components) and heat flux vector. Substituting equation (6.67) into the 20 moment system of equations ((6.52), (6.53), (6.55), (6.56), and (6.60)) leads to the following set of 13 transport equations:

$$\frac{\partial\left(mn_0\right)}{\partial t} + \frac{\partial\left(mn_0\, u_{0,i}\right)}{\partial x_i} = \frac{\delta(mn_0)}{\delta t} \tag{6.68a}$$

$$mn_0\left(\frac{\partial u_{0,k}}{\partial t} + u_{0,i}\frac{\partial u_{0,k}}{\partial x_i}\right) + \frac{\partial p_0}{\partial x_k} - \frac{\partial \tau_{ik}}{\partial x_i} - mn_0\langle a_k\rangle = mn_0\frac{\delta u_{0,k}}{\delta t} \tag{6.68b}$$

$$\frac{\partial p_0}{\partial t} + \frac{\partial\left(p_0 u_{0,i}\right)}{\partial x_i} + \frac{2}{3}p_0\frac{\partial u_{0,i}}{\partial x_i} - \frac{2}{3}\tau_{ih}\frac{\partial u_{0,h}}{\partial x_i} + \frac{2}{3}\frac{\partial h_i}{\partial x_i} = \frac{\delta p_0}{\delta t} \tag{6.68c}$$

$$\frac{\partial \tau_{kl}}{\partial t} + \frac{\partial \left(\tau_{kl} u_{0,i} \right)}{\partial x_i} + \tau_{ik} \frac{\partial u_{0,l}}{\partial x_i} + \tau_{il} \frac{\partial u_{0,k}}{\partial x_i} - \frac{2}{3} \delta_{kl} \tau_{ij} \frac{\partial u_{0,i}}{\partial x_j}$$

$$- p_0 \left(\frac{\partial u_{0,k}}{\partial x_l} + \frac{\partial u_{0,l}}{\partial x_k} - \frac{2}{3} \delta_{kl} \frac{\partial u_{0,i}}{\partial x_i} \right) - \left(\frac{2}{5} \frac{\partial h_l}{\partial x_k} + \frac{2}{5} \frac{\partial h_k}{\partial x_l} - \frac{4}{15} \delta_{kl} \frac{\partial h_i}{\partial x_i} \right)$$

$$+ mn_0 \langle a_k c_l + a_l c_k \rangle = \frac{\delta \tau_{kl}}{\delta t} \tag{6.68d}$$

$$\frac{\partial h_k}{\partial t} + u_{0,i} \frac{\partial h_k}{\partial x_i} + \frac{7}{5} h_k \frac{\partial u_{0,i}}{\partial x_i} + \frac{2}{5} h_i \frac{\partial u_{0,i}}{\partial x_k} + \frac{7}{5} h_i \frac{\partial u_{0,k}}{\partial x_i} + \frac{5}{2} p_0 \frac{\partial}{\partial x_k} \left(\frac{p_0}{mn_0} \right)$$

$$- \frac{5}{2} \tau_{ik} \frac{\partial}{\partial x_i} \left(\frac{p_0}{mn_0} \right) - p_0 \frac{\partial}{\partial x_i} \left(\frac{\tau_{ik}}{mn_0} \right) - \tau_{kl} \frac{\partial}{\partial x_i} \left(\frac{\tau_{il}}{mn_0} \right) + \frac{5}{2} p_0 \langle a_k \rangle - \frac{1}{2} mn_0 \langle a_k c^2 \rangle$$

$$- \tau_{ki} \langle a_i \rangle - mn_0 \langle a_i c_i c_k \rangle = \frac{\delta h_k}{\delta t} - \frac{5}{2} p_0 \frac{\delta u_{0,k}}{\delta t} + \tau_{kh} \frac{\delta u_{0,h}}{\delta t} \tag{6.68e}$$

These equations constitute the governing equations of the 13 moment approximation.

6.4.2 Collision terms for a single species gas

In the 13 moment approximation the Chapman–Enskog distribution function can be written in the following form:

$$F = F_0 \left[1 - \frac{mn_0}{2p_0^2} \tau_{ij} c_i c_j + \frac{(mn_0)^2}{p_0^3} \left(\frac{1}{5} c^2 - \frac{p_0}{mn_0} \right) h_i c_i \right] \tag{6.69}$$

We will use this form of the distribution function to evaluate the collision terms for a single species gas. For the sake of mathematical simplicity we will use the relaxation time approximation in this derivation. Later in this chapter a more accurate (and more complicated) approach will be used to calculate the collision terms for multispecies gases.

In the relaxation time approximation (equation (5.52)) the collision term for the Chapman–Enskog distribution function (6.69) can be written in the following form:

$$\frac{\delta F}{\delta t} = -\frac{F - F_0}{\tau_0} = \frac{F_0}{\tau_0}\left[\frac{mn_0}{2p_0^2}\tau_{ij}c_ic_j - \frac{(mn_0)^2}{p_0^3}\left(\frac{1}{5}c^2 - \frac{p_0}{mn_0}\right)h_ic_i\right] \quad (6.70)$$

where τ_0 is the formally defined relaxation time. Note that the distribution function, F, locally relaxes to F_0, the equilibrium distribution function. The right hand side of Maxwell's equation of change is the appropriate velocity moment of the collision term, therefore one can write

$$\Delta[W] = \iiint_\infty d^3c\, W\frac{F_0}{\tau_0}\left[\frac{mn_0}{2p_0^2}\tau_{ij}c_ic_j - \frac{(mn_0)^2}{p_0^3}\left(\frac{1}{5}c^2 - \frac{p_0}{mn_0}\right)h_ic_i\right] \quad (6.71)$$

Next we evaluate equation (6.71) for the 13 moment approximation. The simplest case is when we take $W=m$. Substituting this into equation (6.71) immediately yields the collision term for the particle mass density:

$$\frac{\delta(mn_0)}{\delta t} = \iiint_\infty d^3c\,\frac{mF_0}{\tau_0}\left[\frac{mn_0}{2p_0^2}\tau_{ij}c_ic_j - \frac{(mn_0)^2}{p_0^3}\left(\frac{1}{5}c^2 - \frac{p_0}{mn_0}\right)h_ic_i\right] = 0 \ (6.72)$$

Here we have used the fact that $\int d^3c\tau_{ij}c_ic_jF_0=0$ since τ_{ij} is traceless ($\tau_{ii}=0$). In the equation of motion the collision term is $\Delta[mc_k]$. In this case the relaxation time approximation of the collision term leads to the following result:

$$mn_0\frac{\delta u_k}{\delta t} = \iiint_\infty d^3c\,\frac{mF_0}{\tau_0}\left[\frac{mn_0}{2p_0^2}\tau_{ij}c_ic_jc_k - \frac{(mn_0)^2}{p_0^3}\left(\frac{1}{5}c^2 - \frac{p_0}{mn_0}\right)h_ic_ic_k\right]$$

$$= -\iiint_\infty d^3c\,\frac{mF_0}{\tau_0}\left[\frac{(mn_0)^2}{p_0^3}\left(\frac{1}{5}c^2 - \frac{p_0}{mn_0}\right)h_ic_ic_k\right] = 0$$

$$\quad (6.73)$$

Note that the final result is achieved by direct integration.

The collision term for the scalar pressure equation can be evaluated with $W=mc^2/3$:

$$\frac{\delta p_0}{\delta t} = \iiint_\infty d^3c\,\frac{1}{2}m\frac{F_0}{\tau_0}\left[\frac{mn_0}{2p_0^2}\tau_{ij}c^2c_ic_j - \frac{(mn_0)^2}{p_0^3}\left(\frac{1}{5}c^4 - \frac{p_0}{mn_0}c^2\right)h_ic_i\right] = 0$$

$$\quad (6.74)$$

where $\tau_{ii}=0$ was used again. Equations (6.72) through (6.74) express the conservation of mass, momentum and energy during the collisional process.

The collision term for the stress tensor can be obtained by taking the multiplier $W=mc_kc_h$:

$$\frac{\delta\tau_{kh}}{\delta t} = -\iiint_{\infty} d^3c\, m \frac{F_0}{\tau_0}\left[\frac{mn_0}{2p_0^2}\tau_{ij}c_ic_jc_kc_h - \frac{(mn_0)^2}{p_0^3}\left(\frac{1}{5}c^2 - \frac{p_0}{mn_0}\right)h_ic_ic_kc_h\right]$$

$$= -\frac{m}{\tau_0}\frac{mn_0}{2p_0^2}\tau_{ij}\iiint_{\infty} d^3c\, c_ic_jc_kc_h F_0 = -\frac{\tau_{kh}}{\tau_0}$$

(6.75)

Note that for τ_{kh} only non-diagonal terms of the tensor are considered, therefore we have explicitly assumed that $k\neq h$. Finally, the collision term of the heat flow vector is

$$\frac{\delta h_k}{\delta t} = \iiint_{\infty} d^3c\, \frac{1}{2}m \frac{F_0}{\tau_0}\left[\frac{mn_0}{2p_0^2}\tau_{ij}c^2c_kc_ic_j - \frac{(mn_0)^2}{p_0^3}\left(\frac{1}{5}c^4 - \frac{p_0}{mn_0}c^2\right)h_ic_ic_k\right]$$

$$= -\frac{1}{2\tau_0}m\frac{(mn_0)^2}{p_0^3}\iiint_{\infty} d^3c\left(\frac{1}{5}c^4 - \frac{p_0}{mn_0}c^2\right)h_ic_ic_k F_0 = -\frac{h_k}{\tau_0}$$

(6.76)

Substituting these collision terms into the 13 moment equations yield the following system of transport equations for single species gases:

$$\frac{\partial(mn_0)}{\partial t} + \frac{\partial(mn_0\,u_{0,i})}{\partial x_i} = 0$$

(6.77a)

$$mn_0\left(\frac{\partial u_{0,k}}{\partial t} + u_{0,i}\frac{\partial u_{0,k}}{\partial x_i}\right) + \frac{\partial p_0}{\partial x_k} - \frac{\partial\tau_{ik}}{\partial x_i} - mn_0\langle a_k\rangle = 0$$

(6.77b)

$$\frac{\partial p_0}{\partial t} + \frac{\partial(p_0 u_{0,i})}{\partial x_i} + \frac{2}{3}p_0\frac{\partial u_{0,i}}{\partial x_i} - \frac{2}{3}\tau_{ih}\frac{\partial u_{0,h}}{\partial x_i} + \frac{2}{3}\frac{\partial h_i}{\partial x_i} = 0$$

(6.77c)

$$\frac{\partial \tau_{kl}}{\partial t} + \frac{\partial\left(\tau_{kl}u_{0,i}\right)}{\partial x_i} + \tau_{ik}\frac{\partial u_{0,l}}{\partial x_i} + \tau_{il}\frac{\partial u_{0,k}}{\partial x_i} - \frac{2}{3}\delta_{kl}\tau_{ij}\frac{\partial u_{0,i}}{\partial x_j}$$

$$-p_0\left(\frac{\partial u_{0,k}}{\partial x_l} + \frac{\partial u_{0,l}}{\partial x_k} - \frac{2}{3}\delta_{kl}\frac{\partial u_{0,i}}{\partial x_i}\right) - \left(\frac{2}{5}\frac{\partial h_l}{\partial x_k} + \frac{2}{5}\frac{\partial h_k}{\partial x_l} - \frac{4}{15}\delta_{kl}\frac{\partial h_i}{\partial x_i}\right)$$

$$+ mn_0\left\langle a_k c_l + a_l c_k\right\rangle = -\frac{\tau_{kl}}{\tau_0} \tag{6.77d}$$

$$\frac{\partial h_k}{\partial t} + u_{0,i}\frac{\partial h_k}{\partial x_i} + \frac{7}{5}h_k\frac{\partial u_{0,i}}{\partial x_i} + \frac{2}{5}h_i\frac{\partial u_{0,i}}{\partial x_k} + \frac{7}{5}h_i\frac{\partial u_{0,k}}{\partial x_i} + \frac{5}{2}p_0\frac{\partial}{\partial x_k}\left(\frac{p_0}{mn_0}\right)$$

$$-\frac{5}{2}\tau_{ik}\frac{\partial}{\partial x_i}\left(\frac{p_0}{mn_0}\right) - p_0\frac{\partial}{\partial x_i}\left(\frac{\tau_{ik}}{mn_0}\right) - \tau_{kl}\frac{\partial}{\partial x_i}\left(\frac{\tau_{il}}{mn_0}\right) + \frac{5}{2}p_0\left\langle a_k\right\rangle - \frac{1}{2}mn_0\left\langle a_k c^2\right\rangle$$

$$-\tau_{ki}\left\langle a_i\right\rangle - mn_0\left\langle a_i c_i c_k\right\rangle = -\frac{h_k}{\tau_0}$$

$$\tag{6.77e}$$

This is a full set of transport equations for the density, bulk flow velocity vector, scalar pressure, stress tensor and heat flow vector.

6.4.3 The Navier–Stokes equation

Equations (6.77) represent a set of 13 simplified equations for the macroscopic flow parameters of a single species gas. However, it is still quite complicated to solve these equations and further physical simplification can be achieved. This physical simplification is based on the relative importance of the coefficients in the Chapman–Enskog form of the distribution function (6.69). It was discussed before how n_0, u_0 and p_0 represent zeroth order quantities in the smallness parameter, while the components of the stress tensor are of second order, and the components of the heat flow vector are of third order. This means that terms containing elements of τ and **h** are usually negligibly small, except when they are multiplied by very large factors. In equations (6.77) this large factor is $1/\tau_0$, the inverse of the relaxation time (or collision frequency) of the molecules.

It has been discussed before how the Chapman–Enskog method can only be applied to gases which are not far from equilibrium. This means that very frequent molecular collisions must nearly eradicate all deviations from equilibrium so that the resulting velocity distribution could satisfy the basic condition of the

Chapman–Enskog solution. In other words the relaxation time (or the mean free time between collisions) must be extremely short, and therefore the coefficient, $1/\tau_0$, must be a very large factor.

Based on this argument we apply the following approximation to the stress and heat flow equations, (6.77d) and (6.77e). All terms containing the stress tensor and the heat flow vector are neglected except the right hand sides of the equations which contain τ_{kl}/τ_0, and h_k/τ_0, respectively. This way we only neglect terms which give small corrections to small quantities. For the sake of mathematical simplicity we also neglect all external forces, i.e. $\mathbf{a}=0$. In this limit the last two equations become the following:

$$\tau_{kl} = \tau_0 p_0 \left(\frac{\partial u_{0,k}}{\partial x_l} + \frac{\partial u_{0,l}}{\partial x_k} - \frac{2}{3} \delta_{kl} \frac{\partial u_{0,i}}{\partial x_i} \right) \tag{6.78}$$

$$h_k = -\frac{5}{2} \tau_0 p_0 \frac{\partial}{\partial x_k} \left(\frac{p_0}{m n_0} \right) = -\frac{5}{2} \frac{k}{m} \tau_0 p_0 \frac{\partial T_0}{\partial x_k} \tag{6.79}$$

This is a fascinating result. The stress and heat flow equations have suddenly simplified to something quite familiar: the stress is proportional to the velocity gradient and the heat flux is proportional to the temperature gradient. These are exactly the well known relations for regular fluids at normal densities. We know that the coefficient between the velocity gradient and the stress tensor is the coefficient of viscosity, η,

$$\eta = \tau_0 p_0 \tag{6.80}$$

while the coefficient between the negative temperature gradient and the heat flux vector is the thermal conductivity, κ:

$$\kappa = \frac{5}{2} \frac{k}{m} \tau_0 \, p_0 \tag{6.81}$$

Equations (6.78) and (6.79) make a small but important contribution to the continuity, momentum and energy equations. We do not neglect these corrections because they represent small contributions to zeroth order quantities in the smallness parameter. Substituting these expressions into equation (6.77a) through (6.77c) we obtain the well known Navier–Stokes equations:

$$\frac{\partial(mn_0)}{\partial t} + \frac{\partial(mn_0 u_{0,i})}{\partial x_i} = 0 \qquad (6.82a)$$

$$mn_0 \left(\frac{\partial u_{0,k}}{\partial t} + u_{0,i} \frac{\partial u_{0,k}}{\partial x_i} \right) + \frac{\partial p_0}{\partial x_k} = \frac{\partial}{\partial x_i} \left[\eta \left(\frac{\partial u_{0,k}}{\partial x_i} + \frac{\partial u_{0,i}}{\partial x_k} - \frac{2}{3} \delta_{ik} \frac{\partial u_{0,j}}{\partial x_j} \right) \right]$$

$$(6.82b)$$

$$\frac{3}{2} \left(\frac{\partial p_0}{\partial t} + u_{0,i} \frac{\partial p_0}{\partial x_i} \right) + \frac{5}{2} p_0 \frac{\partial u_{0,i}}{\partial x_i} = \frac{\partial}{\partial x_i} \left(\kappa \frac{\partial T_0}{\partial x_i} \right)$$

$$+ \eta \left(\frac{\partial u_{0,k}}{\partial x_l} + \frac{\partial u_{0,l}}{\partial x_k} - \frac{2}{3} \delta_{kl} \frac{\partial u_{0,i}}{\partial x_i} \right) \frac{\partial u_{0,k}}{\partial x_l} \qquad (6.82c)$$

The Navier–Stokes equations represent a more realistic description of the macroscopic gas parameters than the simple Euler equations. The reason is that in the derivation of the Euler equations we assumed strict local thermodynamic equilibrium, while the Navier–Stokes equations do allow for small deviations from LTE. These deviations are accounted for by the viscous and the heat conduction terms.

6.5 Collision terms for multispecies gases

In this section we examine the moments of the Boltzmann collision integral for multispecies gases (see equation (5.28)). We will concentrate on the 13 moment approximation, but the results are directly applicable to other approximations as well.

The general moment of the Boltzmann collision term can be written in the following form:

$$\frac{\delta W_s}{\delta t} = \sum_t \iiint_\infty d^3 c_s \iiint_\infty d^3 c_t \int_0^{2\pi} d\varepsilon \int_0^\pi d\chi \, W_s(\mathbf{c}_s) \sin\chi \, S(g,\chi) g \left[F_s{'} F_t{'} - F_s F_t \right]$$

$$(6.83)$$

where $W_s(\mathbf{c}_s)$ is a molecular quantity of the species 's'. Next we repeat a trick used when proving the H-theorem and when deriving Maxwell's equation of change.

The point is that one can interchange primed and unprimed quantities in the multiple integral. Using this trick expression (6.83) can be written as

$$\frac{\delta W_s}{\delta t} = \sum_t \iiint_\infty d^3 c_s \iiint_\infty d^3 c_t \int_0^{2\pi} d\varepsilon \int_0^\pi d\chi (W_s' - W_s) \sin \chi S(g, \chi) g F_s F_t \quad (6.84)$$

In the 13 moment approximation one substitutes the molecular quantities, $W_s = m_s$, $m_s c_{si}$, $m_s c_{si} c_{sj}$, and $m_s c_s^2 c_{si}/2$, respectively. In the simplest case, when $W_s = m_s$, the particle mass is conserved in the collision, therefore

$$\Delta m_s = m_s' - m_s = 0 \quad (6.85)$$

This immediately means that the zeroth moment of the Boltzmann collision integral vanishes:

$$\frac{\delta(m_s n_s)}{\delta t} = 0 \quad (6.86)$$

6.5.1 Collision term for the momentum equation

The momentum exchange term of the momentum equation is much more complicated. In this case the change of the quantity, W_s, can be written as

$$\Delta(m_s c_{si}) = m_s (c_{si}' - c_{si}) \quad (6.87)$$

On the other hand we know that $c_s' - c = v_s' - v_s$ and therefore one can use equation (5.74a) to write expression (6.87) in the following form:

$$\Delta(m_s c_{si}) = m_{st} (g_i' - g_i) \quad (6.88)$$

It has to be recognized that in the integrand of expression (6.84) the only term which depends on the impact azimuth, ε, is $\Delta(m_s c_{si})$. The integral over ε can be evaluated with the help of equation (5.79):

$$\int_0^{2\pi} d\varepsilon (W_s' - W_s) = m_{st} \int_0^{2\pi} d\varepsilon (g_i' - g_i) = -2\pi m_{st} g_i (1 - \cos \chi) \quad (6.89)$$

Substituting this expression into equation (6.84) yields the following:

$$m_s n_s \frac{\delta u_{si}}{\delta t} = -2\pi \sum_t m_{st} \iiint_\infty d^3c_s \iiint_\infty d^3c_t \int_0^\pi d\chi (1 - \cos\chi) \sin\chi \, S(g, \chi) g_i \, g \, F_s F_t$$

(6.90)

Finally, one can use the definition of the momentum transfer cross section (given by equation (3.107)) to obtain a general form of the collision term:

$$m_s n_s \frac{\delta u_{si}}{\delta t} = -\sum_t m_{st} \iiint_\infty d^3c_s \iiint_\infty d^3c_t \, \sigma_1(g) g_i \, g \, F_s F_t$$

(6.91)

For general inverse power interactions the momentum transfer cross section is given by equation (3.109). Substituting this expression into equation (6.91) we obtain the following:

$$m_s n_s \frac{\delta u_{si}}{\delta t} = -\sum_t 2\pi m_{st} \left(\frac{K_a}{m_{st}} \right)^{\frac{2}{a-1}} A_1(a) \iiint_\infty d^3c_s \iiint_\infty d^3c_t \, g_i \, g^{-\frac{a-5}{a-1}} F_s F_t$$

(6.92)

where a is the spectral index of the intermolecular force and $A_1(a)$ is a pure number given in Table 3.1. In general it is very difficult to evaluate the six remaining integrals in expression (6.92), due to the fractional power of the magnitude of the relative velocity, g. However, in the case of Maxwell molecular interaction, when a=5, the integrals can be readily evaluated. In order to demonstrate the fundamental physics and retain mathematical simplicity we consider the case of Maxwell molecules. It should be noted, however, that Maxwell molecules do not exist in reality, so our results should be treated with caution. At the same time these results represent a physically meaningful, good approximation to a wide range of molecular collisions (when the particles are not electrically charged).

For Maxwell molecules a=5 and equation (6.92) simplifies to the following:

$$m_s n_s \frac{\delta u_{si}}{\delta t} = -\sum_t 2\pi \left(\frac{K_5}{m_{st}} \right)^{\frac{1}{2}} A_1(5) m_{st} \iiint_\infty d^3c_s \iiint_\infty d^3c_t \, g_i \, F_s F_t$$

(6.93)

On the other hand we know that $g_i = v_{si} - v_{ti} = c_{si} + u_{si} - c_{ti} - u_{ti}$, therefore one obtains

$$m_s n_s \frac{\delta u_{si}}{\delta t} = -\sum_t 2\pi \left(\frac{K_5}{m_{st}}\right)^{\frac{1}{2}} A_1(5) m_{st} \iiint_\infty d^3 c_s \iiint_\infty d^3 c_t \, (c_{si} - c_{ti}) F_s F_t$$

$$-\sum_t 2\pi \left(\frac{K_5}{m_{st}}\right)^{\frac{1}{2}} A_1(5) m_{st} (u_{si} - u_{ti}) \iiint_\infty d^3 c_s \iiint_\infty d^3 c_t \, F_s F_t \qquad (6.94)$$

The first term in equation (6.94) vanishes because the averages of both c_{si} and c_{ti} are zero. The remaining term can be readily evaluated to obtain

$$m_s n_s \frac{\delta u_{si}}{\delta t} = -\sum_t 2\pi \left(\frac{K_5}{m_{st}}\right)^{\frac{1}{2}} A_1(5) n_s n_t m_{st} (u_{si} - u_{ti}) \qquad (6.95)$$

Finally one can introduce the momentum transfer collision frequency between species 's' and 't', v_{st},

$$v_{st} = 2\pi \left(\frac{K_5}{m_{st}}\right)^{\frac{1}{2}} A_1(5) \frac{m_{st}}{m_s} n_t \qquad (6.96)$$

Using this definition we obtain the final form of the collision term for the momentum equation:

$$m_s n_s \frac{\delta u_{si}}{\delta t} = \sum_t m_s n_s v_{st} (u_{ti} - u_{si}) \qquad (6.97)$$

Equation (6.97) is a truly fascinating result. The physical meaning is that collisions between the molecules of the various gas components create a macroscopic friction which tries to decrease the velocity difference between the species.

6.5.2 Collision term for the pressure tensor

In the case of the energy equation the molecular quantity, W_s, is chosen to be $W_s = m_s c_{si} c_{sj}$. With the help of equation (5.74a) the change of this quantity due to a collision can be written as

$$\Delta(m_s c_{si} c_{sj}) = m_{st} \left[c_{si}(g_j' - g_j) + c_{sj}(g_i' - g_i) \right] + m_s \left(\frac{m_{st}}{m_s} \right)^2 (g_i' - g_i)(g_j' - g_j)$$

$$(6.98)$$

The integral of this quantity over the impact azimuth angle, ε, can be written in the following form:

$$\int_0^{2\pi} d\varepsilon \, \Delta(m_s c_{si} c_{sj}) = \pi \, m_{st} \left(\frac{m_{st}}{m_s} \right) (1 - \cos^2 \chi)(g^2 \delta_{ij} - 3g_i g_j)$$

$$-2\pi \, m_{st} \left(c_{si} g_j + g_i c_{sj} - 2 \frac{m_{st}}{m_s} g_i g_j \right) (1 - \cos\chi) \qquad (6.99)$$

When deriving equation (6.99) we made use of equations (5.79) and (5.80). In the next step one can substitute equation (6.99) into the collision integral, (6.84). Using the definition of the higher order transport cross sections given by equation (3.107) we obtain the following integral for the collision term of the pressure tensor:

$$\frac{\delta P_{sij}}{\delta t} = \sum_t \frac{1}{2} m_{st} \left(\frac{m_{st}}{m_s} \right) \iiint_\infty d^3 c_s \iiint_\infty d^3 c_t \, \sigma_2(g) g (g^2 \delta_{ij} - 3g_i g_j) F_s F_t$$

$$-\sum_t m_{st} \iiint_\infty d^3 c_s \iiint_\infty d^3 c_t \, \sigma_1(g) g \left(c_{si} g_j + g_i c_{sj} - 2 \frac{m_{st}}{m_s} g_i g_j \right) F_s F_t \quad (6.100)$$

For the sake of mathematical simplicity we again consider an inverse power interaction between the molecules with a spectral index of a=5 (Maxwell molecules). For Maxwell molecular collisions this expression greatly simplifies and one obtains:

$$\frac{\delta P_{sij}}{\delta t} = \sum_t \frac{\nu_{st}}{n_t} \frac{1}{2} m_{st} \frac{A_2(5)}{A_1(5)} \iiint_\infty d^3 c_s \iiint_\infty d^3 c_t (g^2 \delta_{ij} - 3g_i g_j) F_s F_t$$

$$-\sum_t \frac{\nu_{st}}{n_t} m_s \iiint_\infty d^3 c_s \iiint_\infty d^3 c_t \left(c_{si} g_j + g_i c_{sj} - 2 \frac{m_{st}}{m_s} g_i g_j \right) F_s F_t \qquad (6.101)$$

where we have used the definition of the momentum transfer collision frequency given by equation (6.96). The next step is to evaluate the velocity integrals. We again use the fact that $g_i = (u_{si} - u_{ti}) + (c_{si} - c_{ti})$ and obtain the following:

$$\frac{\delta P_{sij}}{\delta t} = \sum_t \frac{3 v_{st} m_{st}}{2 m_s} \frac{A_2(5)}{A_1(5)}$$

$$\left[m_s n_s \left(\frac{\delta_{ij}}{3} (u_s - u_t)^2 - (u_{si} - u_{ti})(u_{sj} - u_{tj}) \right) + \tau_{sij} + \frac{m_s n_s}{m_t n_t} \tau_{tij} \right]$$

$$+ \sum_t \frac{2 v_{st} m_s}{m_s + m_t} \left(\frac{n_s}{n_t} P_{tij} - P_{sij} + m_t n_s (u_{si} - u_{ti})(u_{sj} - u_{tj}) \right) \qquad (6.102)$$

This collision term can be readily decomposed into two expressions: one describing the effects of collisions on the scalar pressure, while the other accounts for collisional effects on the stress tensor:

$$\frac{\delta p_s}{\delta t} = \sum_t \frac{2 v_{st} m_s n_s}{m_s + m_t} \left(k(T_t - T_s) + \frac{m_t}{3} (u_s - u_t)^2 \right) \qquad (6.103a)$$

$$\frac{\delta \tau_{sij}}{\delta t} = \sum_t \frac{v_{st} m_s}{m_s + m_t} \left(2 - \frac{3 m_t}{2 m_s} \frac{A_2(5)}{A_1(5)} \right)$$

$$\left[m_s n_s \left(\frac{\delta_{ij}}{3} (u_s - u_t)^2 - (u_{si} - u_{ti})(u_{sj} - u_{tj}) \right) \right]$$

$$+ \sum_t \frac{v_{st} m_s}{m_s + m_t} \left[\left(2 - \frac{3}{2} \frac{A_2(5)}{A_1(5)} \right) \frac{n_s}{n_t} \tau_{tij} - \left(2 + \frac{3 m_t}{2 m_s} \frac{A_2(5)}{A_1(5)} \right) \tau_{sij} \right] \qquad (6.103b)$$

This result tells us that collisions affect the pressure in two ways. First of all collisions try to eliminate temperature differences between the various gas components. This effect is described by the first term in equation (6.103a). Second, the energy of the relative bulk motion of the various components which is dissipated via friction (see equation (6.97)) is heating the gas components. This effect is accounted for by the second term in equation (6.103a).

6.5.3 Collision term for the heat flow vector

In the case of the heat flow vector the molecular quantity to be considered is

$W_s = m_s c_s^2 c_{si}/2$. The change of this quantity as a result of a collision between an 's' and a 't' particle can be written in the following form:

$$\Delta\left(\frac{1}{2}m_s c_s^2 c_{si}\right) = m_{st} c_{sk} c_{si}(g_k{}'-g_k) + \frac{1}{2}m_{st} c_s^2(g_i{}'-g_i)$$

$$+\frac{1}{2}m_{st}\frac{m_{st}}{m_s}c_{si}(g_k{}'-g_k)(g_k{}'-g_k) + m_{st}\frac{m_{st}}{m_s}c_{sk}(g_k{}'-g_k)(g_i{}'-g_i)$$

$$+\frac{1}{2}m_{st}\left(\frac{m_{st}}{m_s}\right)^2(g_k{}'-g_k)(g_k{}'-g_k)(g_i{}'-g_i) \tag{6.104}$$

This expression can be substituted into equation (6.84) to obtain the collision term for the heat flow. The derivation is tedious but straightforward: one has to evaluate a large number of integrals similar to those we evaluated in Subsections 6.5.1 and 6.5.2. Without boring the reader with unnecessary details (students should be encouraged to carry out the detailed calculation) the collision term for Maxwell molecules is given by the following expression:[5]

$$\frac{\delta h_{si}}{\delta t} = -\sum_t v_{st}\left\{\alpha_{st}^{(1)} h_{si} - \frac{m_s n_s}{m_t n_t}\alpha_{st}^{(4)} h_{ti} + (u_{si}-u_{ti})\left(\frac{1}{2}\alpha_{st}^{(2)} P_{sij} + \frac{m_s n_s}{m_t n_t}\alpha_{st}^{(4)} P_{tij}\right)\right.$$

$$\left. +\frac{1}{2}(u_{si}-u_{ti})\left(3\alpha_{st}^{(3)} p_s + 3\alpha_{st}^{(4)}\frac{m_s n_s}{m_t n_t}p_t + m_s n_s \alpha_{st}^{(4)}(u_s-u_t)^2\right)\right\}$$

$$\tag{6.105}$$

where

$$\alpha_{st}^{(1)} = \frac{1}{(m_s+m_t)^2}\left(3m_s^2 + m_t^2 + 2m_s m_t\frac{A_2(5)}{A_1(5)}\right) \tag{6.106a}$$

$$\alpha_{st}^{(2)} = \frac{1}{(m_s+m_t)^2}\left(2(m_s-m_t)^2 + m_t(m_s-3m_t)\frac{A_2(5)}{A_1(5)}\right) \tag{6.106b}$$

$$\alpha_{st}^{(3)} = \frac{1}{(m_s+m_t)^2}\left((m_s-m_t)^2 + m_t(3m_s+m_t)\frac{A_2(5)}{A_1(5)}\right) \tag{6.106c}$$

$$\alpha_{st}^{(4)} = \frac{2m_t^2}{(m_s + m_t)^2}\left(2 - \frac{A_2(5)}{A_1(5)}\right) \tag{6.106d}$$

Equations (6.86), (6.97), (6.103) and (6.105) represent the complete set of collision integrals for composite gases in the 13 moment approximation assuming Maxwell molecular interactions between the molecules. The complete set of 13 moment equations with these collision terms gives a good description of the transport of rarefied neutral gases.

6.6 Simplified sets of transport equations

The 13 moment system of equations (equations (6.68)) gives a good approximation to the transport of neutral gases. In several cases the particle density (and consequently the collision frequency) is high enough to justify additional simplification of the transport equation. One such simplification has already been discussed in Subsection 6.4.3, where the Navier–Stokes equations were derived from the 13 moment transport equation in the case of a single species gas. The derivation of the Navier–Stokes equation was based on the assumption that the physical quantities describing the deviation of the distribution function from local thermodynamic equilibrium (the stress tensor, τ, and the heat flow vector, $\mathbf{h}$) are small and can be neglected except when they are multiplied by the large collision frequency.

In this section we consider several additional simplified sets of transport equations.

6.6.1 The 10 moment approximation

The 10 moment approximation neglects the heat flow altogether, but assumes that the stress tensor cannot be neglected. In this limit we keep 10 moments of the distribution function: density, three components of the velocity vector, scalar pressure, and five elements of the stress tensor. In the 10 moment approximation the transport equations simplify to the following:

$$\frac{\partial(mn)}{\partial t} + \frac{\partial(mn\,u_i)}{\partial x_i} = \frac{\delta(mn)}{\delta t} \tag{6.107a}$$

$$mn\left(\frac{\partial u_k}{\partial t} + u_i\frac{\partial u_k}{\partial x_i}\right) + \frac{\partial p}{\partial x_k} - \frac{\partial \tau_{ik}}{\partial x_i} - mn\langle a_k\rangle = mn\frac{\delta u_k}{\delta t} \tag{6.107b}$$

$$\frac{\partial p}{\partial t} + \frac{\partial(pu_i)}{\partial x_i} + \frac{2}{3}p\frac{\partial u_i}{\partial x_i} - \frac{2}{3}\tau_{ih}\frac{\partial u_h}{\partial x_i} = \frac{\delta p}{\delta t} \tag{6.107c}$$

$$\frac{\partial \tau_{kl}}{\partial t} + \frac{\partial(\tau_{kl}u_i)}{\partial x_i} + \tau_{ik}\frac{\partial u_l}{\partial x_i} + \tau_{il}\frac{\partial u_k}{\partial x_i} - \frac{2}{3}\delta_{kl}\tau_{ij}\frac{\partial u_i}{\partial x_j}$$

$$-p\left(\frac{\partial u_k}{\partial x_l} + \frac{\partial u_l}{\partial x_k} - \frac{2}{3}\delta_{kl}\frac{\partial u_i}{\partial x_i}\right) + mn\langle a_k c_l + a_l c_k\rangle = \frac{\delta \tau_{kl}}{\delta t} \tag{6.107d}$$

These equations refer to a single species. The effects of other gas components are included in the collision term (see Section 6.5).

At first glance this approximation seems to be very reasonable, because it neglects the third order term, while it keeps the second and higher order ones. However, in realistic physical situations the heat flow term is quite often more important than the stress tensor, therefore the application of the 10 moment approximation has to be handled with great care.

6.6.2 The 8 moment approximation

A physically more useful approximation is obtained when one neglects the stress tensor altogether, but keeps the heat flow vector. This is the very widely used 8 moment approximation (sometimes also called Euler equations with heat flow). In this case the 13 moment set of transport equations simplifies to the following 8 equations:

$$\frac{\partial(mn)}{\partial t} + \frac{\partial(mn\,u_i)}{\partial x_i} = \frac{\delta(mn)}{\delta t} \tag{6.108a}$$

$$mn\left(\frac{\partial u_k}{\partial t} + u_i\frac{\partial u_k}{\partial x_i}\right) + \frac{\partial p}{\partial x_k} - mn\langle a_k\rangle = mn\frac{\delta u_k}{\delta t} \tag{6.108b}$$

$$\frac{\partial p}{\partial t} + \frac{\partial(pu_i)}{\partial x_i} + \frac{2}{3}p\frac{\partial u_i}{\partial x_i} + \frac{2}{3}\frac{\partial h_i}{\partial x_i} = \frac{\delta p}{\delta t} \tag{6.108c}$$

$$\frac{\partial h_k}{\partial t} + u_i \frac{\partial h_k}{\partial x_i} + \frac{7}{5} h_k \frac{\partial u_i}{\partial x_i} + \frac{2}{5} h_i \frac{\partial u_i}{\partial x_k} + \frac{7}{5} h_i \frac{\partial u_k}{\partial x_i} + \frac{5}{2} p \frac{\partial}{\partial x_k} \left(\frac{p}{mn} \right)$$

$$+ \frac{5}{2} p \langle a_k \rangle - \frac{1}{2} mn \langle a_k c^2 \rangle - mn \langle a_i c_i c_k \rangle = \frac{\delta h_k}{\delta t} - \frac{5}{2} p \frac{\delta u_k}{\delta t} \qquad (6.108d)$$

6.6.3 *The 5 moment approximation*

In the simplest case we neglect all higher order terms and thus obtain five transport equations for the macroscopic parameters of the local Maxwell–Boltzmann distribution. In the case of a single species gas these equations are the Euler equations discussed earlier. In the case of a multicomponent gas the right hand sides of the equations do not vanish and we obtain the following set of five transport equations (for the sake of mathematical simplicity we assume Maxwell molecular interactions and neglect all external accelerations):

$$\frac{\partial (m_s n_s)}{\partial t} + \frac{\partial (m_s n_s u_{si})}{\partial x_i} = 0 \qquad (6.109a)$$

$$m_s n_s \left(\frac{\partial u_{sk}}{\partial t} + u_{si} \frac{\partial u_{sk}}{\partial x_i} \right) + \frac{\partial p_s}{\partial x_k} = \sum_t m_s n_s v_{st} (u_{tk} - u_{sk}) \qquad (6.109b)$$

$$\frac{\partial p_s}{\partial t} + \frac{\partial (p_s u_{si})}{\partial x_i} + \frac{2}{3} p_s \frac{\partial u_{si}}{\partial x_i} = \sum_t \frac{2 v_{st} m_s n_s}{m_s + m_t} \left(k(T_t - T_s) + \frac{m_t}{3} (u_s - u_t)^2 \right)$$

$$(6.109c)$$

This set of equations is the most frequently used approximation in kinetic theory. One should, however, keep in mind the simplifying assumptions we made along the road and be very careful when applying it to low density gases.

6.7 Problems

6.1 Show that for a charged particle moving in an electromagnetic field the acceleration is divergence free in velocity space.

6.2 Show that for electromagnetic forces $\langle a_i c_i \rangle = 0$.

6.3 Evaluate the 20 moment equations (equations (6.52), (6.53), (6.55), (6.56), and (6.60)) for charged particles moving under the influence of gravity, electric and magnetic fields.

6.4 Evaluate the 13 moment collision terms for Maxwell molecules in the case of a single species gas. Compare the results to the collision terms obtained using the relaxation time approximation.

6.8 References

[1] Grad, H., On the kinetic theory of rarefied gases, *Comm. Pure Appl. Math.*, **2**, 331, 1949.

[2] Chapman, S., On the kinetic theory of a gas. Part II. – A composite monatomic gas: Diffusion, viscosity, and thermal conduction, *Phil. Trans. Royal Soc. London*, **217A**, 115, 1916.

[3] Enskog, D., Kinetische Theorie der Vorgänge in massing verdumten Gasen, Ph.D. Thesis, University of Uppsala, Sweden, 1917.

[4] Burgers, J.M., *Flow equations for composite gases*, Academic Press, New York, 1969.

[5] Schunk, R.W., Transport equations for aeronomy, *Planet. Space Sci.*, **23**, 485, 1975.

7

Free molecular aerodynamics

In free molecular flows, occurring at very low gas densities, the interaction between gas molecules is neglected and only the interaction between the gas molecules and the solid surface is taken into account. In Chapter 4 we discussed the flow of highly rarefied gases in the case of 'internal' flows which are typically associated with vacuum installations. There is, however, a much more interesting situation which arises with respect to high altitude flights at hypersonic velocities. This situation involves the aerodynamics of high speed rarefied gas flow past submerged bodies.

In this chapter we consider the interaction of solid bodies with rarefied gas flows with Knudsen numbers (the ratio of the molecular mean free path and the characteristic size of the body) larger than unity. This regime, which is called free molecular interaction, is quite different from normal gasdynamic flows around solid bodies and therefore very different mathematical methods must be applied.

Typical mean free path values in the terrestrial upper atmosphere are shown in Figure 7.1 (when calculating the mean free path an average cross section value of $\sigma = 10^{-19}$ m^2 was used). Inspection of Figure 7.1 reveals that above approximately 150 km the condition Kn $\gg$ 1 is satisfied (at this altitude $\lambda \sim 100$ m, which is large compared to typical space vehicles). In the h $\geq$ 150 km region the interaction is free molecular but the number density is still high enough to make macroscopic averages meaningful (the atmospheric density profile is shown in Figure 1.2).

In the free molecular interaction regime the molecules which hit the body and are reflected usually travel very far before colliding with other molecules. Consequently, one can neglect the interaction of reflected particles with the incident stream, at least so far as effects on the body itself are concerned. The free molecular interaction of a solid body with rarefied gas is mainly governed by the interaction of individual gas molecules with the surface of the solid body.

The basic assumptions of free molecular flow theory can be summarized as follows.

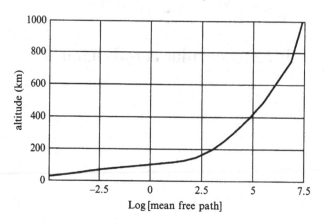

Figure 7.1

(i) In the free molecular interaction regime the mean free path of the molecules is much larger than the largest characteristic dimension of the body which is assumed to be in a gas flow of infinite extent.

(ii) The molecules which hit the surface of the body and are reflected (or re-emitted) usually travel very far from the body before colliding with other molecules. The reflected molecules therefore 'forget' their previous interaction with the body before they hit again. Also, the effect of the reflected molecules on the particle streams incident on the body is assumed to be negligible. In other words the incident flow is assumed to be entirely undisturbed by the presence of the body.

It is an immediate consequence of these basic assumptions that no shock wave is expected to form in the free molecular interaction regime. The 'boundary layer' will be very thick and diffuse. Therefore it has no effect on the incident flow.

From the practical point of view the main assumptions mean that in the free molecular flow approximation the incident and reflected (or re-emitted) molecules can be treated separately. The incoming gas particles are assumed to be in a local thermodynamic equilibrium (LTE), characterized by a Maxwell–Boltzmann velocity distribution. The usual justification is that when Kn $\gg$ 1, there is only a small probability that one of the scattered molecules will interact with the incoming beam in such a way that it will be scattered back to the body, and there is an even smaller probability that it will strike the body before it reaches equilibrium with the incoming beam after many collisions.

The assumption that reflected molecules play only a negligible role in the formation of the incident flux is not always true. Figure 7.2 shows two surface elements: one is concave, the other is convex. It is obvious that our fundamental assumptions are only valid for convex surfaces for which there is no possibility for a reflected particle to immediately hit another surface element. However, in the

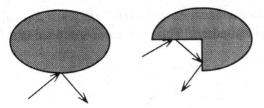

Figure 7.2

case of concave surfaces reflected particles can directly reach another surface element of the body without colliding with other molecules first. It should be remembered that the free molecular interaction of bodies with large concave surfaces is very complicated and the proper incident streams must be taken into account. In this textbook we do not deal with concave bodies; rather we will concentrate on the much simpler case of convex surface bodies and will demonstrate the methods of describing the interaction between the gas flow and the solid body.

A good summary of the basic principles of free molecular aerodynamics was given by Schaaf and Chambré,[1] with an extensive list of the relevant literature. The interested reader is referred to this work.

7.1 Transfer of mass, momentum, and translational energy

7.1.1 Reflection coefficients

The determination of the flux of mass, momentum or energy carried by gas molecules reflected or reemitted from the surface of the body requires a specification of the interaction between the impinging particles and the surface. For a detailed description of this interaction one also has to determine the velocity distribution function of the re-emitted (reflected) molecules as a function of their incident velocity. However, in the free-molecular approximation this detailed knowledge is not necessary; it is sufficient to know certain average parameters characterizing the interaction process. These average quantities are the thermal accommodation and reflection coefficients.

As has been discussed earlier in connection with slip flow (see Subsection 4.4.2) there are two basic types of interaction between a wall and gas molecules: diffuse and elastic. In an elastic collision the molecules are reflected specularly from the wall, while in the case of diffuse reflection they are temporarily absorbed and later re-emitted. The direction of the re-emitted molecules is randomly distributed in 2π solid angle (particles must leave the wall), and their velocity distribution is Maxwellian. In other words diffusely reflected molecules completely forget their pre-collision velocity.

The thermal accommodation coefficient, α, was originally introduced by Smoluchowski[2] and Knudsen.[3] This coefficient is defined in the following way:

$$\alpha = \frac{\Phi_i^{(E)} - \Phi_r^{(E)}}{\Phi_i^{(E)} - \Phi_w^{(E)}} \tag{7.1}$$

where $\Phi_i^{(E)}$ is the incident energy flux (energy per unit surface area per unit time) of the gas molecules reaching a surface element, while $\Phi_r^{(E)}$ is the reflected (re-emitted) energy flux carried by the gas molecules leaving the surface element. The quantity $\Phi_w^{(E)}$ is the energy flux that would have been carried away by the reflected (re-emitted) gas molecules if all incident molecules were fully accommodated (if they were all re-emitted with a Maxwellian distribution corresponding to the surface temperature, T_w).

The coefficient, α, is thus a measure of the degree of accommodation of the mean energy of the molecules to the temperature of the wall with which they are interacting. It is obvious that $\alpha=1$ corresponds to perfect accommodation ($\Phi_r^{(E)}=\Phi_w^{(E)}$, i.e. all molecules are re-emitted with the appropriate Maxwellian velocity distribution), while $\alpha=0$ corresponds to no accommodation at all. In the $\alpha=0$ case $\Phi_r^{(E)}=\Phi_i^{(E)}$, meaning that all molecules undergo perfect elastic reflections. It should be noted that perfect accommodation is identical to the diffuse reflection mechanism discussed in Section 4.4 in connection with slip flow (the slip flow approximation applies to the Kn~1 regime, while the free molecular interaction assumes Kn>>1).

It is implicitly assumed that all forms of energy (translational and internal) associated with those molecular degrees of freedom which participate in the energy exchange with the surface are accommodated to the same degree. Experimental evidence indicates that this implicit assumption is true for translational and rotational energies, while the vibrational energy is affected to a much smaller degree. In principle, one could introduce a different accommodation coefficient for each degree of freedom. In this introductory treatment, however, we use a single value for each material.

Experimental evidence indicates that the thermal accommodation coefficient of many materials is quite close to unity ($0.87 \leq \alpha \leq 0.97$).

In Chapter 4 we briefly discussed the traditional treatment of momentum transfer between the impinging molecules and the surface of a solid body and assumed that a fraction, σ', of the incident particles is reflected 'diffusely', while the rest is reflected specularly. The diffusely reflected fraction contributes to the tangential momentum transfer between the molecules and the wall. The remaining fraction $(1-\sigma')$ is reflected 'specularly' (mirror-like), with a reversal of the normal momentum component. This fraction contributes only to the transfer of the normal momentum component to the wall.

There is some experimental evidence that the situation is more complicated and a single parameter σ' is not sufficient to characterize both the normal and tangential momentum transfer in an adequate manner. In order to adequately specify both the tangential and normal components of the aerodynamic forces acting on the body we introduce two reflection coefficients in the following way:

$$\sigma_t = \frac{\Phi_i^{(mv_t)} - \Phi_r^{(mv_t)}}{\Phi_i^{(mv_t)} - \Phi_w^{(mv_t)}} = \frac{\Phi_i^{(mv_t)} - \Phi_r^{(mv_t)}}{\Phi_i^{(mv_t)}} \tag{7.2a}$$

$$\sigma_n = \frac{\Phi_i^{(mv_n)} - \Phi_r^{(mv_n)}}{\Phi_i^{(mv_n)} - \Phi_w^{(mv_n)}} \tag{7.2b}$$

where $\Phi_i^{(mv_t)}$ and $\Phi_r^{(mv_t)}$ represent the magnitudes of the incident and reflected tangential momentum fluxes. $\Phi_w^{(mv_t)}$ would be the magnitude of the re-emitted tangential momentum flux if all incident particles were re-emitted with a Maxwellian velocity distribution corresponding to the surface temperature, T_w. From this definition it is obvious that $\Phi_w^{(mv_t)}=0$, because the directional distribution is symmetric around the normal vector of the surface element (the tangential momentum components of the released particles cancel each other). Similarly, the fluxes in the expression for the normal reflection coefficient represent the magnitude of the normal component of the incident momentum flux, $\Phi_i^{(mv_n)}$, the magnitude of the normal component of the reflected momentum flux, $\Phi_r^{(mv_n)}$, and the magnitude of the normal component of the hypothetical reflected flux, $\Phi_w^{(mv_n)}$.

It can be easily seen that in the case of specular reflection $\Phi_r^{(mv_t)}=\Phi_i^{(mv_t)}$ and $\Phi_r^{(mv_n)}=\Phi_i^{(mv_n)}$, consequently $\Phi_r^{(E)}=\Phi_i^{(E)}$ and therefore $\alpha=0$, $\sigma_t=0$, and $\sigma_n=0$. In the case of diffuse reflection $\Phi_r^{(mv_t)}=\Phi_w^{(mv_t)}=0$ and $\Phi_r^{(mv_n)}=\Phi_w^{(mv_n)}$, consequently $\Phi_r^{(E)}=\Phi_w^{(E)}$ and therefore $\alpha=1$, $\sigma_t=1$, and $\sigma_n=1$.

Experiments show that most materials have tangential reflection coefficients close to unity, $\sigma_t \approx 1$. The normal reflection coefficient is relatively poorly known, but observations are consistent with $\sigma_n \approx 1$.

It should be emphasized that the quantities α, σ_t, and σ_n are average phenomenological parameters. Additional interaction parameters can be specified as needed. However, for the purpose of this introductory description, these three parameters provide an adequate description of the free molecular interaction between gas molecules and a solid surface.

In this section we will investigate the transfer of mass, momentum and translational energy by gas molecules to and from a surface element, dS, at a temperature, T_w. It is assumed that the incoming molecules follow a Maxwell–Boltzmann velocity distribution in the frame of reference of the gas. For the sake of simplicity it is postulated that molecules are reflected diffusely with a Maxwell–Boltzmann velocity distribution, characterized by a reflection temperature, T_r, which is in general different from the wall temperature and from the temperature of the incident population, T_i.

Since the motions of incident and emergent molecules are assumed to be independent, the transfer of mass, momentum and energy by the incoming and reflected molecules might be considered separately.

The interaction will be described using the following local coordinate system. Let us choose a system of rectangular coordinates (x_1, x_2, x_3) fixed relative to the surface with the x_3 axis perpendicular to the surface and pointing inward (see Figure 7.3). The velocity components of a gas molecule, v_1, v_2, v_3, arise partly from the motion of the surface relative to the gas (u_1, u_2, u_3) and partly from the random motion of the gas molecule (c_1, c_2, c_3): $\mathbf{v=u+c}$.

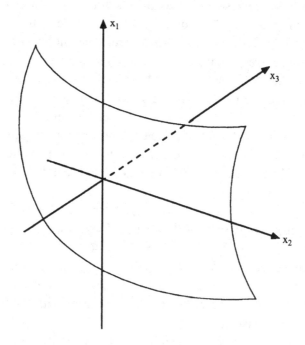

Figure 7.3

7.1.2 Transfer of mass

We start by calculating the mass flux absorbed (or released) by the surface element, dS. This net flux is the difference of the absorbed and re-emitted mass fluxes. Our model assumes that all particles reaching the surface element are temporarily absorbed and then reemitted with a Maxwellian velocity distribution in a random direction. In this model absorbed particle flux is the mass flux (number of molecules per unit time per unit area) reaching the surface:

$$\Phi_i^{(m)} = m \, n_i \left(\frac{\beta_i}{\pi}\right)^{3/2} \int\limits_{-\infty}^{\infty} dv_1 \, e^{-\beta_i c_1^2} \int\limits_{-\infty}^{\infty} dv_2 \, e^{-\beta_i c_2^2} \int\limits_{0}^{\infty} dv_3 \, v_3 \, e^{-\beta_i c_3^2} \tag{7.3}$$

where $\beta_i = m/(2kT_i)$. It is important to note that we integrate over the entire velocity space in the horizontal (tangential) direction, but only over the positive velocities in the x_3 direction. The reason is that only particles with $v_3 > 0$ are moving towards the surface element, therefore only these particles are able to reach it. In the exponential factors, random velocity components are present, because the velocity distribution is a Maxwellian in the frame of reference of the gas. The first two integrals can easily be worked out and equation (7.3) can be written in the following form:

$$\Phi_i^{(m)} = m \, n_i \left(\frac{\beta_i}{\pi}\right)^{1/2} \int\limits_{-u_3}^{\infty} dc_3 \, (u_3 + c_3) \, e^{-\beta_i c_3^2} \tag{7.4}$$

Here a variable transformation has been carried out: the old integration variable, v_3, was replaced by the normal random velocity component, c_3. This variable transformation is reflected in the fact that the lower integration boundary is now $-u_3$, because when the total normal velocity is zero, the random particle velocity must be $-u_3$. The physical meaning of this new integration boundary is that particles with random velocities $-u_3$ are still swept to the surface by the moving gas.

The integrals in equation (7.4) can be evaluated. This process leads to the following expression:

$$\Phi_i^{(m)} = \frac{1}{4} m \, n_i \sqrt{\frac{8kT_i}{\pi m}} \left[e^{-s_3^2} + \sqrt{\pi} \, s_3 \left(1 + \frac{2}{\sqrt{\pi}} \int\limits_{0}^{s_3} dx_3 \, e^{-x_3^2} \right) \right] \tag{7.5}$$

where we have introduced the following dimensionless velocity:

$$s_3 = \beta_i^{1/2} u_3 = u_3 \sqrt{\frac{m}{2kT_i}} \tag{7.6}$$

The remaining integral in equation (7.5) cannot be evaluated analytically, but is a well known special function, the so-called error function, erf. The basic properties of the error function are discussed in Appendix 9.5.3. We also recognize that the square root in equation (7.5) is the average thermal speed of the incident molecules, $\bar{v}_i$. With the introduction of the error function the incident mass flux can be written in the following form:

$$\Phi_i^{(m)} = \frac{1}{4} m\, n_i\, \bar{v}_i \left\{ e^{-s_3^2} + \sqrt{\pi}\, s_3 \left[1 + \mathrm{erf}(s_3) \right] \right\} \qquad (7.7)$$

This expression means that the incident flux absorbed by the surface element is the thermal flux of particles going through the surface element multiplied by a correction factor due to the relative motion of the gas and the body.

The correction factor, $C_i^{(m)} = 4\Phi_i^{(m)}/m n_i \bar{v}_i$, is shown in Figure 7.4 as a function of the normal component of the dimensionless gas flow velocity, s_3. It can be seen that for large negative values the correction factor approaches zero, meaning that only the fastest few particles can reach the surface. For positive values of s_3, however, particles are swept to the wall and the correction factor rapidly increases.

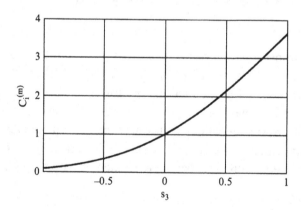

Figure 7.4

The reflected molecules can be regarded as particles coming from a hypothetical gas reservoir on the reverse side of the surface. This gas reservoir is at rest relative to the surface; its molecules follow a Maxwell–Boltzmann velocity distribution at a temperature T_r (it should be emphasized that in general T_r is not equal to the wall temperature). This is the same as the previously studied problem of molecular effusion into vacuum through an orifice (see Section 4.1.2), therefore, by direct analogy, the escaping particle flux is

$$\Phi_r^{(m)} = \frac{1}{4} m n_r \bar{v}_r = m n_r \sqrt{\frac{kT_r}{2\pi m}} \tag{7.8}$$

Under equilibrium conditions it is reasonable to assume that the absorbed and re-emitted particle fluxes are identical, $\Phi_r^{(m)}=\Phi_i^{(m)}$. This relation can immediately be used to obtain a relation between the incident particle density and the density of the reflected particle population:

$$n_r = n_i \sqrt{\frac{T_i}{T_r}} \left\{ e^{-s_3^2} + \sqrt{\pi}\, s_3 \left[1 + \mathrm{erf}(s_3)\right] \right\} \tag{7.9}$$

It should be noted that n_r is the particle density in the hypothetical leaky particle reservoir behind the surface element, but n_r is not the physical number density of reflected particles.

7.1.3 Transfer of perpendicular momentum

The pressure on the surface is defined as the perpendicular momentum transfer flux from the gas to the surface. In our particular model the pressure arises from two independent components. The first component is due to the flux of perpendicular momentum carried to the wall by the absorbed incident gas molecules. The second component arises from the perpendicular momentum flux carried away by the re-emitted molecules. In the present model the two components are independent of each other and therefore one can treat them separately.

We start by calculating the flux of perpendicular momentum carried by the incident particles to the wall. We calculated these types of fluxes repeatedly in Chapter 4 (see equation (4.23)), therefore we can write

$$\Phi_i^{(mv_n)} = m n_i \left(\frac{\beta_i}{\pi}\right)^{3/2} \int_{-\infty}^{\infty} dv_1\, e^{-\beta_i c_1^2} \int_{-\infty}^{\infty} dv_2\, e^{-\beta_i c_2^2} \int_0^{\infty} dv_3\, v_3^2\, e^{-\beta_i c_3^2} \tag{7.10}$$

The integrals in the v_1 and v_2 directions can easily be carried out and the remaining integral can be transformed to get the following:

$$\Phi_i^{(mv_n)} = m n_i \left(\frac{\beta_i}{\pi}\right)^{1/2} \int_{-u_3}^{\infty} dc_3\, (u_3 + c_3)^2\, e^{-\beta_i c_3^2} \tag{7.11}$$

The integral can be evaluated to obtain the following expression for the incident flux of perpendicular momentum:

$$\Phi_i^{(mv_n)} = n_i \, k \, T_i \left\{ \frac{s_3}{\sqrt{\pi}} e^{-s_3^2} + \left(\frac{1}{2} + s_3^2 \right) \left[1 + \mathrm{erf}(s_3) \right] \right\} \qquad (7.12)$$

Equation (7.12) tells us that the contribution of the incident molecules to the transfer of perpendicular momentum to the wall can be obtained by multiplying the external gas pressure by a correction factor. This correction factor, given by $C_i^{(mv_n)} = \Phi_i^{(mv_n)}/p_i$, is plotted in Figure 7.5. Inspection of Figure 7.5 shows that when there is no relative motion in the perpendicular direction the incident momentum flux on the surface is identical to half the gas pressure. This is of course not a surprising result, because in the classical case the momentum change of specularly reflected molecules is twice the perpendicular incident momentum of the incident particles. This question has been discussed in considerable detail in Subsection 2.4.1.

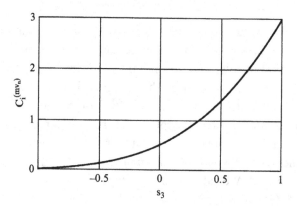

Figure 7.5

The flux of perpendicular momentum carried away by the reflected (re-emitted) molecules can be calculated fairly readily. Our assumption of perfect accommodation means that reflected particles can be approximated by a population of particles escaping from a hypothetical gas reservoir attached to the back of the wall. In this approximation we obtain

$$\Phi_r^{(mv_n)} = m \, n_r \left(\frac{\beta_r}{\pi} \right)^{3/2} \int\limits_{-\infty}^{\infty} dc_1 \, e^{-\beta_r c_1^2} \int\limits_{-\infty}^{\infty} dc_2 \, e^{-\beta_r c_2^2} \int\limits_{-\infty}^{0} dc_3 \, c_3^2 \, e^{-\beta_r c_3^2} \qquad (7.13)$$

The three integrals can be worked out to obtain

$$\Phi_r^{(mv_n)} = \frac{mn_r}{4\beta_r} = \frac{1}{2} n_r k T_r \qquad (7.14)$$

This result indicates that the flux of perpendicular momentum carried away by the re-emitted particles is half the pressure in the hypothetical gas reservoir. Substituting expression (7.9) for the reservoir density, n_r, we obtain the following:

$$\Phi_r^{(mv_n)} = \frac{1}{2} n_i k T_i \sqrt{\frac{T_r}{T_i}} \left\{ e^{-s_3^2} + \sqrt{\pi} s_3 \left[1 + \mathrm{erf}(s_3) \right] \right\} \qquad (7.15)$$

The total gas pressure acting on the surface element is the sum of the incident and reflected fluxes of perpendicular momentum:

$$p = \Phi_i^{(mv_n)} + \Phi_r^{(mv_n)} = p_i \left\{ g(s_3) e^{-s_3^2} + \left(\frac{1}{2} + \sqrt{\pi} s_3 g(s_3) \right) \left[1 + \mathrm{erf}(s_3) \right] \right\} \quad (7.16)$$

where $p_i = n_i k T_i$ is the pressure of the incident gas and

$$g(s_3) = \frac{s_3}{\sqrt{\pi}} + \frac{1}{2} \sqrt{\frac{T_r}{T_i}} \qquad (7.17)$$

In the simplest case when $s_3 = 0$ ($u_3 = 0$) and $T_i = T_r$ expression (7.16) reduces to the classical result, $p = p_i$. The dependence of the pressure correction factor (the expression in the curly bracket in equation (7.16)) is plotted in Figure 7.6 for identical incident and reflection temperatures, $T_i = T_r$.

It is important to mention that T_r is still an unknown quantity which can be determined by making appropriate assumptions about the reflection parameters. This issue will be addressed in detail in Section 7.2.

7.1.4 Transfer of tangential momentum

Next we calculate the transfer of parallel or tangential momentum from the dilute gas to the surface of a solid body. Since in our model the incident molecules are temporarily trapped by the surface they lose all their tangential momentum on

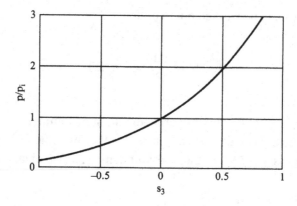

Figure 7.6

contact, so the incident flux of parallel momentum is given by the following integral:

$$\Phi_i^{(mv_t)} = m\,n_i \left(\frac{\beta_i}{\pi}\right)^{3/2} \int_{-\infty}^{\infty} dv_1 \int_{-\infty}^{\infty} dv_2 \left(v_t\, e^{-\beta_i c_1^2} e^{-\beta_i c_2^2}\right) \int_{0}^{\infty} dv_3\, v_3\, e^{-\beta_i c_3^2} \qquad (7.18)$$

where v_t denotes the projection of the particle velocity to the (x_1, x_2) plane (which is tangential to the surface element). The integrals over the variables v_1 and v_2 can be readily evaluated:

$$\Phi_i^{(mv_t)} = m\,n_i \left(\frac{\beta_i}{\pi}\right)^{1/2} u_t \int_{-\infty}^{0} dc_3\, v_3\, e^{-\beta_i c_3^2} \qquad (7.19)$$

It should be recognized that the remaining integral is related to the incident mass flux given by equation (7.4). Using this result one can obtain the following expression for the magnitude of the incident flux of parallel momentum:

$$\Phi_i^{(mv_t)} = u_t\, \Phi_i^{(m)} \qquad (7.20)$$

Substituting expression (7.7) into equation (7.20) yields the following for the incident flux of parallel momentum:

$$\Phi_i^{(mv_t)} = p_i \left\{ s_t \frac{e^{-s_3^2}}{\sqrt{\pi}} + s_t\, s_3 \left[1 + \mathrm{erf}(s_3)\right] \right\} \qquad (7.21)$$

where we have introduced the dimensionless tangential gas velocity:

$$s_t = u_t \sqrt{\frac{m}{2kT_i}} \qquad (7.22)$$

In the limit $s_3 \to 0$ equation (7.21) becomes the following:

$$\lim_{s_3 \to 0} \Phi_i^{(mv_t)} = \frac{p_i s_t}{\sqrt{\pi}} \qquad (7.23)$$

The reflected molecules are re-emitted isotropically, therefore the reflected flux of tangential momentum must vanish:

$$\Phi_r^{(mv_t)} = m\,n_r \left(\frac{\beta_r}{\pi}\right)^{3/2} \int_{-\infty}^{\infty} dc_1 \int_{-\infty}^{\infty} dc_2 \left(c_t\, e^{-\beta_r c_1^2} e^{-\beta_r c_2^2}\right) \int_0^{\infty} dc_3 c_3\, e^{-\beta_r c_3^2} = 0 \quad (7.24)$$

The shearing stress acting on the surface is the difference of the incident and reflected fluxes of tangential momentum. In this particular case the reflected flux is zero; therefore one can conclude that

$$\tau_t = \Phi_i^{(mv_t)} = p_i s_t \left\{ \frac{e^{-s_3^2}}{\sqrt{\pi}} + s_3 \left[1 + \mathrm{erf}(s_3)\right] \right\} \qquad (7.25)$$

7.1.5 Transfer of translational energy

Finally we calculate the transfer of translational energy from the gas to the wall. As before, the flux of translational energy can be considered as the sum of the incident and reflected fluxes, which in turn can be calculated independently of each other. We start with the incident translational energy flux, which can be written as

$$\Phi_i^{(\frac{1}{2}mv^2)} = \frac{1}{2} m\,n_i \left(\frac{\beta_i}{\pi}\right)^{3/2} \int_{-\infty}^{\infty} dv_1 \int_{-\infty}^{\infty} dv_2 \int_0^{\infty} dv_3 v_3\, v^2\, e^{-\beta_i c^2} \qquad (7.26)$$

By expanding v^2 into its components the integral can be separated into three terms:

$$\Phi_i^{(\frac{1}{2}mv^2)} = \frac{1}{2}mn_i\left(\frac{\beta_i}{\pi}\right)\int\limits_{-\infty}^{\infty}dv_1\,v_1^2\,e^{-\beta_ic_1^2}\int\limits_0^{\infty}dv_3v_3\,e^{-\beta_ic_3^2}$$

$$+\frac{1}{2}mn_i\left(\frac{\beta_i}{\pi}\right)\int\limits_{-\infty}^{\infty}dv_2\,v_2^2\,e^{-\beta_ic_2^2}\int\limits_0^{\infty}dv_3v_3\,e^{-\beta_ic_3^2}+\frac{1}{2}mn_i\left(\frac{\beta_i}{\pi}\right)^{1/2}\int\limits_0^{\infty}dv_3\,v_3^3\,e^{-\beta_ic_3^2}$$

$$(7.27)$$

The integrals can be evaluated in a straightforward but somewhat tedious way. This process leads to the following result:

$$\Phi_i^{(\frac{1}{2}mv^2)} = \frac{\Phi_i^{(m)}}{m}\left(\frac{1}{2}mu^2 + \frac{5}{2}kT_i\right) - \frac{1}{8}n_i\,\bar{v}_i\,kT_i\,e^{-s_3^2} \qquad (7.28)$$

In the limit $s_3 \to 0$ this expression becomes

$$\lim_{s_3\to 0}\Phi_i^{(\frac{1}{2}mv^2)} = \frac{1}{4}n_i\,\bar{v}\left(\frac{1}{2}mu^2 + 2kT_i\right) \qquad (7.29)$$

This is a very interesting result which tells us that the translational energy flux incident on the surface has two components: one is due to the organized (bulk) motion of the gas with respect of the solid body, while the other is due to the random (thermal) motion of the gas molecules. An interesting aspect of the result is that in the case of no relative bulk motion (u=0), the average thermal energy of the molecules hitting the surface element is $2kT_i$, which is larger that the average thermal energy of the gas molecules, $3kT_i/2$. The physical reason is the same as it was in the case of molecular effusion through an orifice (see Subsection 4.1.3 and equation (4.27)), namely, that in unit time faster particles can reach the surface element from a larger distance, therefore the distribution of particles hitting the wall is biased towards faster molecules.

Next we calculate the reflected energy flux. The re-emitted molecules transport the following translational energy flux away from the surface:

$$\Phi_r^{(\frac{1}{2}mv^2)} = -\frac{1}{2}mn_r\left(\frac{\beta_r}{\pi}\right)^{3/2}\int\limits_{-\infty}^{\infty}dc_1\int\limits_{-\infty}^{\infty}dc_2\int\limits_{-\infty}^{0}dc_3c_3\,c^2\,e^{-\beta_rc^2} \qquad (7.30)$$

The minus sign in this expression reflects the fact that the value of the integral is negative, because the flux is in the $-x_3$ direction. Recall that Φ is defined as the magnitude of the flux, therefore one has to compensate for the negative sign. The integrals can be evaluated to obtain

$$\Phi_r^{(\frac{1}{2}mv^2)} = \sqrt{\frac{2}{\pi}} \, mn_r \left(\frac{kT_r}{m}\right)^{3/2} = 2kT_r \frac{\Phi_r^{(m)}}{m} \qquad (7.31)$$

This result means that the average energy carried away by re-emitted molecules is $2kT_r$. The explanation of this result is the same as was given above (see the discussion about the incident energy flux). Under steady-state conditions the incident and reflected mass fluxes are identical, therefore one obtains

$$\Phi_r^{(\frac{1}{2}mv^2)} = 2kT_r \frac{\Phi_i^{(m)}}{m} \qquad (7.32)$$

7.1.6 Limiting cases

Next we examine the free molecular transfer of mass, momentum and energy to a surface element in two limits. The first is the low speed limit, when $\mathbf{u} \to 0$, but the temperature remains finite. The second limit is the so-called hypersonic limit, when the Mach number of the free molecular flow (with respect to the body) is much larger than unity, $M \gg 1$. The Mach number is closely related to our dimensionless velocity, s, because

$$s = \sqrt{\frac{mu^2}{2kT_i}} = \sqrt{\frac{\gamma}{2}} \sqrt{\frac{mu^2}{\gamma kT_i}} = \sqrt{\frac{\gamma}{2}} M \qquad (7.33)$$

First we consider the limit $s \to 0$. In this case we obtain the following expressions for the various free molecular fluxes:

$$\lim_{s \to 0} \Phi_i^{(m)} = \frac{1}{4} m n_i \bar{v}_i \qquad (7.34)$$

$$\lim_{s \to 0} n_r = n_i \sqrt{\frac{T_i}{T_r}} \qquad (7.35)$$

$$\lim_{s \to 0} \Phi_i^{(mv_n)} = \frac{1}{2} n_i kT_i \qquad (7.36)$$

$$\lim_{s \to 0} \Phi_r^{(mv_n)} = \frac{1}{2} n_i \, k \, T_i \sqrt{\frac{T_r}{T_i}} \qquad (7.37)$$

$$\lim_{s \to 0} p = \Phi_i^{(mv_n)} + \Phi_r^{(mv_n)} = \frac{1}{2} p_i \left(1 + \sqrt{\frac{T_r}{T_i}} \right) \qquad (7.38)$$

$$\lim_{s \to 0} \tau_t = \Phi_i^{(mv_t)} = 0 \qquad (7.39)$$

$$\lim_{s \to 0} \Phi_i^{(\frac{1}{2}mv^2)} = \frac{1}{4} n_i \, \bar{v}_i \, (2 \, k \, T_i) \qquad (7.40)$$

$$\lim_{s \to 0} \Phi_r^{(\frac{1}{2}mv^2)} = \frac{1}{4} n_i \, \bar{v}_i \, (2 \, k \, T_r) \qquad (7.41)$$

These results do not differ from the hydrodynamic results, because there is no relative motion between the gas and the solid body. In the hypersonic limit (when $s \to \infty$ and all components of the normalized relative velocity, s_t, and s_3, approach infinity, as well) the same quantities become very different:

$$\lim_{s \to \infty} \Phi_i^{(m)} = m \, n_i \, u_3 \qquad (7.42)$$

$$\lim_{s \to \infty} n_r = 2\sqrt{\pi} \, s_3 \, n_i \qquad (7.43)$$

$$\lim_{s \to \infty} \Phi_i^{(mv_n)} = m \, n_i \, u_3^2 \qquad (7.44)$$

$$\lim_{s \to \infty} \Phi_r^{(mv_n)} = \frac{\pi}{4} m \, n_i \, \bar{v}_r \, u_3 \qquad (7.45)$$

$$\lim_{s \to \infty} p = \Phi_i^{(mv_n)} + \Phi_r^{(mv_n)} = m n_i \, u_3^2 \qquad (7.46)$$

$$\lim_{s \to \infty} \tau_t = \Phi_i^{(mv_t)} = m \, n_i \, u_3 \, u_t \qquad (7.47)$$

$$\lim_{s \to \infty} \Phi_i^{(\frac{1}{2}mv^2)} = \frac{1}{2} m \, n_i \, u^2 \, u_3 \qquad (7.48)$$

$$\lim_{s \to \infty} \Phi_r^{(\frac{1}{2}mv^2)} = 2 \, m \, n_i \, u_3 \, k \, T_r \qquad (7.49)$$

It is interesting to note that in the hypersonic case even the reflected fluxes depend on the normal component of the dimensionless relative velocity, s_3. The reason is that the density in the hypothetical reservoir of reflected particles, n_r, is a function of s_3. This density is directly related to the reflected fluxes.

7.2 Free molecular heat transfer

Next, we consider the convective heat transfer to (and from) a surface element in a steady, uniform, free molecular flow. The results obtained in the previous section will be extensively used in the following derivations. In this discussion we will take into account the energy stored in and transferred from internal degrees of freedom of the gas molecules.

Let us first consider the contribution of the internal energy of the gas molecules to their total energy. It has been discussed in Chapter 2 how the relation between the specific heat ratio of the gas, γ, and the total number of internal degrees of freedom of the gas molecules, ν, is the following (see equation (2.111)):

$$\nu = \frac{5 - 3\gamma}{\gamma - 1} \qquad (7.50)$$

We also recall the equipartition theorem and realize that each internal degree of freedom carries an average energy of $kT_i/2$ per molecule. Assuming equilibrium between the translational and internal degrees of freedom the total incident and reflected energy fluxes can be written as

$$\Phi_i^{(E)} = \Phi_i^{(\frac{1}{2}mv^2)} + \frac{\nu}{2} k \, T_i \, \frac{\Phi_i^{(m)}}{m} = \frac{\Phi_i^{(m)}}{m} \left(\frac{1}{2} m u^2 + \frac{\gamma}{\gamma - 1} k \, T_i \right) - \frac{1}{8} n_i \, \bar{v}_i \, k \, T_i \, e^{-s_3^2}$$

$$(7.51)$$

$$\Phi_r^{(E)} = \Phi_r^{(\frac{1}{2}mv^2)} + \frac{\nu}{2} k \, T_r \, \frac{\Phi_r^{(m)}}{m} = \frac{\gamma + 1}{2(\gamma - 1)} k \, T_r \, \frac{\Phi_r^{(m)}}{m} \qquad (7.52)$$

The net energy transfer rate to unit wall area is the difference of the incident and reflected energy fluxes:

$$\Delta\Phi^{(E)} = \Phi_i^{(E)} - \Phi_r^{(E)} = \frac{\Phi_i^{(m)}}{m}\left(\frac{1}{2}mu^2 + \frac{2\gamma kT_i - (\gamma+1)kT_r}{2(\gamma-1)}\right) - \frac{1}{8}n_i\,\bar{v}_i\,kT_i\,e^{-s_3^2}$$

(7.53)

This is a very nice result which has only one problem: we have no idea of the value of the reflected temperature, T_r. This problem can be circumvented by making use of the thermal accommodation coefficient, α, introduced by equation (7.1). If α is not zero (i.e. the free molecular interaction is not entirely specular) the energy transfer rate can also be expressed as

$$\Delta\Phi^{(E)} = \alpha\left(\Phi_i^{(E)} - \Phi_w^{(E)}\right)$$

(7.54)

It should be noted that the mathematical form of $\Phi_w^{(E)}$ is the same as that of the reflected flux, $\Phi_r^{(E)}$, with the difference that T_r must be replaced by the known wall temperature, T_w. This means that the total energy transfer rate can be expressed in terms of the empirical parameter, α, and known quantities, such as the gas and wall temperatures, and the magnitude of the dimensionless relative velocity between the gas and the body, s:

$$\Delta\Phi^{(E)} = \frac{\alpha}{4}n_i\,\bar{v}_i\,kT_i\left\{e^{-s_3^2}\left(s^2 + \frac{\gamma+1}{2(\gamma-1)} - \frac{(\gamma+1)}{2(\gamma-1)}\frac{T_w}{T_i}\right)\right.$$
$$\left. +\sqrt{\pi}\,s_3\left[1+\mathrm{erf}(s_3)\right]\left(s^2 + \frac{\gamma}{(\gamma-1)} - \frac{(\gamma+1)}{2(\gamma-1)}\frac{T_w}{T_i}\right)\right\}$$

(7.55)

7.2.1 Recovery temperature, Stanton number and thermal recovery factor

In this section we introduce three functions, which are instrumental in describing the free molecular heat transfer between a solid body and the surrounding low density gas.

Equation (7.55) describes the net energy transfer rate between the gas and a surface element. When the gas and the surface reach equilibrium there is no net heat transfer between them and $\Delta\Phi^{(E)}=0$ (assuming that heat conduction effects can be neglected). The wall temperature under equilibrium conditions is called the recovery temperature, T_{rec}, (this is the temperature the surface element would reach if it was left in the gas for an infinitely long time). The recovery temperature can be obtained by solving the equation, $\Delta\Phi^{(E)}=0$. The solution is the following:

$$T_{rec} = T_i \frac{\gamma}{(\gamma+1)}\left[2\left(\frac{\gamma-1}{\gamma}s^2+1\right)-\frac{\gamma-1}{\gamma}\frac{e^{-s_3^2}}{e^{-s_3^2}+\sqrt{\pi}s_3\left[1+\mathrm{erf}(s_3)\right]}\right] \quad (7.56)$$

The stagnation state of a gas is defined as the state in which the gas flow velocity is zero. The stagnation energy is the energy of the gas when it is decelerated to zero speed by an adiabatic process (one with no external work). It follows from this definition that the stagnation temperature of the gas is defined in the following way:

$$\frac{\gamma}{\gamma-1}kT_{stag} = \frac{1}{2}mu^2 + \frac{\gamma}{\gamma-1}kT_i \quad (7.57)$$

Using equation (7.57) one can easily obtain the ratio of the gas temperature to the stagnation temperature:

$$\frac{T_{stag}}{T_i} = \frac{\gamma-1}{\gamma}s^2+1 \quad (7.58)$$

Substituting this result into equation (7.56) yields the following:

$$T_{rec} = T_i \frac{\gamma}{(\gamma+1)}\left[2\frac{T_{stag}}{T_i}-\frac{\gamma-1}{\gamma}\frac{e^{-s_3^2}}{e^{-s_3^2}+\sqrt{\pi}s_3\left[1+\mathrm{erf}(s_3)\right]}\right] \quad (7.59)$$

It is important to note that the local recovery temperature is independent of the current temperature of the surface element, T_w: it is only a function of gas parameters. It is interesting to explore the limiting values of the recovery temperature. In the low speed limit, when $s\to0$, the stagnation temperature, T_{stag}, approaches the gas temperature, T_i, and the recovery temperature becomes the gas temperature:

$$\lim_{s\to0}T_{rec} = T_i \quad (7.60)$$

In the other limiting case, when s>>1, the recovery temperature becomes

$$\lim_{s\to\infty} T_{rec} = \frac{\gamma-1}{\gamma+1}\frac{m}{k}u^2 \tag{7.61}$$

These two limiting cases are achieved by different physical processes: in the hypersonic case (s>>1) the recovery temperature reflects the 'ram' energy of the incident gas, while in the low speed case the recovery temperature results from thermal equilibrium between the random energies of the gas molecules and the body.

With the help of the stagnation and recovery temperatures one can define the modified thermal recovery factor, r', in the following way:

$$r'=\frac{\gamma+1}{\gamma}\frac{T_{rec}-T_i}{T_{stag}-T_i} \tag{7.62}$$

The modified thermal recovery factor characterizes the difference between the steady-state wall temperature and the stagnation temperature of the gas. The modified Stanton number, St', characterizes the ratio of the free molecular heat transfer rate (to the surface element) to the hydrodynamic heat transfer rate:

$$St'=\frac{1}{\alpha}\frac{\gamma}{\gamma+1}\frac{\Delta\Phi^{(E)}}{m\,n_i\,u\,C_p\,(T_{rec}-T_w)} \tag{7.63}$$

where C_p is the specific heat of the gas at constant pressure.

The recovery (equilibrium) temperature and the heat transfer rate can be readily calculated for a surface element if one knows the gas parameters and has a specific expression for the modified thermal recovery factor, r', and the modified Stanton number, St'. In general, the heat transfer properties of the body itself are characterized by these two parameters. In the case of real bodies the components of the relative velocity vector, **u**, vary with surface location (because the normal vector of the surface elements changes) and the calculation of the recovery temperature and free molecular heat transfer rate becomes much more complicated. In the case of bodies with large heat conductivities it is still possible to find a single recovery temperature and an average heat transfer rate, but the solution of the problem becomes much more complicated. Some simple cases will be discussed in the following section.

7.2.2 Free molecular heat transfer to specific bodies

Let us start with the simplest possible body shape and consider a flat conducting plate with a perfect insulator covering its back side. This means that the flow velocity vector, **u**, is the same for all surface elements. We will calculate r' and St' characterizing the heat transfer properties of the front side of the body.

In this case the recovery temperature is given by equation (7.59). The modified thermal recovery factor, r', can now be obtained using expression (7.62):

$$r' = \frac{1}{s^2}\left[2s^2 + 1 - \frac{e^{-s_3^2}}{e^{-s_3^2} + \sqrt{\pi}\, s_3\left[1 + \mathrm{erf}(s_3)\right]} \right] \tag{7.64}$$

The modified Stanton number, St', can also be calculated using equations (7.55), (7.59) and (7.63):

$$St' = \frac{1}{4\sqrt{\pi}\, s}\left(e^{-s_3^2} + \sqrt{\pi}\, s_3\left[1 + \mathrm{erf}(s_3)\right] \right) \tag{7.65}$$

Once we know the functions r' and St', the recovery temperature and heat transfer rate can be easily obtained by the following relations:

$$T_{rec} = T_i\left(1 + \frac{\gamma - 1}{\gamma + 1}s^2 r' \right) \tag{7.66a}$$

$$\Delta\Phi^{(E)} = \alpha\,\frac{\gamma + 1}{\gamma - 1}p_i\, u\left(1 + \frac{\gamma - 1}{\gamma + 1}s^2 r' - \frac{T_w}{T_i} \right)St' \tag{7.66b}$$

In general the functions r' and St' can be derived by integration for any body shape. In order to get the recovery temperature one has to solve the following integral equation:

$$\oint_{\text{surface}} dA\,\Delta\Phi^{(E)}(s, T_i, T_{rec}) = 0 \tag{7.67}$$

With the help of the recovery temperature one can calculate the modified recovery factor, r', for the given body shape. Similarly, the heat transfer rate can be obtained by evaluating the following integral:

$$\Delta\Phi^{(E)} = \oint_{\text{surface}} dA \, \Delta\Phi^{(E)}(s, T_i, T_w) \tag{7.68}$$

Once we know the total heat transfer rate, we can calculate the modified Stanton number, St'.

Below we present functional forms of the modified recovery factor and the modified Stanton number for several simple body shapes.

In the case of a *flat conducting plate* with front and rear surfaces in perfect thermal contact we obtain the following:

$$r' = \frac{1}{s^2}\left[2s^2 + 1 - \frac{e^{-s_3^2}}{e^{-s_3^2} + \sqrt{\pi}\, s_3 \, \mathrm{erf}(s_3)} \right] \tag{7.69a}$$

$$St' = \frac{1}{4\sqrt{\pi}\, s}\left(e^{-s_3^2} + \sqrt{\pi}\, s_3 \, \mathrm{erf}(s_3) \right) \tag{7.69b}$$

In the two limiting cases these quantities become (assuming a perpendicular incident flow, $s_3 = s$)

$$\lim_{s\to 0} r' = 4 \qquad \lim_{s\to 0} St' = \frac{1}{4\sqrt{\pi}\, s} \tag{7.70a}$$

$$\lim_{s\to\infty} r' = 2 \qquad \lim_{s\to\infty} St' = \frac{1}{4} \tag{7.70b}$$

In the case of a *right circular cylinder* with its axis perpendicular to the flow direction we obtain

$$r' = \frac{(2s^2 + 3)I_0(\tfrac{1}{2}s^2) + (2s^2 + 1)I_1(\tfrac{1}{2}s^2)}{(s^2 + 1)I_0(\tfrac{1}{2}s^2) + s^2\, I_1(\tfrac{1}{2}s^2)} \tag{7.71a}$$

$$St' = \frac{\exp(-\tfrac{1}{2}s^2)}{4\sqrt{\pi}}\left(\frac{s^2 + 1}{s}I_0(\tfrac{1}{2}s^2) + s\, I_1(\tfrac{1}{2}s^2) \right) \tag{7.71b}$$

where $I_0(x)$ and $I_1(x)$ are modified Bessel function of integer order (see Appendix 9.5.4). In the two limiting cases these expressions become

$$\lim_{s\to 0} r' = 3 \qquad \lim_{s\to 0} St' = \frac{1}{4\sqrt{\pi}\,s} \qquad\qquad (7.72a)$$

$$\lim_{s\to \infty} r' = 2 \qquad \lim_{s\to \infty} St' = \frac{1}{2\pi} \qquad\qquad (7.72b)$$

In the case of a *sphere* we get the following expression:

$$r' = \frac{\left(2s+\dfrac{1}{s}\right)\dfrac{e^{-s^2}}{\sqrt{\pi}} + \left(2s^2+2-\dfrac{1}{2s^2}\right)\mathrm{erf}(s)}{s\dfrac{e^{-s^2}}{\sqrt{\pi}} + \left(s^2+\dfrac{1}{2}\right)\mathrm{erf}(s)} \qquad\qquad (7.73a)$$

$$St' = \frac{e^{-s^2}}{8s\sqrt{\pi}} + \frac{1}{8}\left(1+\frac{1}{2s^2}\right)\mathrm{erf}(s) \qquad\qquad (7.73b)$$

In the two limiting cases these quantities become

$$\lim_{s\to 0} r' = \frac{8}{3} \qquad \lim_{s\to 0} St' = \frac{1}{4\sqrt{\pi}\,s} \qquad\qquad (7.74a)$$

$$\lim_{s\to \infty} r' = 2 \qquad \lim_{s\to \infty} St' = \frac{1}{8} \qquad\qquad (7.74b)$$

The functions, $r'(s)$ and $St'(s)$, are shown in Figures 7.7 and 7.8 for the various simple body shapes discussed in the text (assuming perpendicular flow, i.e. $s_3=s$). Inspection of these figures reveals that the thermal properties of an insulated plate are very different than those for other body shapes. The reason is that for body shapes other than the insulated plate the angle between the normal vector of a surface element and the incident flow varies over a wide range and therefore the heat transfer is 'averaged' over these incident angles. In the case of the insulated plate the impact angle is the same for all surface elements.

7.2.3 *Heat transfer between two plates*

Let us consider two infinite plates (with front and back faces thermally insulated) facing each other and calculate the free molecular heat exchange between them. It

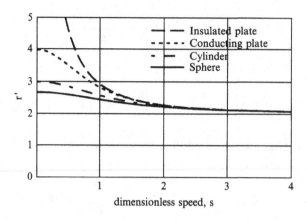

Figure 7.7

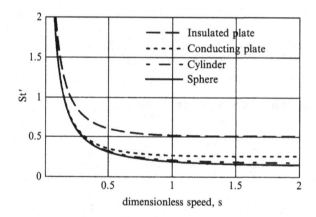

Figure 7.8

is assumed that the distance between the plates is much smaller than the mean free path of interparticle collisions, therefore the flow between the two planes is free molecular.

We assume that each plate emits particles with equilibrium velocity distributions characterized by the plate temperatures, T_1 and T_2. It is obvious from Figure 7.9 that all particles emitted from plate 1 have $v_3 < 0$, while all particles emitted from plate 2 have $v_3 > 0$. In this approximation the emitted velocity distribution functions are half Maxwellians:

$$F_1 = \begin{cases} 0 & \text{for } v_3 \geq 0 \\ 2n_1 \left(\dfrac{m}{2\pi kT_1} \right)^{3/2} \exp\left(-\dfrac{mv^2}{2kT_1} \right) & \text{for } v_3 < 0 \end{cases} \qquad (7.75a)$$

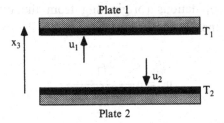

Figure 7.9

$$F_2 = \begin{cases} 2n_2 \left(\dfrac{m}{2\pi kT_2} \right)^{3/2} \exp\left(-\dfrac{mv^2}{2kT_2} \right) & \text{for } v_3 \geq 0 \\ 0 & \text{for } v_3 < 0 \end{cases} \qquad (7.75b)$$

where n_1 and n_2 represent the number densities of the particles leaving plate 1 and plate 2, respectively. The factor of 2 in the normalization comes from the fact that F_1 and F_2 are normalized to n_1 and n_2, respectively.

It is important to note that from the point of view of the absorbing surfaces, plate 2 and plate 1, the incident particles behave as full Maxwellian velocity distributions with spatial densities of $2n_1$ and $2n_2$, respectively. The reason the densities appear doubled is that the absorbing surface would never 'see' the missing part of the velocity distribution (these are the molecules which move away from the absorbing surface). This means that the absorption takes place as if the incident particles had zero flow velocity in the x_3 direction, so $u_1 = u_2 = 0$.

In this zero flow velocity limit the recovery temperatures are identical to the temperatures of the other plate, $T_{rec}^{(1)} = T_2$ and $T_{rec}^{(2)} = T_1$. One can substitute this result into equation (7.55) to obtain the heat transfer rate from plate 2 to plate 1:

$$\Delta\Phi_1^{(E)} = \frac{\alpha}{4} n_2 \, \bar{v}_2 \, k \, \frac{\gamma+1}{(\gamma-1)} \left(T_2 - T_1 \right) \qquad (7.76)$$

Under equilibrium conditions the absorbed and re-emitted particle fluxes are identical for each plate, therefore the following relation must be satisfied:

$$n_1 \sqrt{T_1} = n_2 \sqrt{T_2} \qquad (7.77)$$

The particle density between the two plates is n, which is the sum of the number densities of the two populations (originating from the two plates), therefore $n=n_1+n_2$. One can combine these two relations to obtain the following:

$$n_2 = n \frac{\sqrt{T_1}}{\sqrt{T_1} + \sqrt{T_2}} \qquad (7.78)$$

Substituting this result into equation (7.76) yields the following expression for the heat transfer rate to plate 1:

$$\Delta\Phi_1^{(E)} = \alpha\, n\, k \sqrt{\frac{k}{2\pi m}} \frac{\gamma+1}{(\gamma-1)} \frac{\sqrt{T_2 T_1}}{\sqrt{T_1} + \sqrt{T_2}} (T_2 - T_1) \qquad (7.79)$$

Finally, let us compare this result with the hydrodynamic (Kn<<1) case. In the Navier—Stokes approximation the heat flux was found to be the following (see equation (6.79)):

$$\Phi_{NS}^{(E)} = -\frac{5}{8}\frac{k}{\sigma} \sqrt{\frac{\pi k}{m}} \sqrt{T} \frac{dT}{dx_3} \qquad (7.80)$$

where we have approximated the average collision time as the ratio of the collisional mean free path and the average thermal velocity. If a high density gas fills the gap between the two infinite plates this expression yields the following heat transfer rate to plate 1:

$$\Delta\Phi_{NS,1}^{(E)} = -\frac{5}{8}\frac{k}{\sigma} \sqrt{\frac{\pi k}{m}} \frac{\sqrt{T_1} + \sqrt{T_2}}{2} \frac{T_1 - T_2}{L} \qquad (7.81)$$

where we have assumed that the temperature gradient was constant between the two plates separated by a distance L. Also, the temperature dependent heat conductivity coefficient was approximated by its average value.

The ratio of the hydrodynamic and free molecular heat transfer rates is the following:

$$\frac{\Delta\Phi_{NS,1}^{(E)}}{\Delta\Phi_{\text{free mol,1}}^{(E)}} = \frac{5}{8}\frac{\pi}{\alpha} \frac{\gamma-1}{\gamma+1} \frac{\left(\sqrt{T_1} + \sqrt{T_2}\right)^2}{\sqrt{T_2 T_1}} \frac{\lambda}{L} \qquad (7.82)$$

where λ is the mean free path of molecular collisions. In the free molecular limit $\lambda \gg L$, therefore the improper use of hydrodynamic formulas in free molecular situations might lead to gross overestimations of the heat transfer rate.

7.3 Free molecular aerodynamic forces

Aerodynamic forces act upon solid objects moving in fluids. In this section we will consider two types of forces: drag and lift. The drag force is parallel to the relative velocity vector between the body and the gas, **u**, and it represents the 'resistance' of the medium, which tends to eliminate the relative motion. The lift force is perpendicular to the velocity vector, and tends to change the 'course' of the body. Both of these aerodynamic forces depend on the reflection characteristics and the shape of the solid body. These dependencies will be explored in this section.

7.3.1 *Pressure and shearing stress*

Figure 7.10 shows a schematic representation of the interaction geometry and the coordinate system for a simple rectangular body. The x_3 axis is perpendicular to the surface and points inward to the body. The angle between the surface normal vector and the direction of the incident flow is denoted by θ, which is identical to the polar angle in our coordinate system. The angle between the surface and the incident flow velocity, **u**, is denoted by α_i. It is obvious from the schematics that $\theta + \alpha_i = \pi/2$. Using these angles one can express the normal and tangential components of the normalized velocity vector:

$$s_3 = s \sin \alpha_i \qquad s_t = s \cos \alpha_i \qquad (7.83)$$

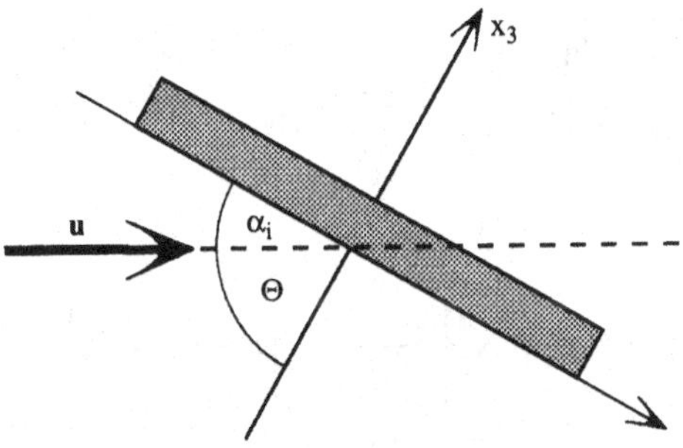

Figure 7.10

where s_t represents the tangential component of the normalized relative velocity vector.

Next we calculate the pressure and shearing stress acting on a surface element using the results of the previous section. Here, however, we do not assume perfect accommodation, but will make use of the reflection coefficients, σ_t and σ_n, given by equations (7.2a) and (7.2b). Using these two equations one can express the contribution of reflected molecules to the pressure and shearing stress:

$$p_r = (1-\sigma_n)p_i + \sigma_n p_w \tag{7.84}$$

$$\tau_r = (1-\sigma_t)\tau_i \tag{7.85}$$

where we have made use of the fact that the pressure (and shearing stress) is defined as the momentum transfer rate per unit surface area. With the help of these two expressions one can easily obtain the total pressure and shearing stress acting on the surface element:

$$p = p_i + p_r = (2-\sigma_n)p_i + \sigma_n p_w \tag{7.86}$$

$$\tau = \tau_i - \tau_r = \sigma_t \tau_i \tag{7.87}$$

The negative sign for τ_r comes from the fact that only specularly reflected molecules contribute to τ_r, and they represent a shear with opposite sign to τ_i.

Next we substitute expressions (7.12), (7.15) and (7.21) into equations (7.86) and (7.87). After some elementary manipulations one obtains the following results for the total pressure and shearing stress acting on the surface:

$$p = p_i \left\{ \left(\frac{2-\sigma_n}{\sqrt{\pi}} s_3 + \frac{\sigma_n}{2}\sqrt{\frac{T_w}{T_i}} \right) \left(e^{-s_3^2} + \sqrt{\pi}\, s_3 \left[1 + \mathrm{erf}(s_3) \right] \right) + \frac{2-\sigma_n}{2} \left[1 + \mathrm{erf}(s_3) \right] \right\} \tag{7.88}$$

$$\tau = \sigma_t p_i s_t \left\{ \frac{e^{-s_3^2}}{\sqrt{\pi}} + s_3 \left[1 + \mathrm{erf}(s_3) \right] \right\} \tag{7.89}$$

The total force acting on a particular body can be obtained by integrating the appropriate projections of the local pressures and shearing stresses over the entire surface. In general, the net force components depend on the ambient gas pressure, p_i, the angle of attack, α_i, the normalized relative speed, s, and the surface temperature, T_w (and hence the accommodation coefficient α), as well as the reflection coefficients, σ_t and σ_n.

7.3.2 *Lift and drag forces*

Figure 7.11 shows a schematic representation of the lift and drag forces (F_L and F_D, respectively) acting on a surface element. The lift and drag force per unit surface area (dF_L/dA and dF_D/dA) can be expressed with the help of the pressure and shearing stress (these represent the normal and tangential force components) in the following way:

$$\frac{dF_L}{dA} = p\cos\alpha_i - \tau\sin\alpha_i \tag{7.90}$$

$$\frac{dF_D}{dA} = p\sin\alpha_i + \tau\cos\alpha_i \tag{7.91}$$

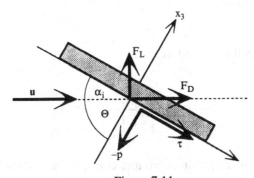

Figure 7.11

The total aerodynamic forces can be obtained by integrating dF_L/dA and dF_D/dA over the entire surface. After we obtain the total lift and drag force, F_L and F_D, the results can be expressed in terms of lift and drag coefficients, C_L and C_D, which characterize the particular body shape:

$$C_L = \frac{F_L}{\frac{1}{2}mn_i u^2 A} \tag{7.92}$$

$$C_D = \frac{F_D}{\frac{1}{2}mn_i u^2 A} \tag{7.93}$$

where A is the reference area of the body which must be specified in each case.

7.3.3 Free molecular aerodynamic coefficients for specific bodies

Finally, we summarize some aerodynamic coefficient functions for simple body shapes.

First we consider a flat plate at an angle of attack α_i and surface area A (here A is the area of one side of the plate). We present the aerodynamic coefficients for the two limiting cases (diffuse and specular reflections). In the case of diffuse reflection ($\sigma_t=1$ and $\sigma_n=1$) these coefficients become

$$C_{D,\text{diff}} = \frac{2}{\sqrt{\pi}} \left[\frac{e^{-s_3^2}}{s} + \sqrt{\pi} \sin\alpha_i \left(1 + \frac{1}{2s^2}\right) \text{erf}(s_3) + \pi \frac{\sin^2\alpha_i}{s_w} \right] \qquad (7.94)$$

$$C_{L,\text{diff}} = \cos\alpha_i \left[\frac{\text{erf}(s_3)}{s^2} + \sqrt{\pi} \frac{\sin\alpha_i}{s_w} \right] \qquad (7.95)$$

where we have introduced the notation

$$s_w = \sqrt{\frac{m\,u^2}{2\,k\,T_w}} \qquad (7.96)$$

In the case of specular reflection ($\sigma_t=0$ and $\sigma_n=0$) the aerodynamic coefficients become the following:

$$C_{D,\text{spec}} = \frac{4}{\sqrt{\pi}} \left[\frac{\sin^2\alpha_i}{s} e^{-s_3^2} + \sqrt{\pi} \sin\alpha_i \left(\frac{1}{2\,s^2} + \sin^2\alpha_i\right) \text{erf}(s_3) \right] \qquad (7.97)$$

$$C_{L,\text{spec}} = \cot\alpha_i\, C_{L,\text{diff}} \qquad (7.98)$$

Next let us consider the behavior of these coefficients in limiting cases. In the case of low speed flow (when T_i remains finite but $\mathbf{u}\to 0$, and consequently $s\to 0$) the aerodynamic lift and drag coefficients for a flat plate become the following:

$$\lim_{u \to 0} C_{D,\text{diff}} = \frac{2}{\sqrt{\pi}\, s}\left[1 + \sin^2 \alpha_i \left(1 + \pi\sqrt{\frac{T_w}{T_i}}\right)\right] \tag{7.99}$$

$$\lim_{u \to 0} C_{D,\text{spec}} = \frac{8}{\sqrt{\pi}}\frac{\sin^2 \alpha_i}{s} \tag{7.100}$$

$$\lim_{u \to 0} C_{L,\text{diff}} = \frac{\sin \alpha_i \cos \alpha_i}{\sqrt{\pi}\, s}\left(2 + \pi\sqrt{\frac{T_w}{T_i}}\right) \tag{7.101}$$

$$\lim_{u \to 0} C_{L,\text{spec}} = \frac{\cos^2 \alpha_i}{\sqrt{\pi}\, s}\left(2 + \pi\sqrt{\frac{T_w}{T_i}}\right) \tag{7.102}$$

It is interesting to note that equations (7.99) through (7.102) all have an s^{-1} type singularity as $s \to 0$. This does not mean that the drag or lift forces become infinite in the low speed limit. For instance, the limiting value of the drag force is the following:

$$\lim_{u \to 0} F_D = \lim_{u \to 0} \frac{1}{2} m n_i\, u^2 A\, C_D = \lim_{u \to 0} p\, s^2 A\, C_D \propto \lim_{u \to 0} s = 0 \tag{7.103}$$

The second limiting case is that of the hypersonic flow, in which both s and s_3 are much larger than unity, $s \gg 1$, and $s_3 \gg 1$. In this limit the aerodynamic coefficients become the following:

$$\lim_{s,s_3 \gg 1} C_{D,\text{diff}} = 2 \sin \alpha_i \tag{7.104}$$

$$\lim_{s,s_3 \gg 1} C_{D,\text{spec}} = 4 \sin^3 \alpha_i \tag{7.105}$$

$$\lim_{s,s_3 \gg 1} C_{L,\text{diff}} = \frac{\sin \alpha_i \cos \alpha_i}{s}\sqrt{\pi \frac{T_w}{T_i}} \tag{7.106}$$

$$\lim_{s,s_3 \gg 1} C_{L,\text{spec}} = \frac{\cos^2 \alpha_i}{s}\sqrt{\pi \frac{T_w}{T_i}} \tag{7.107}$$

The hypersonic behavior of the drag and lift forces can be obtained using the limiting values derived above:

$$\lim_{s,s_3 \gg 1} F_D = \lim_{s,s_3 \gg 1} \frac{1}{2} mn_i \, u^2 A C_D \lim_{s,s_3 \gg 1} p \, s^2 A C_D \propto s^2 \qquad (7.108)$$

$$\lim_{s,s_3 \gg 1} F_L = \lim_{s,s_3 \gg 1} \frac{1}{2} mn_i \, u^2 A C_L \lim_{s,s_3 \gg 1} p \, s^2 A C_L \propto s \qquad (7.109)$$

Next, we consider a right circular cylinder with its axis normal to the flow and with projected area A=2ah (where a and h are the radius and height of the cylinder, respectively). In this case the lift force (and consequently the lift coefficient) is zero because of the symmetric geometry. The drag coefficients are the following:

$$C_{D,\text{diff}} = \frac{\sqrt{\pi}}{s} \exp\left(-\frac{s^2}{2}\right)\left[\left(s^2+\frac{3}{2}\right)I_0\left(\tfrac{1}{2}s^2\right)+\left(s^2+\frac{1}{2}\right)I_1\left(\tfrac{1}{2}s^2\right)\right]+\frac{\sqrt{\pi^3}}{4\,s_w}$$

$$(7.110)$$

$$C_{D,\text{spec}} = \frac{4\sqrt{\pi}}{3s} \exp\left(-\frac{s^2}{2}\right)\left[\left(s^2+\frac{3}{2}\right)I_0\left(\tfrac{1}{2}s^2\right)+\left(s^2+\frac{1}{2}\right)I_1\left(\tfrac{1}{2}s^2\right)\right] \qquad (7.111)$$

where I_0 and I_1 refer to the modified Bessel functions.

In the case of low speed flow (when T_i remains finite but $u \to 0$, and consequently $s \to 0$) the aerodynamic drag coefficients for the right circular cylinder become the following:

$$\lim_{s \to 0} C_{D,\text{diff}} = \frac{\sqrt{\pi}}{s}\left(\frac{3}{2}+\frac{\pi}{4}\sqrt{\frac{T_w}{T_i}}\right) \qquad (7.112)$$

$$\lim_{s \to 0} C_{D,\text{spec}} = \frac{2\sqrt{\pi}}{s} \qquad (7.113)$$

In the case of hypersonic flow ($s \gg 1$, and $s_3 \gg 1$) the drag coefficients become the following:

$$\lim_{s \gg 1} C_{D,\text{diff}} = 2 \qquad (7.114)$$

$$\lim_{s \gg 1} C_{D,\text{spec}} = \frac{8}{3} \tag{7.115}$$

Finally, we consider a *sphere* with projected area $A = r^2\pi$ (where r is the radius of the sphere). In this case there is no lift and the drag coefficients are the following:

$$C_{D,\text{diff}} = \frac{1 + 2s^2}{s^3\sqrt{\pi}} e^{-s^2} + \frac{4s^4 + 4s^2 - 1}{2s^4} \text{erf}(s) + \frac{2\sqrt{\pi}}{3s_w} \tag{7.116}$$

$$C_{D,\text{spec}} = \frac{1 + 2s^2}{s^3\sqrt{\pi}} e^{-s^2} + \frac{4s^4 + 4s^2 - 1}{2s^4} \text{erf}(s) \tag{7.117}$$

In the low speed limit the aerodynamic drag coefficients for the sphere become the following:

$$\lim_{s \to 0} C_{D,\text{diff}} = \frac{16}{3} \frac{1}{\sqrt{\pi} s} + \frac{2\sqrt{\pi}}{3s} \sqrt{\frac{T_w}{T_i}} \tag{7.118}$$

$$\lim_{s \to 0} C_{D,\text{spec}} = \frac{16}{3} \frac{1}{\sqrt{\pi} s} \tag{7.119}$$

When deriving these limits we had to use a fifth order expansion for the error function to account for all first order terms (because there is an s^4 term in the denominator). The hypersonic limit of the drag coefficients gives the following result:

$$\lim_{s \gg 1} C_{D,\text{diff}} = 2 \tag{7.120}$$

$$\lim_{s \gg 1} C_{D,\text{spec}} = 2 \tag{7.121}$$

Figure 7.12 shows the free molecular drag coefficients (assuming specular reflection) as a function of the dimensionless speed, s, for a plate (with 45° angle of attack), a cylinder, and a sphere.

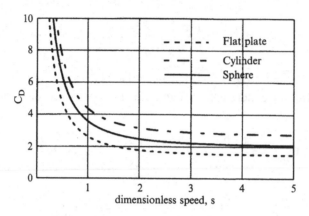

Figure 7.12

7.4 Problems

7.1 Two infinite plates are set at a distance d apart in a gas at very low pressure with mean free path $\lambda \gg d$. The temperatures of the plates are T_1 and T_2, respectively. The temperature of the surrounding gas is equal to T_2, while n is the number density of the gas. Find the force of repulsion per unit area between the plates. Assume that edge effects can be neglected and that diffuse reflection with $\alpha = \sigma_t = \sigma_n = 1$ occurs at the plates.

7.2 A small spherical chamber is immersed in a low density stream, where the mean free path is long compared to the sphere diameter. Derive the steady-state density of the gas within the chamber. The temperature of the sphere is T_0, the incident gas density, bulk velocity and temperature are n, u, and T, respectively. The angle between the flow direction and the normal vector of the orifice (pointing into the spherical chamber) is $0°$.

7.3 A very thin plate with a radius of 0.1 m and mass 0.01 kg levitates 100 m above the surface in the atmosphere of an earth size planet of unknown mass. The plate consists of two well conducting metals separated by an insulating layer. The lower part of the plate is kept at a temperature $T_1 = 10\,000$ K, while the upper part is in thermal equilibrium with the atmosphere at $T_0 = 200$ K. Assume perfect accommodation and calculate the mass ratio between the unknown planet and earth if the atmospheric density is $n = 10^{19}$ m^{-3} and the collision cross section of the molecules is $\sigma = 10^{-20}$ m^2.

7.4 A spherical spacecraft orbits the earth along a circular trajectory with an orbital velocity of 7.8 km/s. The number density in the high altitude atmosphere can be approximated by a barometric formula:

$$n(z) = n_0 \exp\left(-\frac{z - z_0}{H}\right)$$

where z_0=200 km, n_0=8×10^{15} m^{-3} and the scale height, H, is H=15 km. Assume that the temperature of the upper atmosphere is T=1000 K at all altitudes, the mean molecular mass and the specific heat ratio are 22 amu (m=3.67×10^{-26} kg) and 7/5, respectively. (a) Assume that the wall temperature of the spacecraft is uniform and calculate the heat transfer rate to the satellite. Show that to a very good approximation the heat transfer rate is independent of the spacecraft temperature. (b) Assume that the only energy loss of the spacecraft is the black body radiation (black body energy loss per unit time per unit spacecraft area = $\sigma_{SB} T_{sc}^4$, where σ_{SB} is the Stefan–Boltzmann constant, σ_{SB}=5.67×10^{-8} joule m^{-2} K^{-4} s^{-1}). Calculate the equilibrium spacecraft temperature at z=100 km and at z=200 km. Can a spacecraft survive at these altitudes?

7.5 References

[1] Shaaf, S.A., and Chambré, P.L., *Flow of rarefied gases*, Princeton Aeronautical Paperbacks, Princeton University Press, Princeton, NJ, 1961.
[2] Smoluchowski, M.V., *Wied. Ann.*, **64**, 101, 1898.
[3] Knudsen, M., *Ann Phys.*, **34**, 593, 1911.

8

Shock waves

It has been observed under certain conditions that a compressible fluid can experience an abrupt change of macroscopic parameters. Examples are detonation waves, explosions, wave systems formed at the nose of projectiles moving with supersonic speeds, etc. In all these cases the wave front is very steep and there is a large pressure rise in traversing the wave, which is called a shock wave.

In the perfect gas approximation shock waves are discontinuity surfaces separating two distinct gas states. In higher order approximations (such as the Navier–Stokes equation) the shock wave is a region where physical quantities change smoothly but rapidly. In this case the shock has a finite thickness, generally of the order of the mean free path.

Since the shock wave is a more or less instantaneous compression of the gas, it cannot be a reversible process. The energy for compressing the gas flowing through the shock wave is derived from the kinetic energy of the bulk flow upstream of the shock wave. It can be shown that a shock wave is not an isentropic phenomenon: the gas experiences an increase in its entropy.

The simplest case for studying shock waves is a normal plane shock wave, when the gas flows parallel to the x axis and all physical quantities depend only on the x coordinate. In this case the normal vector of the shock surface is parallel to the flow direction. In this chapter we compare hydrodynamic and simple kinetic descriptions of a steady normal plane shock wave.

8.1 Hydrodynamic description

8.1.1 Normal shock waves in perfect gases[1,2]

First we consider the hydrodynamic description of shock waves in perfect gases. For the sake of mathematical simplicity let us consider a steady-state one dimensional flow in the x direction with constant flow area and neglect the effects of external forces. Such a flow occurs for instance in a very long tube. Also, we assume that the shock wave is perpendicular to the flow direction. The direction of shock waves is usually characterized by the shock normal vector. Shock waves with normal vectors parallel to the flow direction are usually referred to as normal shock waves (sometimes they are also called parallel shocks).

It is assumed that far upstream and downstream of the shock the gas is very close to thermodynamic equilibrium. This means that in these two regions the macroscopic conservation equations can be obtained using the five moment approximation (Euler equations). In the present one dimensional, steady-state (time independent) case these equations simplify to the following:

$$\frac{d(n\,u)}{dx} = 0 \tag{8.1}$$

$$m\,n\,u\frac{du}{dx} + \frac{dp}{dx} = 0 \tag{8.2}$$

$$\frac{3}{2}u\frac{dp}{dx} + \frac{5}{2}p\frac{du}{dx} = 0 \tag{8.3}$$

where $n(x)$, $u(x)$ and $p(x)$ represent the particle number density, flow velocity and pressure, respectively.

The first two equations can immediately be integrated to yield

$$n_1\,u_1 = n_2\,u_2 \tag{8.4}$$

$$mn_1\,u_1^2 + p_1 = mn_2\,u_2^2 + p_2 \tag{8.5}$$

where the subscripts 1 and 2 refer to far upstream and far downstream conditions, respectively. The energy equation can be rewritten into a total derivative using the momentum equation:

$$\frac{3}{2}u\frac{dp}{dx} + \frac{5}{2}p\frac{du}{dx} = \frac{5}{2}\frac{d(up)}{dx} - u\frac{dp}{dx} = \frac{d}{dx}\left(\frac{5}{2}up + \frac{1}{2}mnu^3\right) \tag{8.6}$$

This form of the energy equation can easily be integrated to yield the following conservation relation:

$$\frac{\gamma}{\gamma-1} u_1 p_1 + \frac{1}{2} mn_1 u_1^3 = \frac{\gamma}{\gamma-1} u_2 p_2 + \frac{1}{2} mn_2 u_2^3 \qquad (8.7)$$

where γ is the specific heat ratio (in our case naturally $\gamma=5/3$).

Next we want to rewrite the conservation equations (8.4), (8.5) and (8.7) in terms of a single characteristic parameter, the upstream Mach number, M_1. The Mach number is the ratio of the flow speed to the local sound speed and it can be expressed as

$$M^2 = \frac{mnu^2}{\gamma p} \qquad (8.8)$$

With the help of this expression the integral form of the momentum and energy equations (equations (8.5) and (8.7)) can be rewritten as

$$p_1 \left(1 + \gamma M_1^2\right) = p_2 \left(1 + \gamma M_2^2\right) \qquad (8.9)$$

$$u_1 p_1 \left(1 + \frac{\gamma-1}{2} M_1^2\right) = u_2 p_2 \left(1 + \frac{\gamma-1}{2} M_2^2\right) \qquad (8.10)$$

With the help of the continuity and energy equations one can immediately express the ratio of the upstream and downstream temperatures:

$$\frac{T_1}{T_2} = \frac{n_1 u_1}{n_2 u_2} \frac{T_1}{T_2} = \frac{p_1 u_1}{p_2 u_2} = \frac{2 + (\gamma-1) M_2^2}{2 + (\gamma-1) M_1^2} \qquad (8.11)$$

It follows from the definition of the Mach number (equation (8.8)) that the gas velocity, u, is proportional to the Mach number times the square root of the temperature, $u \propto MT^{1/2}$. With the help of this relation and of equation (8.11) one obtains the following expression for the velocity ratio:

$$\frac{u_1}{u_2} = \frac{M_1}{M_2} \sqrt{\frac{T_1}{T_2}} = \frac{M_1}{M_2} \sqrt{\frac{2 + (\gamma-1) M_2^2}{2 + (\gamma-1) M_1^2}} \qquad (8.12)$$

The same velocity ratio can also be expressed independently with the help of equations (8.4), (8.9) and (8.11):

$$\frac{u_1}{u_2} = \frac{1+\gamma M_1^2}{1+\gamma M_2^2}\frac{2+(\gamma-1)M_2^2}{2+(\gamma-1)M_1^2} \tag{8.13}$$

Equations (8.12) and (8.13) can be combined to obtain an algebraic equation connecting the upstream and downstream Mach numbers:

$$\frac{1+\gamma M_1^2}{1+\gamma M_2^2} = \frac{M_1}{M_2}\sqrt{\frac{2+(\gamma-1)M_1^2}{2+(\gamma-1)M_2^2}} \tag{8.14}$$

This equation has two solutions. The first solution describes no change in the gas parameters, $M_2 = M_1$. This trivial solution, of course, is not a shock. The second solution is non-trivial: it relates the upstream and downstream Mach numbers for a normal shock wave:

$$M_2 = \sqrt{\frac{2+(\gamma-1)M_1^2}{2\gamma M_1^2 - (\gamma-1)}} \tag{8.15}$$

It is important to note that this solution has a singularity at $M_1{}^2 = (\gamma-1)/2\gamma$. Also, the limiting value of $M_2{}^2$ as M_1 goes to infinity turns out to be the same value, $(\gamma-1)/2\gamma$. This function is shown in Figure 8.1. Inspection of Figure 8.1 reveals that for supersonic upstream flows, $M_1 > 1$, the downstream Mach number is subsonic. This is a physically meaningful solution describing a normal shock wave. For subsonic upstream Mach numbers, however, the solution becomes unphysical with supersonic downstream Mach numbers. This result essentially means that normal shocks can only be formed in supersonic flows, where the condition $M_1 > 1$ is fulfilled.

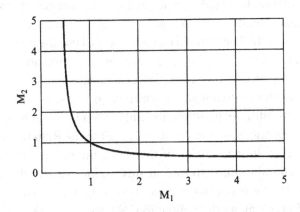

Figure 8.1

The normal shock wave is a sharp (infinitely thin) discontinuity in the gas parameters. Equation (8.15) gives a unique relation between the upstream and downstream Mach numbers. With the help of this equation one can express the change of macroscopic physical quantities through a normal shock wave:

$$\frac{T_2}{T_1} = \frac{\left(2 + (\gamma-1)M_1^2\right)\left(2\gamma M_1^2 - (\gamma-1)\right)}{(\gamma+1)^2 M_1^2} \tag{8.16}$$

$$\frac{n_2}{n_1} = \frac{u_1}{u_2} = \frac{(\gamma+1)M_1^2}{2 + (\gamma-1)M_1^2} \tag{8.17}$$

$$\frac{p_2}{p_1} = \frac{2\gamma M_1^2 - (\gamma-1)}{\gamma+1} \tag{8.18}$$

Equations (8.16) through (8.18) are the well known Rankine–Hugoniot relations for normal shock waves in perfect gases.

8.1.2 *The structure of shock waves in viscous gases*

So far in this chapter, shock waves were regarded as surfaces of zero thickness. Next we consider the thickness of these discontinuities and we shall see that shock waves in reality are transition layers with finite thickness. In perfect gases shock waves are idealized discontinuities with no internal structure. In order to obtain shocks with finite thickness one has to introduce some kind of localized dissipation mechanism (such as viscosity) near the shock. This dissipation converts bulk energy to random energy and thus initiates the shock transition. The upstream gas parameters must still satisfy the Rankine–Hugoniot jump conditions, but the transition will take place over a region of finite thickness. In this section we will use the Navier–Stokes transport equation set (corresponding to the 13 moment approximation) which is capable of describing dissipation due to viscosity and heat conduction.

It is assumed that far upstream and downstream of the shock region, the gas is in thermodynamic equilibrium, therefore the Euler equations can be used to describe the conservation equations. These equations result in the Rankine–Hugoniot jump conditions. However, in the vicinity of the shock the gas can deviate from local thermodynamic equilibrium, therefore in this region we use the Navier–Stokes equations. Consider the Navier–Stokes equations for a steady-state gas flow of a monatomic perfect gas along the x direction. Neglecting mass addition, the continuity equation is the same as it was before, therefore equations (8.1) and (8.4) are valid throughout the shock region. This means that the particle flux is conserved everywhere:

$$j_1 = n_1 u_1 = n(x) u(x) \tag{8.19}$$

where the subscripts 1 and 2 again refer to far upstream and downstream conditions, respectively.

The steady-state, one dimensional momentum equation in the Navier–Stokes approximation can be written as

$$mn\, u \frac{du}{dx} + \frac{dp}{dx} - \frac{d}{dx}\left(\frac{4}{3}\eta \frac{du}{dx} \right) = 0 \tag{8.20}$$

This equation can readily be integrated to yield the following algebraic conservation relation:

$$mn\, u^2 + p - \frac{4}{3}\eta \frac{du}{dx} = mn_1\, u_1^2 + p_1 \tag{8.21}$$

Here quantities without subscripts refer to conditions inside the dissipation region. This form of the momentum conservation equation reflects the fact that dissipation processes are neglected in the far upstream region.

The energy equation in the steady-state, one dimensional Navier–Stokes approximation can be written in the following form:

$$\frac{3}{2} u \frac{dp}{dx} + \frac{5}{2} p \frac{du}{dx} - \frac{4}{3}\eta \left(\frac{du}{dx} \right)^2 - \frac{d}{dx}\left(\kappa \frac{dT}{dx} \right) = 0 \tag{8.22}$$

Multiplying the momentum equation by u and adding the result to the energy equation yields

$$\frac{5}{2}\frac{d(u\,p)}{dx} + \frac{1}{2}\frac{d(mn\,u^3)}{dx} - \frac{d}{dx}\left(\frac{2}{3}\eta \frac{du^2}{dx} \right) - \frac{d}{dx}\left(\kappa \frac{dT}{dx} \right) = 0 \tag{8.23}$$

This form of the energy equation can easily be integrated:

$$\frac{5}{2} u\,p + \frac{1}{2} mn\, u^3 - \frac{2}{3}\eta \frac{du^2}{dx} - \kappa \frac{dT}{dx} = \frac{5}{2} u_1\, p_1 + \frac{1}{2} mn_1\, u_1^3 \tag{8.24}$$

This form of the energy equation can be further manipulated to yield the following result:

$$\frac{1}{2}(u^2 - u_1^2) + C_p(T - T_1) = \frac{\kappa}{m\,j_1\,C_p}\left[\frac{4}{3}Pr\frac{d}{dx}\left(\frac{1}{2}u^2\right) + C_p\frac{dT}{dx}\right] \quad (8.25)$$

where $C_p = 5k/2m$ is the specific heat of a monatomic gas (considered here) at constant pressure, and Pr is the Prandtl number (given by equation (4.65)).

There is no known general analytic solution to this differential equation system. However, the conservation equations can be integrated for Pr=3/4, which is a good approximation of the actual value over a wide temperature range. Assuming Pr=3/4 the energy conservation equation becomes the following:

$$\frac{1}{2}(u^2 - u_1^2) + C_p(T - T_1) = \frac{4\eta}{3m\,j_1}\frac{d}{dx}\left[\frac{1}{2}(u^2 - u_1^2) + C_p(T - T_1)\right] \quad (8.26)$$

The solution of this equation is the following.[3]

$$\frac{1}{2}(u^2 - u_1^2) + C_p(T - T_1) = C_1\exp\left(\frac{4}{3m\,j_1}\int_{-\infty}^{x}dx\,\eta\right) \quad (8.27)$$

where C_1 is an integration constant. It can be seen that the only physical solution is the case $C_1=0$, otherwise the total energy density of the gas would change exponentially with location. In this case the energy conservation equation reduces to

$$\frac{1}{2}(u^2 - u_1^2) + \frac{5}{2}\frac{k}{m}(T - T_1) = 0 \quad (8.28)$$

With the help of the continuity equation the momentum equation can be expressed as follows:

$$mj_1(u - u_1) + (p - p_1) = \frac{4}{3}\eta\frac{du}{dx} \quad (8.29)$$

With the help of equation (8.28) this can be rewritten as

$$\frac{5}{3}\frac{\eta}{m\,j_1}V\frac{dV}{dx}-(V-1)(V-\alpha)=0 \tag{8.30}$$

where $V=u/u_1$ and the dimensionless constant, α, is defined as

$$\alpha = \frac{1}{4}\left(1+\frac{3}{M_1^2}\right) \tag{8.31}$$

In general the viscosity is a temperature dependent physical quantity. Equation (8.30) can be solved analytically for an arbitrary power law dependence.[3] However, for the sake of mathematical simplicity (and without loss of physical generality) we consider the simplest problem, when the approximation $\eta = \eta_1$ can be applied. In this case one can use the simple expression obtained for the coefficient of viscosity with the help of the mean free path method (equation (4.54)) to obtain

$$\frac{5}{3}\frac{\eta}{m\,j_1}=\frac{5}{9}\frac{v_1}{u_1}\lambda_1=\frac{2}{9}\sqrt{\frac{30}{\pi}}\frac{1}{M_1}\lambda_1 \tag{8.32}$$

With the help of this expression the differential equation (8.30) can be written in the following form:

$$\frac{V}{(V-1)(V-\alpha)}dV=\frac{9}{2}\sqrt{\frac{\pi}{30}}M_1\,dX \tag{8.33}$$

where the dimensionless variable, X, is defined as $X=x/\lambda_1$. Equation (8.33) can be immediately integrated to yield the following implicit equation for V:

$$\frac{1-V}{(V-\alpha)^\alpha}=C_2\exp\left(\frac{9}{2}\sqrt{\frac{\pi}{30}}(1-\alpha)M_1\,X\right) \tag{8.34}$$

where C_2 is an integration constant.

The boundary condition far upstream is that $u=u_1$, therefore $V=1$ as X approaches $-\infty$. This condition can be satisfied only if $\alpha<1$, because otherwise the exponential term becomes infinite as $X\to-\infty$. It can be immediately seen from equation (8.31) that this condition, in turn, implies that the upstream Mach number must be supersonic, $M_1 > 1$. Thus we again conclude that the only physical solution for a shock transition involves supersonic upstream flow.

For values of $C_2 < 0$, equation (8.34) describes a monotonically increasing velocity with $V \to \infty$ as $X \to \infty$. This is again an unphysical case. We conclude that physically meaningful solutions are constrained to supersonic upstream flows and positive values of C_2.

The asymptotic values of the normalized velocity, V, are $V=1$ for $x \to -\infty$ and $V = \alpha$ for $x \to +\infty$. It can also be shown that the V(X) function has an inflection point (characterized by $d^2V/dX^2=0$) at $V = \alpha^{1/2}$. The actual value of C_2 determines the physical location of the shock. We choose its value in a way that the inflection point may be found at the origin of our coordinate system, $V(X=0) = \alpha^{1/2}$. With this condition equation (8.34) becomes the following:

$$\frac{1-V}{(V-\alpha)^\alpha} = \frac{1-\sqrt{\alpha}}{(\sqrt{\alpha}-\alpha)^\alpha} \exp\left(\frac{9}{2}\sqrt{\frac{\pi}{30}}(1-\alpha)M_1 X\right) \qquad (8.35)$$

This implicit equation describes the change of velocity through the shock wave. The function V(X) is plotted in Figure 8.2 for the case of $M_1=2$. Inspection of Figure 8.2 reveals that most of the velocity change takes place in the immediate vicinity of the inflection point. This means that the shock wave thickness is of the order of magnitude of the mean free path of the gas molecules. It can also be shown that the shock thickness decreases with increasing upstream Mach number. The important point to remember is that the thickness of the shock wave is very small, and most of the shock transition takes place within a region with the extent of a couple of mean free paths.

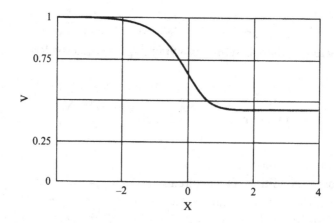

Figure 8.2

Once we know the normalized velocity, V, as a function of location we can easily calculate the remaining gas parameters. From equations (8.19) and (8.28) we get the density and the temperature as a function of V:

$$\frac{n}{n_1} = \frac{1}{V} \tag{8.36}$$

$$\frac{T}{T_1} = 1 + \frac{M_1^2}{3}(1 - V^2) \tag{8.37}$$

The pressure can be obtained by combining these two expressions:

$$\frac{p}{p_1} = \frac{1}{V}\left(1 + \frac{M_1^2}{3}(1 - V^2)\right) \tag{8.38}$$

Finally, it is instructive to examine the change of entropy through the shock wave. For a perfect gas the change of entropy is defined in the following way:[1,2]

$$S = \frac{s - s_1}{C_v} = \frac{5}{3}\ln\left(\frac{T}{T_1}\right) - \frac{2}{3}\ln\left(\frac{p}{p_1}\right) \tag{8.39}$$

where S is the dimensionless entropy change. With the help of expressions (8.37) and (8.38) one can express S in terms of the normalized velocity, V:

$$S = \ln\left[V^{2/3}\left(1 + \frac{M_1^2}{3}(1 - V^2)\right)\right] \tag{8.40}$$

The entropy profile for an upstream Mach number of 2 is shown in Figure 8.3. This profile is very interesting and it warrants a brief discussion. Inspection of Figure 8.3 reveals that the entropy profile has a maximum at the inflection point, X=0. This means that the gas recovers some bulk energy from the randomized motion on the downstream side of the shock wave. At first glance this result seems to violate the second law of thermodynamics. However, one should remember that the second law only applies to the entire system and it permits energy recovery (decrease of entropy) in localized regions. The negative entropy gradient means that an important physical assumption is violated inside the shock: the gas is not near thermodynamic equilibrium in this transition region.

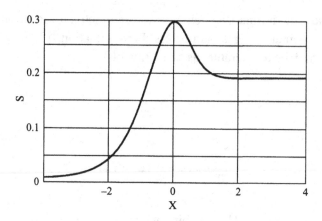

Figure 8.3

If we compare the physical solutions far upstream and far downstream from the shock we can conclude that the gas parameters satisfy the Rankine–Hugoniot jump conditions and that the entropy of the gas is larger in the downstream region than far upstream. The transition takes place within a region with spatial extent $\sim\lambda$: within this shock layer the gas is not in equilibrium.

8.2 Kinetic description of shocks: the Mott-Smith model[4]

The structure of shock waves can also be investigated with the help of kinetic theory. An intuitive approximate solution of the shock structure was proposed by Mott-Smith in 1951. This intuitive solution is based on the fundamental assumption that far upstream and far downstream of the shock the velocity distributions are Maxwell–Boltzmann distributions:

$$\lim_{x\to-\infty} F(x,\mathbf{v}) = n_1 f_1(\mathbf{v}) = n_1 \left(\frac{m}{2\pi k T_1}\right)^{3/2} \exp\left\{-\frac{m}{2 k T_1}\left[\left(v_x - u_1\right)^2 + v_y^2 + v_z^2\right]\right\}$$

$$(8.41)$$

$$\lim_{x\to+\infty} F(x,\mathbf{v}) = n_2 f_2(\mathbf{v}) = n_2 \left(\frac{m}{2\pi k T_2}\right)^{3/2} \exp\left\{-\frac{m}{2 k T_2}\left[\left(v_x - u_2\right)^2 + v_y^2 + v_z^2\right]\right\}$$

$$(8.42)$$

Mott-Smith's idea was to approximate the distribution function inside the shock as a linear combination of the upstream and downstream solutions:

$$F(x, v) = a_1(x) n_1 f_1(v) + a_2(x) n_2 f_2(v) \qquad (8.43)$$

where a_1 and a_2 are to be determined by suitable moment equations. It is assumed that the far upstream and far downstream gas parameters satisfy the Rankine–Hugoniot jump conditions given by equations (8.16) through (8.18). It is assumed that the mass flux is constant everywhere, therefore we immediately conclude that

$$a_1(x) + a_2(x) = 1 \qquad (8.44)$$

This means that it is enough to determine the function $a_1(x)$. Far upstream the distribution is the upstream Maxwellian, therefore we know that $a_1(-\infty)=1$. Similarly, we know that far downstream the distribution approaches the downstream Maxwellian, therefore $a_1(+\infty)=0$.

In order to determine the function $a_1(x)$ one needs a suitable moment equation which is independent of the basic conservation laws resulting in the Rankine–Hugoniot jump conditions. Basic conservation laws (equivalent to those obtained from the Euler equations) cannot describe a smooth transition from the upstream to the downstream distribution function, therefore one has to use a different moment equation to obtain the function $a_1(x)$.

We shall make use of Maxwell's equation of change (given by equation (6.38)) to determine the function $a_1(x)$. In our steady-state one dimensional case this equation becomes

$$\frac{d}{dx}\left(n\langle v_x W\rangle\right) = \Delta[W] \qquad (8.45)$$

We have considerable freedom in selecting the molecular quantity $W(v)$. It should be noted that the resulting function $a_1(x)$ is slightly different for different choices of the function W; however, the difference is quantitatively unimportant. We choose the simplest meaningful function, the momentum flux in the x direction. In this case $W = m v_x^2$. Substituting this expression into equation (8.45) we obtain

$$\frac{d}{dx}\left(mn\langle v_x^3\rangle\right) = m\Delta[v_x^2] \qquad (8.46)$$

Substituting our distribution function yields the following expression for the left hand side of the equation:

$$j_1\left(3kT_1 - 3kT_2 + mu_1^2 - mu_2^2\right)\frac{da_1}{dx} = m\Delta[v_x^2] \tag{8.47}$$

One can use equation (8.5) to obtain the following differential equation for $a_1(x)$:

$$\frac{da_1}{dx} = -\frac{m\Delta[v_x^2]}{2kj_1(T_2 - T_1)} \tag{8.48}$$

Next we evaluate the right hand side of Maxwell's equation of change. The integral on the right hand side can be written as

$$\Delta[v_x^2] = \frac{1}{2}\iiint_\infty d^3v_a \iiint_\infty d^3v_b \int_0^{2\pi} d\varepsilon \int_0^{\pi} d\chi \sin\chi \, Sg(v_{a,x}'^2 + v_{b,x}'^2 - v_{a,x}^2 - v_{b,x}^2)F_a F_b \tag{8.49}$$

where the subscripts a and b refer to two colliding gas molecules. Substituting the Mott-Smith distribution function yields the following:

$$\Delta[v_x^2] = \frac{a_1^2(x)}{2}\iiint_\infty d^3v_a \iiint_\infty d^3v_b \int_{4\pi} d\Omega Sg(v_{a,x}'^2 + v_{b,x}'^2 - v_{a,x}^2 - v_{b,x}^2)F_{1,a}F_{1,b}$$

$$+\frac{(1-a_1(x))^2}{2}\iiint_\infty d^3v_a \iiint_\infty d^3v_b \int_{4\pi} d\Omega Sg(v_{a,x}'^2 + v_{b,x}'^2 - v_{a,x}^2 - v_{b,x}^2)F_{2,a}F_{2,b}$$

$$+a_1(x)(1-a_1(x))\iiint_\infty d^3v_a \iiint_\infty d^3v_b \int_{4\pi} d\Omega Sg(v_{a,x}'^2 + v_{b,x}'^2 - v_{a,x}^2 - v_{b,x}^2)F_{1,a}F_{2,b} \tag{8.50}$$

The last term is a combination of two integrals, because the colliding particles can be interchanged in the collision integral. It is obvious that the first two integrals vanish, because both particles belong to the same Maxwellian population. The only non-vanishing term comes from the third integral which describes the interaction between the upstream and downstream particle populations. This means that equation (8.50) can be rewritten in the following form:

$$\Delta[v_x^2] = a_1(x)(1-a_1(x))\beta \tag{8.51}$$

where the location independent quantity β is defined as

$$\beta = \iiint_{\infty} d^3v_a \iiint_{\infty} d^3v_b \int_{4\pi} d\Omega\, S\, g(v_{a,x}'^2 + v_{b,x}'^2 - v_{a,x}^2 - v_{b,x}^2) F_{1,a} F_{2,b} \quad (8.52)$$

Substituting expression (8.51) into equation (8.48) yields the following differential equation for the function $a_1(x)$:

$$\frac{da_1}{dx} = -\alpha\, a_1(1-a_1) \quad (8.53)$$

where α is defined as

$$\alpha = \frac{m\beta}{2k\, j_1(T_2 - T_1)} \quad (8.54)$$

Equation (8.53) can be readily integrated to obtain the following solution for the function $a_1(x)$:

$$a_1(x) = \frac{1}{1 + e^{\alpha x}} \quad (8.55)$$

For the sake of convenience we set an integration constant (which controls the location of the shock) to zero.[5] It is obvious that this solution satisfies the necessary boundary limiting conditions at $\pm\infty$.

With the help of the function $a_1(x)$, we can obtain the gas number density, velocity and temperature profiles:

$$n(x) = \iiint_{\infty} d^3v\, F(x, \mathbf{v}) = a_1 n_1 + (1-a_1)n_2 = \frac{n_1}{1 + e^{\alpha x}}\left(1 + \frac{4M_1^2}{3 + M_1^2} e^{\alpha x}\right) \quad (8.56)$$

$$u(x) = \frac{n_1 u_1}{n(x)} = u_1 \frac{(3 + M_1^2)(1 + e^{\alpha x})}{3 + M_1^2(1 + 4e^{\alpha x})} \quad (8.57)$$

$$T(x) = \frac{1}{3}\frac{m}{n(x)k}\iiint_{\infty} d^3v\,(\mathbf{v} - \mathbf{u})^2\, F(x, \mathbf{v})$$

$$= T_1\left\{a_1(x)\frac{u(x)}{u_1}\left(1 + \frac{5}{9}M_1^2\right) - \frac{5}{9}M_1^2\left(\frac{u(x)}{u_1}\right)^2 + (1 - a_1(x))\frac{u(x)}{u_1}\frac{25M_1^2 + 3}{18}\right\}$$

$$(8.58)$$

Finally, we evaluate the quantity β. First of all let us examine the expression inside the parenthesis in equation (8.52). In the present case the masses of the colliding particles are identical, therefore one obtains with the help of equation (5.59) the following:

$$v_{a,x}'^2 + v_{b,x}'^2 - v_{a,x}^2 - v_{b,x}^2 = \frac{1}{2}\left(g_x'^2 - g_x^2\right) \tag{8.59}$$

Substituting this into equation (8.52) yields the following integral:

$$\beta = \frac{1}{2}\iiint d^3v_a \iiint_\infty d^3v_b \int_0^\pi d\chi \sin\chi\, S(g,\chi)g \int_0^{2\pi} d\varepsilon \left(g_x'^2 - g_x^2\right) F_{1,a}F_{2,b} \tag{8.60}$$

The integral over the impact azimuth can be evaluated with the help of equation (5.80). Substituting this result into our integral yields the following:

$$\beta = \frac{\pi}{2}\iiint d^3v_a \iiint_\infty d^3v_b \int_0^\pi d\chi \sin\chi(1-\cos^2\chi)S(g,\chi)g(g^2-3g_x^2)F_{1,a}F_{2,b} \tag{8.61}$$

The integral over the deflection angle, χ, yields the second order transport cross section, σ_2 (see the definition given by equation (3.88)). With the help of this cross section the integral becomes the following:

$$\beta = \frac{1}{4}\iiint d^3v_a \iiint_\infty d^3v_b \,\sigma_2(g)g(g^2-3g_x^2)F_{1,a}F_{2,b} \tag{8.62}$$

The evaluation of equation (8.62) is very complicated except in the case of Maxwell molecules. In the case of Maxwell molecules the transport cross section is proportional to $1/g$, therefore the integral can readily be evaluated. In order to demonstrate the physical meaning of the Mott-Smith solution let us consider the case of Maxwell molecules and evaluate β. For Maxwell molecules σ_2 is given by equation (3.109) with spectral index of 5. In this case the parameter β can be expressed as follows:

$$\beta = \frac{\pi}{2}\left(\frac{2K_5}{m}\right)^{1/2} A_2(5)\iiint d^3v_a \iiint_\infty d^3v_b\,(g^2-3g_x^2)F_{1,a}F_{2,b} \tag{8.63}$$

where $A_2(5)$ is a pure number (for Maxwell molecules $A_2=0.436\,19$). On the other hand we know that the functions F_1 and F_2 are Maxwellian and that

$$g^2 - g_x^2 = v_a^2 + v_b^2 - 2v_a \cdot v_b - v_{a,x}^2 - v_{b,x}^2 + 2v_{a,x}v_{b,x} \tag{8.64}$$

Therefore we immediately obtain the following result:

$$\beta = \pi \left(\frac{2K_5}{m}\right)^{1/2} A_2(5)\,n_1\,n_2\,\frac{k}{m}(T_1+T_2) \tag{8.65}$$

A physically more revealing form of this result can be obtained by introducing the mean free path and collision frequency with the help of equations (3.73) and (6.96):

$$\beta = \frac{A_2(5)}{A_1(5)}\,n_2\,\frac{k}{m}(T_1+T_2)\frac{\bar{v}_1}{\lambda_1} \tag{8.66}$$

where λ_1 is the upstream mean free path of the gas molecules. Substituting this result into the expression for the parameter α (defined by equation (8.54)) yields the following result:

$$\alpha = \frac{\alpha_0}{\lambda_1} \tag{8.67}$$

where

$$\alpha_0 = \frac{A_2(5)}{A_1(5)}\sqrt{\frac{24}{5\pi}}\,\frac{4M_1}{3+M_1^2}\,\frac{(3+M_1^2)(5M_1^2-1)+16M_1^2}{(3+M_1^2)(5M_1^2-1)-16M_1^2} \tag{8.68}$$

The dimensionless parameter α_0 characterizes the thickness of the shock as a function of the upstream Mach number, M_1. It is obvious that α_0 has a singularity at $M_1=1$: this means that for sonic flow the shock is infinitely broad (i.e. there is no shock). The function $\alpha_0(M_1)$ is shown in Figure 8.4 for supersonic upstream flows. Inspection of Figure 8.4 reveals that the shock rapidly narrows with increasing upstream Mach numbers and for hypersonic flows ($M_1 \gg 1$) its thickness is much less than the upstream mean free path.

Now we can substitute our results and express the Mott-Smith distribution function (given by equation (8.43)) in terms of the normalized location, $X=x/\lambda_1$, and

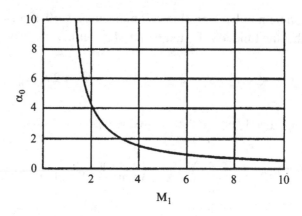

Figure 8.4

velocity, $\xi=v/u_1$. After some manipulation we obtain the following expression for the distribution function:

$$F(X,\xi) = \frac{n_1}{1+e^{\alpha_0 X}}\left(\frac{m}{2kT_1}\right)^{3/2}\left\{\exp\left(-\frac{5M_1^2}{6}\left[(\xi_x - 1)^2 + \xi_y^2 + \xi_z^2\right]\right)\right.$$

$$\left. + \psi\, e^{\alpha_0 X}\exp\left(-\frac{5(3+M_1^2)^3}{96\,M_1^4(5M_1^2 - 1)}\left[\left(\xi_x - \frac{3+M_1^2}{4M_1^2}\right)^2 + \xi_y^2 + \xi_z^2\right]\right)\right\} \qquad (8.69)$$

where

$$\psi = \frac{256\,M_1^5}{(3+M_1^2)^{3/2}(5M_1^2 - 1)^{5/2}} \qquad (8.70)$$

The macroscopic gas parameters, n, u, and T, are given by equations (8.56) through (8.58). Figure 8.5 shows the variation of the normalized bulk velocity, $V(X)=u(X)/u_1$, across the shock. It can be seen from Figure 8.5 that the velocity profile is quantitatively very similar to that obtained with the help of Navier–Stokes equations (see Figure 8.2). The physical reason for these similarities is that the collision integral incorporated into the α_0 parameter describes the collisional (viscous) interaction between the upstream and downstream particle distributions.

Finally, one can also calculate the entropy density profile across the shock. As mentioned previously (see Section 5.2), the entropy density is proportional to $-k\int d^3v\, F\ln(F)$. This integral can be numerically evaluated with the help of equation

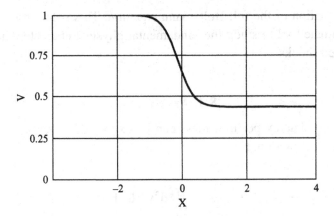

Figure 8.5

(8.69) and the resulting entropy profile can be plotted as a function of location. A much simpler way to get the entropy profile is to make use of equation (8.39) expressing the entropy in terms of the macroscopic gas parameters, T and p. Substituting expressions (8.56) and (8.58) into equation (8.39) also gives us the entropy profile through the shock. This profile is shown in Figure 8.6. Inspection of Figure 8.6 reveals that the entropy again peaks inside the shock, pretty much the same way as it did in the case of the Navier–Stokes solution. There are some minor differences between the two solutions, but qualitatively the results are the same.

It should be kept in mind that the Mott-Smith solution is only an approximate one: it satisfies the continuity, momentum and energy conservation laws and is based on one equation of change, but it does not satisfy all moments of the

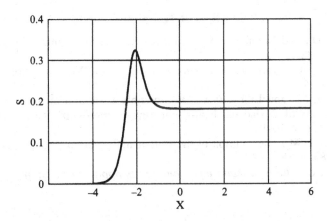

Figure 8.6

Boltzmann equation or the full Boltzmann equation. This solution is, however, a good approximate tool to study the fundamental physical characteristics of collision dominated shocks.

8.3 Problems

8.1 The entropy density per unit mass can be expressed in terms of the phase-space distribution function, F:

$$s = -\frac{k}{m\,n} \iiint_{\infty} d^3v \, F \ln(F)$$

Consider a monatomic perfect gas in an infinitely long tube with constant cross section. There is a steady-state flow through the tube with a shock wave somewhere in the tube. Far upstream from the shock the gas is Maxwellian and hypersonic ($M_1 \gg 1$) with density n_1 and temperature T_1. Far downstream the gas is Maxwellian and subsonic. Show that in the upstream region the entropy is smaller than in the downstream region.

8.2 Consider a 1 dimensional perfect gas flow with a normal shock. The upstream Mach number is M_1, the mean free path is λ_1, and the number density is n_1. Plot the velocity profile through the shock using the Navier–Stokes and the Mott-Smith approximations and compare the results.

8.3 Calculate the increase of entropy ($s_2 - s_1$) through a hypersonic ($M_1 \gg 1$) normal shock using the Navier–Stokes and Mott-Smith methods. Compare the results.

8.4 References

[1] Landau, L.D., and Lifshitz, E.M., *Fluid mechanics*, 2nd edition, Pergamon Press, New York, 1987.

[2] Zucrow, M.J. and Hoffman, J.D., *Gas dynamics*, John Wiley and Sons, New York, 1976.

[3] Morduchow, M., and Libby, P.A., On a complete solution of the one-dimensional flow equations of a viscous, heat-conducting, compressible gas, *J. Aeronautical Sciences*, **16**, 674, 1949.

[4] Mott-Smith, H.M., The solution of the Boltzmann equation for a shock wave, *Phys. Rev.*, **82**, 885, 1951.

[5] Cercignani, C., *The Boltzmann equation and its applications*, Springer, New York, 1988.

9

Appendices

9.1 Physical constants[1]

Avogadro's number	N_A	6.0221×10^{23}	mole^{-1}
Universal gas constant	R	8.3145	joule K^{-1} mole^{-1}
Boltzmann's constant	k	1.3807×10^{-23}	joule K^{-1}
Electron mass	m_e	9.1094×10^{-31}	kg
Proton mass	m_p	1.6726×10^{-27}	kg
Atomic mass unit	m_u	1.6605×10^{-27}	kg
Gravitational constant	G	6.6726×10^{-11}	m^3 s^{-1} kg^{-1}
Gravitational acceleration at the surface of the earth	g	9.8067	m s^{-2}
Planck's constant	h	6.6261×10^{-34}	Joule s
Stefan–Boltzmann constant	σ	5.6705×10^{-8}	W m^{-2} K^{-4}
Speed of light in vacuum	c	2.9979×10^{8}	m s^{-1}
Elementary charge	e	1.6022×10^{-19}	coulomb

9.2 Vector and tensor identities[1]

Below are some vector and tensor identities resulting in scalar or vector quantities. In these relations f represents a scalar function, f=f(**r**), while **A**(**r**), **B**(**r**), **C**(**r**), and **D**(**r**) are vector quantities. The symbol ∇ represents the configuration space differential operator.

$$\mathbf{A}\cdot(\mathbf{B}\times\mathbf{C}) = (\mathbf{A}\times\mathbf{B})\cdot\mathbf{C} = \mathbf{B}\cdot(\mathbf{C}\times\mathbf{A}) = (\mathbf{B}\times\mathbf{C})\cdot\mathbf{A} = \mathbf{C}\cdot(\mathbf{A}\times\mathbf{B}) = (\mathbf{C}\times\mathbf{A})\cdot\mathbf{B}$$

$$\mathbf{A}\times(\mathbf{B}\times\mathbf{C}) = \mathbf{C}\times(\mathbf{B}\times\mathbf{A}) = (\mathbf{A}\cdot\mathbf{C})\mathbf{B} - (\mathbf{A}\cdot\mathbf{B})\mathbf{C}$$

$$(\mathbf{A}\times\mathbf{B})\cdot(\mathbf{C}\times\mathbf{D}) = (\mathbf{A}\cdot\mathbf{C})(\mathbf{B}\cdot\mathbf{D}) - (\mathbf{A}\cdot\mathbf{D})(\mathbf{B}\cdot\mathbf{C})\mathbf{B}$$

$$(\mathbf{A}\times\mathbf{B})\times(\mathbf{C}\times\mathbf{D}) = [(\mathbf{A}\times\mathbf{B})\cdot\mathbf{D}]\mathbf{C} - [(\mathbf{A}\times\mathbf{B})\cdot\mathbf{C}]\mathbf{D}$$

$$\nabla\cdot(f\mathbf{A}) = f\nabla\cdot\mathbf{A} + \mathbf{A}\cdot\nabla f$$

$$\nabla\times(f\mathbf{A}) = f\nabla\times\mathbf{A} + \nabla f\times\mathbf{A}$$

$$\nabla\cdot(\mathbf{A}\times\mathbf{B}) = \mathbf{B}\cdot(\nabla\times\mathbf{A}) - \mathbf{A}\cdot(\nabla\times\mathbf{B})$$

$$(\mathbf{A}\times\mathbf{B})\times(\mathbf{C}\times\mathbf{D}) = [(\mathbf{A}\times\mathbf{B})\cdot\mathbf{D}]\mathbf{C} - [(\mathbf{A}\times\mathbf{B})\cdot\mathbf{C}]\mathbf{D}$$

$$\nabla\cdot(f\mathbf{A}) = f(\nabla\cdot\mathbf{A}) + (\mathbf{A}\cdot\nabla)f$$

$$\nabla\times(f\mathbf{A}) = f(\nabla\times\mathbf{A}) + (\nabla f)\times\mathbf{A}$$

$$\nabla\cdot(\mathbf{A}\times\mathbf{B}) = \mathbf{B}\cdot(\nabla\cdot\mathbf{A}) - \mathbf{A}\cdot(\nabla\cdot\mathbf{B})$$

$$\nabla\times(\mathbf{A}\times\mathbf{B}) = \mathbf{A}(\nabla\cdot\mathbf{B}) - \mathbf{B}(\nabla\cdot\mathbf{A}) + (\mathbf{B}\cdot\nabla)\mathbf{A} - (\mathbf{A}\cdot\nabla)\mathbf{B}$$

$$\mathbf{A}\times(\nabla\times\mathbf{B}) = (\nabla\mathbf{B})\cdot\mathbf{A} - (\mathbf{A}\cdot\nabla)\mathbf{B}$$

$$\nabla(\mathbf{A}\cdot\mathbf{B}) = \mathbf{A}\times(\nabla\times\mathbf{B}) + \mathbf{B}\times(\nabla\times\mathbf{A}) + (\mathbf{A}\cdot\nabla)\mathbf{B} + (\mathbf{B}\cdot\nabla)\mathbf{A}$$

$$\nabla^2\mathbf{A} = \nabla(\nabla\cdot\mathbf{A}) - \nabla\times(\nabla\times\mathbf{A})$$

$$\nabla\times\nabla f = 0$$

$$\nabla\cdot(\nabla\times\mathbf{B}) = 0$$

$$\nabla\cdot(\mathbf{A}\mathbf{B}) = (\nabla\cdot\mathbf{A})\mathbf{B} + (\mathbf{A}\cdot\nabla)\mathbf{B}$$

9.3 Differential operators in spherical and cylindrical coordinates

9.3.1 *Spherical coordinates*

In spherical coordinates the vector, $\mathbf{A}$, is given by three independent coordinates, A_r, A_θ, and A_ϕ. The vector itself is expressed as $\mathbf{A}=A_r\mathbf{e}_r+A_\theta\mathbf{e}_\theta+A_\phi\mathbf{e}_\phi$, where $\mathbf{e}_r$, $\mathbf{e}_\theta$ and $\mathbf{e}_\phi$ are orthonormal unit vectors in the r, θ and ϕ directions. In these relations f represents a scalar function, $f=f(\mathbf{r})$, while $\mathbf{A}(\mathbf{r})$ and $\mathbf{B}(\mathbf{r})$ are vector quantities. The symbol, T, refers to a tensor with two indices, T_{ij}. Finally, ∇ represents the configuration space differential operator.

Divergence of a vector:

$$\nabla\cdot\mathbf{A} = \frac{1}{r^2}\frac{\partial(r^2 A_r)}{\partial r} + \frac{1}{r\sin\theta}\frac{\partial(\sin\theta\, A_\theta)}{\partial\theta} + \frac{1}{r\sin\theta}\frac{\partial A_\phi}{\partial\phi}$$

Gradient of a scalar quantity:

$$\nabla f = \frac{\partial f}{\partial r}\mathbf{e}_r + \frac{1}{r}\frac{\partial f}{\partial\theta}\mathbf{e}_\theta + \frac{1}{r\sin\theta}\frac{\partial f}{\partial\phi}\mathbf{e}_\phi$$

Curl of a vector:

$$\nabla \times \mathbf{A} = \left[\frac{1}{r\sin\theta} \frac{\partial(\sin\theta\, A_\phi)}{\partial\theta} - \frac{1}{r\sin\theta} \frac{\partial A_\Theta}{\partial\phi} \right] \mathbf{e}_r$$

$$+ \left[\frac{1}{r\sin\theta} \frac{\partial A_r}{\partial\phi} - \frac{1}{r} \frac{\partial(rA_\phi)}{\partial r} \right] \mathbf{e}_\Theta + \left[\frac{1}{r} \frac{\partial(rA_\Theta)}{\partial r} - \frac{1}{r} \frac{\partial A_r}{\partial\theta} \right] \mathbf{e}_\phi$$

Laplacian of a scalar:

$$\nabla^2 f = \frac{1}{r^2} \frac{\partial}{\partial r}\left(r^2 \frac{\partial f}{\partial r} \right) + \frac{1}{r^2 \sin\theta} \frac{\partial}{\partial\theta}\left(\sin\theta \frac{\partial f}{\partial\theta} \right) + \frac{1}{r^2 \sin^2\theta} \frac{\partial^2 f}{\partial\phi^2}$$

Laplacian of a vector:

$$\nabla^2 \mathbf{A} = \left[\nabla^2 A_r - \frac{2A_r}{r^2} - \frac{2}{r^2} \frac{\partial A_\Theta}{\partial\theta} - \frac{2\cot\theta\, A\Theta}{r^2} - \frac{2}{r^2 \sin\theta} \frac{\partial A_\phi}{\partial\phi} \right] \mathbf{e}_r$$

$$+ \left[\nabla^2 A_\theta + \frac{2}{r^2} \frac{\partial A_r}{\partial\theta} - \frac{A_\Theta}{r^2 \sin^2\theta} - \frac{2\cos\theta}{r^2 \sin^2\theta} \frac{\partial A_\phi}{\partial\phi} \right] \mathbf{e}_\theta$$

$$+ \left[\nabla^2 A_\phi - \frac{A_\phi}{r^2 \sin^2\theta} + \frac{2}{r^2 \sin\theta} \frac{\partial A_r}{\partial\phi} - \frac{2\cos\theta}{r^2 \sin^2\theta} \frac{\partial A_\Theta}{\partial\phi} \right] \mathbf{e}_\phi$$

Components of $(\mathbf{A}\cdot\nabla)\mathbf{B}$:

$$(\mathbf{A}\cdot\nabla)\mathbf{B} = \left[A_r \frac{\partial B_r}{\partial r} + \frac{A_\theta}{r} \frac{\partial B_r}{\partial\theta} + \frac{A_\phi}{r\sin\theta} \frac{\partial B_r}{\partial\phi} - \frac{A_\theta B_\theta + A_\phi B_\phi}{r} \right] \mathbf{e}_r$$

$$+ \left[A_r \frac{\partial B_\theta}{\partial r} + \frac{A_\theta}{r} \frac{\partial B_\theta}{\partial\theta} + \frac{A_\phi}{r\sin\theta} \frac{\partial B_\theta}{\partial\phi} + \frac{A_\theta B_r - \cot\theta\, A_\phi B_\phi}{r} \right] \mathbf{e}_\theta$$

$$+ \left[A_r \frac{\partial B_\phi}{\partial r} + \frac{A_\theta}{r} \frac{\partial B_\phi}{\partial\theta} + \frac{A_\phi}{r\sin\theta} \frac{\partial B_\phi}{\partial\phi} + \frac{A_\phi B_r + \cot\theta\, A_\phi B_\theta}{r} \right] \mathbf{e}_\phi$$

Divergence of a tensor:

$$\nabla \cdot \mathbf{T} = \left[\frac{1}{r^2} \frac{\partial (r^2 T_{rr})}{\partial r} + \frac{1}{r \sin \theta} \frac{\partial (\sin \theta \, T_{\theta r})}{\partial \theta} + \frac{1}{r \sin \theta} \frac{\partial T_{\phi r}}{\partial \phi} - \frac{T_{\theta\theta} + T_{\phi\phi}}{r} \right] \mathbf{e}_r$$

$$+ \left[\frac{1}{r^2} \frac{\partial (r^2 T_{r\theta})}{\partial r} + \frac{1}{r \sin \theta} \frac{\partial (\sin \theta \, T_{\theta\theta})}{\partial \theta} + \frac{1}{r \sin \theta} \frac{\partial T_{\phi\theta}}{\partial \phi} + \frac{T_{\theta r} - \cot \theta \, T_{\phi\phi}}{r} \right] \mathbf{e}_\theta$$

$$+ \left[\frac{1}{r^2} \frac{\partial (r^2 T_{r\phi})}{\partial r} + \frac{1}{r \sin \theta} \frac{\partial (\sin \theta \, T_{\theta\phi})}{\partial \theta} + \frac{1}{r \sin \theta} \frac{\partial T_{\phi\phi}}{\partial \phi} + \frac{T_{\phi r} + \cot \theta \, T_{\phi\theta}}{r} \right] \mathbf{e}_\phi$$

9.3.2 Cylindrical coordinates

In cylindrical coordinates the vector, $\mathbf{A}$, is given by three independent coordinates, A_ρ, A_ϕ, and A_z. The vector itself is expressed as $\mathbf{A} = A_\rho \mathbf{e}_\rho + A_\phi \mathbf{e}_\phi + A_z \mathbf{e}_z$, where $\mathbf{e}_\rho$, $\mathbf{e}_\phi$ and $\mathbf{e}_z$ are orthonormal unit vectors in the r, ϕ and z directions. In these relations f represents a scalar function, $f = f(\mathbf{r})$, while $\mathbf{A}(\mathbf{r})$ and $\mathbf{B}(\mathbf{r})$ are vector quantities. The symbol, T, refers to a tensor with two indices, T_{ij}. Finally, ∇ represents the configuration space differential operator.

Divergence of a vector:

$$\nabla \cdot \mathbf{A} = \frac{1}{\rho} \frac{\partial (\rho A_\rho)}{\partial \rho} + \frac{1}{\rho} \frac{\partial A_\phi}{\partial \phi} + \frac{\partial A_z}{\partial z}$$

Gradient of a scalar quantity:

$$\nabla f = \frac{\partial f}{\partial \rho} \mathbf{e}_\rho + \frac{1}{\rho} \frac{\partial f}{\partial \phi} \mathbf{e}_\phi + \frac{\partial f}{\partial z} \mathbf{e}_z$$

Curl of a vector:

$$\nabla \times \mathbf{A} = \left[\frac{1}{\rho} \frac{\partial A_z}{\partial \phi} - \frac{\partial A_\phi}{\partial z} \right] \mathbf{e}_\rho + \left[\frac{\partial A_\rho}{\partial z} - \frac{\partial A_z}{\partial \rho} \right] \mathbf{e}_\phi + \left[\frac{1}{\rho} \frac{\partial (\rho A_\phi)}{\partial \rho} - \frac{1}{\rho} \frac{\partial A_\rho}{\partial \phi} \right] \mathbf{e}_z$$

Laplacian of a scalar:

$$\nabla^2 f = \frac{1}{\rho} \frac{\partial}{\partial \rho} \left(\rho \frac{\partial f}{\partial \rho} \right) + \frac{1}{\rho^2} \frac{\partial^2 f}{\partial \phi^2} + \frac{\partial^2 f}{\partial z^2}$$

Laplacian of a vector:

$$\nabla^2 \mathbf{A} = \left[\nabla^2 A_\rho - \frac{2}{\rho^2} \frac{\partial A_\phi}{\partial \phi} - \frac{A_r}{\rho^2} \right] \mathbf{e}_\rho + \left[\nabla^2 A_\phi + \frac{2}{\rho^2} \frac{\partial A_r}{\partial \phi} - \frac{A_\phi}{\rho^2} \right] \mathbf{e}_\phi + \nabla^2 A_z \mathbf{e}_z$$

Components of $(\mathbf{A} \cdot \nabla)\mathbf{B}$:

$$(\mathbf{A} \cdot \nabla)\mathbf{B} = \left[A_\rho \frac{\partial B_\rho}{\partial \rho} + \frac{A_\phi}{\rho} \frac{\partial B_\rho}{\partial \phi} + A_z \frac{\partial B_\rho}{\partial z} - \frac{A_\phi B_\phi}{\rho} \right] \mathbf{e}_\rho$$

$$+ \left[A_\rho \frac{\partial B_\phi}{\partial \rho} + \frac{A_\phi}{\rho} \frac{\partial B_\phi}{\partial \phi} + A_z \frac{\partial B_\phi}{\partial z} + \frac{A_\phi B_\rho}{\rho} \right] \mathbf{e}_\phi$$

$$+ \left[A_\rho \frac{\partial B_z}{\partial \rho} + \frac{A_\phi}{\rho} \frac{\partial B_z}{\partial \phi} + A_z \frac{\partial B_z}{\partial z} \right] \mathbf{e}_z$$

Divergence of a tensor:

$$\nabla \cdot T = \left[\frac{1}{\rho} \frac{\partial (\rho T_{\rho\rho})}{\partial \rho} + \frac{1}{\rho} \frac{\partial T_{\phi\rho}}{\partial \phi} + \frac{\partial T_{z\rho}}{\partial z} - \frac{T_{\phi\phi}}{\rho} \right] \mathbf{e}_\rho$$

$$+ \left[\frac{1}{\rho} \frac{\partial (\rho T_{\rho\phi})}{\partial \rho} + \frac{1}{\rho} \frac{\partial T_{\phi\phi}}{\partial \phi} + \frac{\partial T_{z\phi}}{\partial z} + \frac{T_{\phi\rho}}{\rho} \right] \mathbf{e}_\phi$$

$$+ \left[\frac{1}{\rho} \frac{\partial (\rho T_{\rho z})}{\partial \rho} + \frac{1}{r} \frac{\partial T_{\phi z}}{\partial \phi} + \frac{\partial T_{zz}}{\partial z} \right] \mathbf{e}_z$$

9.4 Some fundamental integrals

A definite integral that frequently appears in kinetic theory is the following:

$$L_n(\beta) = \int_0^\infty dt \, t^n \, e^{-\beta t^2} \qquad\qquad \beta > 0, \quad n = 0, 1, 2, 3, \cdots$$

where $\beta > 0$ and n is a non-negative integer. This integral can be evaluated for even and odd values of n:

$$L_{2n}(\beta) = \int_0^\infty dt\, t^{2n}\, e^{-\beta t^2} = \frac{(2n-1)(2n-3)\cdots 1}{2^{n+1}\beta^{(n+1/2)}}\sqrt{\pi} \qquad\qquad \beta > 0, \quad n = 0,1,2,3,\cdots$$

$$L_{2n+1}(\beta) = \int_0^\infty dt\, t^{2n+1}\, e^{-\beta t^2} = \frac{n!}{2\beta^{(n+1)}} \qquad\qquad \beta > 0, \quad n = 0,1,2,3,\cdots$$

For specific values of n one obtains

$$L_0(\beta) = \frac{1}{2}\sqrt{\frac{\pi}{\beta}} \qquad\qquad\qquad L_1(\beta) = \frac{1}{2\beta}$$

$$L_2(\beta) = \frac{\sqrt{\pi}}{4\beta^{3/2}} \qquad\qquad\qquad L_3(\beta) = \frac{1}{2\beta^2}$$

$$L_4(\beta) = \frac{3\sqrt{\pi}}{8\beta^{5/2}} \qquad\qquad\qquad L_5(\beta) = \frac{1}{\beta^3}$$

$$L_6(\beta) = \frac{15\sqrt{\pi}}{16\beta^{7/2}} \qquad\qquad\qquad L_7(\beta) = \frac{3}{\beta^4}$$

$$L_8(\beta) = \frac{105\sqrt{\pi}}{32\beta^{9/2}} \qquad\qquad\qquad L_9(\beta) = \frac{12}{\beta^5}$$

One can also obtain the following recursive formula for the evaluation of the definite integral, L_n:

$$L_{n+2}(\beta) = -\frac{dL_n(\beta)}{d\beta}$$

The function $L_n(\beta)$ can also be expressed in terms of the gamma function:

$$L_n(\beta) = \frac{\Gamma\left(\dfrac{n+1}{2}\right)}{2\beta^{\frac{n+1}{2}}}$$

Some frequently used integrals involving the error function are the following (all these integrals are only defined for values $\beta > 0$):

$$\int_0^x dt \ e^{-\beta t^2} = \frac{1}{2} \sqrt{\frac{\pi}{\beta}} \ erf\left(\sqrt{\beta} \ x\right)$$

$$\int_0^x dt \ t \ e^{-\beta t^2} = \frac{1}{2\beta} \left(1 - e^{-\beta x^2}\right)$$

$$\int_0^x dt \ t^2 \ e^{-\beta t^2} = \frac{1}{2\beta} \left[\frac{1}{2} \sqrt{\frac{\pi}{\beta}} \ erf\left(\sqrt{\beta} \ x\right) - x \ e^{-\beta x^2}\right]$$

$$\int_0^x dt \ t^{n+2} \ e^{-\beta t^2} = -\frac{d}{d\beta} \int_0^x dt \ t^n \ e^{-\beta t^2}$$

9.5 Some special functions[2]

9.5.1 *The Dirac delta function,* $\delta(x)$

The Dirac delta function, $\delta(x)$, is a very convenient mathematical concept (strictly speaking it is not a function) which singles out a particular value of the variable, x. The delta function is characterized by the basic properties

$$\delta(x - x_0) = \begin{cases} 0 & \text{for } x \neq x_0 \\ \infty & \text{for } x = x_0 \end{cases}$$

but in such a way that the integral of the delta function satisfies the following relation:

$$\int_{x_0-\varepsilon}^{x_0+\varepsilon} dx \ \delta(x - x_0) = 1$$

for any value $\varepsilon > 0$. This means that the delta function has a very sharp peak at $x = x_0$, but in such a way that the area under the peak is unity. It follows from the above definition that for any smooth function, $f(x)$,

$$\int_a^b dx \ f(x) \delta(x - x_0) = \begin{cases} f(x_0) & \text{if } a < x_0 < b \\ 0 & \text{otherwise} \end{cases}$$

The derivative of the delta function, $\delta'(x)$, can be interpreted the following way:

$$\int_a^b dx\, f(x)\delta'(x-x_0) = \int_a^b dx\, \frac{d}{dx}\big[f(x)\delta(x-x_0)\big] - \int_a^b dx\, f'(x)\delta(x-x_0)$$

$$= \big[f(x)\delta(x-x_0)\big]_{x=a}^{x=b} - f'(x_0) = -f'(x_0)$$

for $a<x_0<b$. Some other fundamental properties of the delta function are the following:

$$\delta(-x) = \delta(x)$$

$$\delta'(-x) = -\delta'(x)$$

$$x\,\delta(x) = 0$$

$$x\,\delta'(x) = -\delta(x)$$

$$\delta(a\,x) = \frac{1}{a}\delta(x)$$

$$x\,\delta(x-y) = y\,\delta(x-y)$$

The following are various examples of analytical representations of the Dirac δ function. In these examples the positive parameter, ε, is taken in the limit $\varepsilon \to 0$.

$$\delta(x) = \begin{cases} \lim\limits_{\varepsilon \to 0} \dfrac{1}{\varepsilon} & \text{for } -\dfrac{\varepsilon}{2} < x < \dfrac{\varepsilon}{2} \\ 0 & \text{otherwise} \end{cases}$$

$$\delta(x) = \lim_{\varepsilon \to 0} \frac{1}{\pi}\frac{\varepsilon}{\varepsilon^2 + x^2}$$

$$\delta(x) = \lim_{\varepsilon \to 0} \frac{1}{\varepsilon\sqrt{2\pi}} \exp\left(-\frac{x^2}{2\varepsilon^2}\right)$$

$$\delta(x) = \frac{1}{2\pi} \int_{-\infty}^{\infty} dk \, e^{ikx}$$

9.5.2 The Heaviside step function, H(x)

The Heaviside step function (named after Oliver Heaviside, who first introduced it), H(x), is defined as

$$H(x) = \begin{cases} 0 & \text{for } x < 0 \\ 1 & \text{for } x > 0 \end{cases}$$

H(x) is piecewise continuous in $-\infty < x < \infty$ with a single jump discontinuity at x=0, where H(x) is undefined. The Heaviside step function is sometimes also called the unit-step function, because it represents the turn-on action of an ideal switch.

The derivative of the Heaviside step function is the delta function:

$$H'(x) = \delta(x)$$

The symmetry relation of the Heaviside step function is the following:

$$H(-x) = 1 - H(x)$$

9.5.3 The error function, erf(x)

The error function, erf(x), is defined by the following relation:

$$\text{erf}(x) = \frac{2}{\sqrt{\pi}} \int_{0}^{x} dt \, e^{-t^2}$$

This integral can be evaluated numerically and is tabulated in Table 9.1.

Table 9.1 *Tabulated values of the error function*

x	erf(x)	x	erf(x)	x	erf(x)
0.00	0.000 000	0.85	0.770 668	1.70	0.983 790
0.05	0.056 372	0.90	0.796 908	1.75	0.986 672
0.10	0.112 463	0.95	0.820 891	1.80	0.989 091
0.15	0.167 996	1.00	0.842 701	1.85	0.991 111
0.20	0.222 703	1.05	0.862 436	1.90	0.992 790
0.25	0.276 326	1.10	0.880 205	1.95	0.994 179
0.30	0.328 627	1.15	0.896 124	2.00	0.995 322
0.35	0.379 382	1.20	0.910 314	2.05	0.996 258
0.40	0.428 392	1.25	0.922 900	2.10	0.997 021
0.45	0.475 482	1.30	0.934 008	2.15	0.997 639
0.50	0.520 500	1.35	0.943 762	2.20	0.998 137
0.55	0.563 323	1.40	0.952 285	2.25	0.998 537
0.60	0.603 856	1.45	0.959 695	2.30	0.998 857
0.65	0.642 029	1.50	0.966 105	2.35	0.999 111
0.70	0.677 801	1.55	0.971 623	2.40	0.999 311
0.75	0.711 156	1.60	0.976 348	2.45	0.999 469
0.80	0.742 101	1.65	0.980 376	2.50	0.999 593

The error function is a monotonically increasing function of x. Obviously erf(0)=0, and it can be shown that

$$\lim_{x \to \infty} \operatorname{erf}(x) = 1$$

The behavior of the error function is illustrated in Figure 9.1.

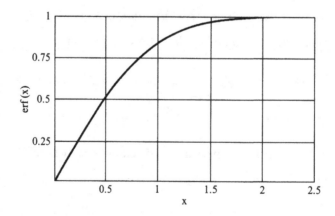

Figure 9.1

It follows from its definition that the error function satisfies the following symmetry relation:

$$\text{erf}(-x) = -\text{erf}(x)$$

The derivative of the error function is an exponential:

$$\frac{d}{dx}\text{erf}(x) = \frac{2}{\sqrt{\pi}}e^{-x^2}$$

The general series expansion of the error function can be written in the following form:

$$\text{erf}(x) = \frac{2e^{-x^2}}{\sqrt{\pi}}\sum_{k=0}^{\infty}\frac{2^k}{1\cdot 3\cdot\ldots\cdot(2k+1)}x^{2k+1}$$

For very large argument values ($x \gg 1$) another, much simpler series expansion can be found:

$$\text{erf}(x) = 1 - \frac{e^{-x^2}}{x\sqrt{\pi}}\left(1 - \frac{1}{2x^2} - \ldots\right) \qquad x \gg 1$$

In the other limiting case, when the argument is much smaller than unity, the series expansion simplifies to the following:

$$\text{erf}(x) = \frac{2}{\sqrt{\pi}}\left(x - \frac{1}{3}x^3 + \frac{1}{10}x^5 - \ldots\right) \qquad x \ll 1$$

A fairly accurate numerical approximation of the error function is given for arbitrary values of x by the following expression:

$$\text{erf}(x) = 1 - (a_1 t + a_2 t^2 + a_3 t^3 + a_4 t^4 + a_5 t^5)e^{-x^2} + \varepsilon(x)$$

where

$$t = \frac{1}{1 + 0.327\,5911x}$$

The magnitude of the numerical error, $\varepsilon(x)$, is smaller than 1.5×10^{-7}, $|\varepsilon(x)| \leq 1.5\times10^{-7}$, for any value of x. The values of the numerical constants are the following: $a_1=0.254\ 829\ 592$, $a_2=-0.284\ 496\ 736$, $a_3=1.421\ 413\ 741$, $a_4=-1.453\ 152\ 027$, and $a_5=1.061\ 405\ 429$.

9.5.4 The modified Bessel function, $I_n(x)$

The modified Bessel function of integer order, $I_n(x)$, is defined by the following differential equation:

$$x^2\frac{d^2I_n(x)}{dx^2}+x\frac{dI_n(x)}{dx}-(x^2+n^2)I_n(x)=0$$

where n=0,1,2,3,....

The integral representation of the function, $I_n(x)$, is given by

$$I_n(x)=\frac{1}{\pi}\int_0^\pi d\theta\, e^{x\cos\theta}\cos(n\theta)$$

The function, $I_n(x)$, satisfies the following symmetry relation:

$$I_{-n}(x)=I_n(x)$$

A recurrence relation for the function, $I_n(x)$, is the following:

$$I_{n+1}(x)=-\frac{2n}{x}I_n(x)+I_{n-1}(x)$$

for n>1.

The derivative (with respect to x) of the modified Bessel function, $I_n(x)$, is given by the following recurrence relation:

$$I_n'(x)=-\frac{n}{x}I_n(x)+I_{n-1}(x)$$

The series expansion of $I_n(x)$ is the following:

$$I_n(x) = \left(\frac{1}{2}x\right)^n \sum_{k=0}^{\infty} \frac{\left(\frac{1}{2}x\right)^{2k}}{k!(n+k)!}$$

Some associated series are the following:

$$1 = I_0(x) - 2I_2(x) + 2I_4(x) - 2I_6(x) + \ldots$$

$$e^x = I_0(x) + 2I_1(x) + 2I_2(x) + 2I_3(x) + \ldots$$

$$e^{-x} = I_0(x) - 2I_1(x) + 2I_2(x) - 2I_3(x) + \ldots$$

$$\cosh(x) = I_0(x) + 2I_2(x) + 2I_4(x) + 2I_6(x) + \ldots$$

$$\sinh(x) = 2I_1(x) + 2I_3(x) + 2I_5(x) + \ldots$$

$$e^{x\cos\theta} = I_0(x) + 2\sum_{k=1}^{\infty} I_k(x)\cos(k\theta)$$

9.5.5 *The complete elliptic integral of the second kind,* E(x)

The complete elliptic integral of the second kind, E(x), is defined in the following way:

$$E(x) = \int_0^{\pi/2} d\theta \sqrt{1 - x^2 \sin^2\theta} = \int_0^1 dt \frac{\sqrt{1 - x^2 t^2}}{\sqrt{1 - t^2}}$$

E(x) is defined for the interval $-1 \le x \le 1$. The limiting values of the complete elliptic integral of the second kind are E(0)=π/2 and E(1)=1. The function E(x) is shown in Figure 9.2.

The derivative of E(x) is given by

$$E'(x) = E\left(\sqrt{1 - x^2}\right)$$

An infinite series expansion of the complete elliptic integral of the second kind, E(x), is the following:

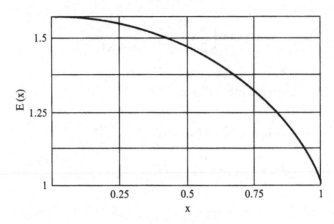

Figure 9.2

$$E(x) = \frac{1}{2}\pi\left[1 - \left(\frac{1}{2}\right)^2\frac{x^2}{1} - \left(\frac{1}{2}\frac{3}{4}\right)^2\frac{x^4}{3} - \left(\frac{1}{2}\frac{3}{4}\frac{5}{6}\right)^2\frac{x^6}{5} - \ldots\right]$$

9.6 References

[1] Book, D.L., *NRL plasma formulary*, Naval Research Laboratory, Washington, D.C., 1990.

[2] Abramowitz, M., and Stegun, I.A., *Handbook of mathematical functions*, Dover, New York, 1965.

Index

Printed in the United States
By Bookmasters